THOMAS KUHN'S "LINGUISTIC TURN" AND THE LEGACY OF LOGICAL EMPIRICISM

Presenting a critical history of the philosophy of science in the twentieth century, focusing on the transition from logical positivism in its first half to the 'new philosophy of science' in its second, Stefano Gattei examines the influence of several key figures, but the main focus of the book are Thomas Kuhn and Karl Popper.

Kuhn as the central figure of the new philosophy of science, and Popper as a key philosopher of the time who stands outside both traditions. Gattei makes two important claims about the development of the philosophy of science in the twentieth century; that Kuhn is much closer to positivism than many have supposed, failing to solve the crisis of neopostivism, and that Popper, in responding to the deeper crisis of foundationalism that spans the whole of the Western philosophical tradition, ultimately shows what is untenable in Kuhn's view.

Gattei has written a very detailed and fine grained, yet accessible discussion making exceptionally interesting use of archive materials.

ASHGATE NEW CRITICAL THINKING IN PHILOSOPHY

The *Ashgate New Critical Thinking in Philosophy* series brings high quality research monograph publishing into focus for authors, the international library market, and student, academic and research readers. Headed by an international editorial advisory board of acclaimed scholars from across the philosophical spectrum, this monograph series presents cutting-edge research from established as well as exciting new authors in the field. Spanning the breadth of philosophy and related disciplinary and interdisciplinary perspectives *Ashgate New Critical Thinking in Philosophy* takes contemporary philosophical research into new directions and debate.

Thomas Kuhn's "Linguistic Turn" and the Legacy of Logical Empiricism

Incommensurability, Rationality and the Search for Truth

STEFANO GATTEI
University of Pisa, Italy

ASHGATE

Published by
Ashgate Publishing Limited
Gower House
Croft Road
Aldershot
Hampshire GU11 3HR
England

Ashgate Publishing Company
Suite 420
101 Cherry Street
Burlington, VT 05401-4405
USA

www.ashgate.com

British Library Cataloguing in Publication Data
Gattei, Stefano
Thomas Kuhn's "linguistic turn" and the legacy of logical empiricism : incommensurability, rationality and the search for truth. – (Ashgate new critical thinking in philosophy)
1. Kuhn, Thomas S. 2. Logical positivism
I. Title
146.4'2

Library of Congress Cataloging-in-Publication Data
Gattei, Stefano.
Thomas Kuhn's "linguistic turn" and the legacy of logical empiricism : incommensurability, rationality and the search for truth / Stefano Gattei.
p. cm. — (Ashgate new critical thinking in philosophy)
Includes bibliographical references.
ISBN 978-0-7546-6160-3 (hardcover : alk. paper)
1. Kuhn, Thomas S. 2. Popper, Karl Raimund, Sir, 1902–1994. 3. Science—Philosophy. 4. Science—History. 5. Logical positivism. I. Title.

Q175.G335 2008
501—dc22

2008006974

ISBN 978-0-7546-6160-3

Mixed Sources
Product group from well-managed forests and other controlled sources
www.fsc.org Cert no. SGS-COC-2482
© 1996 Forest Stewardship Council
FSC

Printed and bound in Great Britain by
TJ International Ltd, Padstow, Cornwall

For my father and mother,
with gratitude and love

Contents

Note

The initial motto is from Imre Lakatos, "Falsification and the Methodology of Scientific Research Programmes", in Imre Lakatos, Alan Musgrave (eds), *Criticism and the Growth of Knowledge*, Cambridge: Cambridge University Press, 1970, pp. 91–195: p. 93. The sources of the mottoes at the beginning of the preface and of each chapter are the following: Pierre Bayle, *Various Thoughts on the Occasion of a Comet*, translated and edited by Robert C. Bartlett, Albany: State University of New York, 2000 [1682, 1683^2], n. 22; Immanuel Kant, *Critique of Pure Reason*, translated by Norman Kemp Smith, London: Macmillan, 1929, 1933^2 [1781, 1787^2], A IX; Bertolt Brecht, *The Life of Galileo*, translated by John Willett, London: Methuen, 1980 [1940], section 9; Friedrich von Hardenbergh (Novalis), *Das allgemeine Brouillon*, in *Schriften*, vol. 3: *Das philosophische Werk II*, edited by Richard Samuel together with Hans-Joachim Mähl and Gerhard Schulz, Stuttgart: W. Kohlhammer Verlag, 1960 [1798-1799], n. 622; Jorge L. Borges, "La luna", in *El hacedor*, Buenos Aires: Emecé, 1960; Sextus Empiricus, *Outlines of Pyrrhonism*, translated by R.G. Bury, London-Cambridge, Massachusetts: Harvard University Press, 1933, Book I, chapter I, 1–2.

Preface

> [N]o single man is without the right to ask that he be listened to when he speaks in favour of his ideas, even if he were to be the only one to hold them, while making an allowance for those who will listen to him to defend themselves, not by prescription or by the prejudice of their number, but by an examination of the core of the matter.
>
> *Pierre Bayle*

From the epistemological point of view, the twentieth century was characterized by two quite different approaches to scientific methodology. On the one hand, in the first three decades of the century philosophers of science were chiefly concerned with logic and the philosophical analysis of language: science was regarded as paradigmatic of empirical knowledge and scientific language was correspondingly regarded as the characteristic element of any language purporting to describe the world. On the other hand, in the second half of the twentieth century the concern of the philosophy of science shifted considerably, differentiating itself from that of the philosophy of language. It got increasingly interested in the dynamics of theories, in the change of scientific categories and in the great intellectual revolutions, thus looking at the history of science as the acid test of rival methodologies.

This fact is extremely significant, not only from the purely philosophical point of view, but also from the wider cultural perspective. And while more than one philosopher contributed to this important shift of focus, Thomas Kuhn undoubtedly played a major role. From the historical point of view this mere fact makes Kuhn one of the most significant philosophers of the past century, and if we think of his influence on such diverse and far-away fields, our consideration of his contribution grows still further. Indeed, few philosophers of science have influenced as many readers as Kuhn: whether one agrees or disagrees with him, no one can deny that the key notions of his philosophy ("normal science", "revolution", or "incommensurability", for instance) and some of the terms he introduced (most notably, "paradigm" and its derivatives, such as "paradigm shift") have been at the very centre of the heated philosophical controversies which characterized the last decades of the past century. Kuhn's 1962 seminal work, *The Structure of Scientific Revolutions*, has become a modern classic, used (and misused) by diverse people in different contexts as the token in some ongoing disputes. Providing a common reference for cross-disciplinary discussions, it has affected debates across fields as different as historiography, sociology, politics, economics, psychology, theology, literature, feminism, cultural studies, art, education and more. Nearly half a century after the publication of *The Structure of Scientific Revolutions*, Kuhn's shadow hangs over almost every field of intellectual inquiry.

In the eyes of many, Kuhn's major result was to undermine a whole philosophical tradition, that of Logical Positivism (or Neo-Positivism, or Logical Empiricism). This is the received view – as we may well call it – of twentieth-century philosophy of science. However, in the past few years, a number of scholars have distanced themselves from such a view, deeming it reductive and, at best, partial. In particular, the assessment of both the differences and the elements of continuity between Kuhn and Logical Positivism has become quite controversial.

Orthodoxy, especially in the light of the enormous impact of Kuhn's ideas, presents us with a picture of a sharp break, of a thorough revolution. I do not think it was. The main thrust of the present work is that from many and often fundamental points of view Kuhn did not manage to break entirely with the preceding philosophical tradition: his works are laden with principles belonging to that very empiricist philosophy he was determined to reject. Furthermore, I shall argue that only a partial challenge of positivism and empiricism can actually account for the genesis of Kuhn's philosophical perspective – incommensurability, the notion of progress, the rejection of the concepts of truth and verisimilitude, and the very thesis of "world change" (one of the theses deemed most radical and characteristic of Kuhn's philosophical stance) are all consequences of the empiricist elements that his philosophy retains. Appearances to the contrary notwithstanding, the implicit presuppositions and the stated principles of Kuhn's philosophy are not very different from those of the logical positivists or logical empiricists he was determined to reject.

The crisis of Neopositivism betrays the deeper crisis of foundationalism, an approach that spans the whole of Western philosophical tradition. Kuhn was unable to offer a viable alternative: in spite of his attempts, the later phase of his philosophy and his vain efforts to finish his last book reveal a failure. By contrast, a concrete response to the collapse of the foundational approach was offered, I suggest, by Karl Popper. Far from being a mere "boundary" philosophy between logical positivists and new philosophers of science, only Popper's critical rationalism in its original and disruptive version (without the later emphasis on the positive role of corroborations for the growth of scientific knowledge, that is) constitutes a sound reaction to the crisis of foundationalism that characterizes philosophy in the past century.

Kuhn's contribution to the philosophy of science grows from his attempt to do history of science from a theoretical point of view. In so doing, he triggered a revolution. He said that revolutions are often started by outsiders, and his own career – that of "a physicist who became a historian for philosophical purposes" – represents a particularly interesting case. However, as Kuhn himself stressed, revolutions are not often total revisions of the system of beliefs from which they originate. Again, Kuhn's case is an exemplary one: the revolution he triggered retained many aspects of the logical empiricist tradition against which he wished to react. In order to find a viable response to the crisis of foundationalism of the twentieth century, we have to acknowledge Kuhn's results, realize the failure of his approach and move on, away from him.

Acknowledgments

This book is the outcome of several years of study and research on the incommensurability thesis and the issues connected to it. As for other works, I owe a great deal to many people for their help, support and encouragement. I have had the chance to mention specific contributions elsewhere. Here I would like to mention Karl Popper and Thomas Kuhn, who not only developed the ideas which form the basis of the present work, but also helped me, with unusual kindness and disposability, to deepen them during several years of intense critical exchange. Popper's thought, in particular, has been the focus of my researches for many years and informed my very approach to philosophy. In concluding this work, the memory of our long conversations and walks in Kenley come to my mind stronger than ever.

Particular gratitude I owe to my Ph.D. supervisor, Alexander Bird, both for offering me the opportunity to do my doctoral studies with him in Bristol and for overseeing my dissertation, on which this work is based. I thank Pierluigi Barrotta, Roberta Corvi, Donald Gillies, Malachi Hacohen, John Heilbron, Mark Notturno, John Preston, Andrew Pyle, William Shea and Ferdinando Vidoni: our conversations about several historical and philosophical topics related to this work have enriched it greatly. Special thanks to Joseph Agassi: I discussed with him nearly every issue I deal with in this work (and much more), and I feel I owe him more than I can express with words. A warm thank you also to Anthea Lockley, who carefully read the manuscript, making many stylistic suggestions.

Finally, my warmest thank you goes to my parents, who have helped and supported me throughout these years: without their care and generosity I would not have been able to write this. Without your love, mum and dad, I could not have been me: thank you, with all my heart.

List of Abbreviations

During my researches I had access to unpublished materials held in various archives and libraries: *The Karl Popper Archive* (Hoover Institution on War, Revolution and Peace, Stanford University), *Thomas S. Kuhn Papers* (MIT Institute Archives and Special Collections), *The Archive of Professor Imre Lakatos* (British Library of Political and Economic Science, London School of Economics and Political Science), *Nachlaß Paul K. Feyerabends* (Philosophisches Archiv, Universität Konstanz) and *Archives for Scientific Philosophy* (University of Pittsburgh), where I consulted the papers of Rudolf Carnap, Herbert Feigl, Otto Neurath, Frank P. Ramsey, Moritz Schlick and Ludwig Wittgenstein. I also had access to the personal and working libraries of Karl R. Popper, Thomas S. Kuhn, Imre Lakatos and Paul K. Feyerabend, which are kept together with the respective archives or held by their heirs.

References to some of Kuhn's works (unpublished papers, interviews or video-recording) are made by adding a prefix (U, I and V, respectively); as to the material from the archives of Popper and Lakatos, I will refer to every item by stating the archive and, within brackets, the corresponding folder and file, divided by a dot; in the case of Feyerabend, I shall indicate three numbers, separated by a dash; finally, to refer to items from the Archive for Scientific Philosophy of Pittsburgh University, I shall use the initials of the author, followed by a number. For example: Kuhn (U-1987), Popper Archive (120.11), Lakatos Archive (6.6), Feyerabend Archive (5-6-2), RC 082-03-01. In agreement with Kuhn's heirs and literary executors, I will not quote from his unpublished material.

All other references are to the bibliography at the end of this work. If an item is referred to with more than one number separated with a comma, the former indicates the first edition, while the latter the edition I am actually referring to or quoting from: for example, Kuhn (1962a, 1970). Otherwise, if the two numbers are separated with a slash, the latter refers to a translation, while the former to the first, original edition: for example, Galileo (1632/1953). All known English translations of foreign works I refer to are listed in the bibliography.

All English translations not explicitly mentioned in the bibliography are mine. I may have occasionally changed some of the translations I refer to for reasons of uniformity.

The clash between Popper and Kuhn is not about a mere technical point in epistemology. It concerns our central intellectual values, and has implications not only for theoretical physics but also for the underdeveloped social sciences and even for moral and political philosophy. If even in science there is no other way of judging a theory but by assessing the number, faith and vocal energy of its supporters, then this must be even more so in the social sciences: truth lies in power.

Imre Lakatos

Chapter 1

Two Revolutions in Twentieth-Century Philosophy of Science

> Her government, under the administration of the *dogmatists*, was at first *despotic*. But inasmuch as the legislation still bore traces of the ancient barbarism, her empire gradually through intestine wars gave way to complete anarchy; and the *sceptics*, a species of nomads, despising all settled modes of life, broke up from time to time all civil society.
>
> *Immanuel Kant*

The notion of incommensurability between scientific theories is one of the most controversial theses to have emerged during the epistemological debate in the twentieth century. The controversy dates back to 1962, when the incommensurability thesis was first advanced by its major advocates, Thomas S. Kuhn and Paul K. Feyerabend. However, despite usual references to this year, the transforming process within the philosophy of science had been under way for a long period.

Indeed, from the epistemological point of view, the past century witnessed two major revolutions, one in the 1920s and one in the early 1960s. In between them, the counter-revolution of critical rationalism. However, while the first revolution – that of Logical Positivism – aimed at re-establishing science in its role as reliable knowledge, after the progress made in mathematics and physics during the early decades of the twentieth century shook its foundations,[1] the second – that of the so-called "new philosophy of science" – had the effect of undermining the privileged position science had been occupying since Francis Bacon's time.

---

[1] I am thinking, in physics, of the birth of quantum mechanics (marked by the black-body radiation theory and the quantum discontinuity discovered by Max Planck in 1900, and developed in the following three decades by, among others, Albert Einstein, Niels Bohr, Werner Heisenberg, Wolfgang Pauli, Louis de Broglie, Paul Dirac and Erwin Schrödinger) and of Einstein's theory of relativity (1905 and 1916); in mathematics, of the "crisis of foundations" (officially opened in 1902 by Bertrand Russell's discovery of a fundamental antinomy in Cantor's set theory, which put an end to Gottlob Frege's ambitious programme); and, in logic, of Kurt Gödel's incompleteness and undecidability theorems (1930–1931). It is no accident that Karl Popper referred to the thick manuscript he was working on in the early 1930s as a "child of his time, a child of crisis – which is, above all, a crisis in *physics*. It affirms the *persistence of crisis* and, if it is right, crisis is the permanent condition of a highly developed rational science" (letter to Egon Friedell, 30 June 1932, quoted in Popper (1979, 1994), p. 443, n. 5).

The Idol of Certainty

From the seventeenth century onwards, until a few decades ago, science enjoyed the greatest intellectual authority as the best form of knowledge, and the highest social consideration as the most appropriate and reliable instrument for the solution of our problems and the cure of our diseases. As a consequence, philosophers had been engaged in inquiring into "*the reason why* science has to be regarded as the supreme and most reliable form of knowledge. *That* it was, it was never actually called into question."[2] In the 1960s, however, philosophers of science raised the problem in these very terms, "causing an unprecedented storm in a relatively well-sheltered region of philosophical reflection".[3]

Modern philosophy, following Bacon and Descartes, equated science and rationality. In the nineteenth century, such a view was reinforced by Positivism which, by acknowledging its certitude and incontrovertibility, granted science the hallmark of *episteme*. This view of solidity and linear progress was undermined by the discovery of non-Euclidean geometries and, later on, by the use Albert Einstein made of them in constructing his general theory of relativity. The turn thus imprinted in physics was tantamount to admitting that science is revisable – even if, as Joseph Agassi has noted, "it is not so much the occurrence of revolutions in science, the fact that science is in flux, that created the major change in the philosophical scene; rather, what has happened is that suddenly the fact that science is in flux ceased to be a secret".[4]

The early version of Positivism was proven wrong and the foundations of classical physics were shaken. In the early 1920s a group of philosophers and scientists undertook the task of winning back science's status, regaining its character of *episteme*. At the roots of their reflections on science they assumed classical empiricism and the tools provided by symbolic logic. The origins of this school as an organized philosophical movement can be traced to the roughly concurrent constitution, in Vienna, of the *Wiener Kreis* (Vienna Circle),[5] which grouped around Moritz Schlick,[6] and, in Berlin, of the *Gesellschaft für empirische Philosophie*

---

2 Pera (1984), p. vii.

3 Corvi (1992), p. 17. The storm actually did not concern science itself, rather, a certain view of science, a certain way to look at it that had propagated from Positivism onwards.

4 Agassi (1968), p. 12.

5 The name was invented and suggested by Otto Neurath.

6 On the basis of his (1917), which went through four editions between 1917 and 1922, and was enthusiastically endorsed by Einstein, Schlick was appointed the chair for the Philosophy of the Inductive Sciences (previously held by Ernst Mach and Ludwig Boltzmann, and which would be offered, decades later, to Karl Popper) at the University of Vienna in 1922. At first he organized a seminar for a restricted group of invited people, which constituted the original core of the Vienna Circle. In order to propagate a scientifically oriented philosophy, in 1928 the group founded the *Verein Ernst Mach* (Ernst Mach Association) – a name which emphasizes the great influence wielded on the newborn movement by the legacy of Machian philosophy, a real intellectual bridge between nineteenth-century positivism and the neo-positivism of the past century. Among the most regular participants in the discussions and initiatives of the group were Otto Neurath, a sociologist and economist, the physicist Philipp

(Society for Empirical Philosophy), fostered by the physicist and philosopher Hans Reichenbach.[7] Another fruitful connection was established with the Polish school of logic (Jan Łukasiewicz, Tadeusz Kotarbińsky, Kazimierz Ajdukiewicz, Alfred Tarski, Stanisław Leśniewski).[8]

The Circle aimed at forming an *Einheitwissenschaft*, that is, a "unified science", empirically connoted and comprising all the knowledge deriving from single scientific specialties.[9] Unification was to be gained by adopting a precise method, that of the

---

Frank, the mathematician Hans Hahn and the philosophers Herbert Feigl and Friedrich Waismann, both pupils of Schlick's. Later additions to the group included Victor Kraft, the mathematicians Karl Menger and Gustav Bergmann, and the logician Kurt Gödel. In 1926, due to Schlick's intervention, the philosopher and physicist Rudolf Carnap was appointed at the University of Vienna (from Prague) and joined the Circle, soon to become one of its leading members. See Barone (1953, 1986), Kraft (1950), Menger (1994) and Stadler (1995) and (1997).

7 In Berlin the scientific philosophical tradition of Ernst Mach and Richard Avenarius had been kept alive by Josef Petzoldt (a Machian empirio-criticist philosopher and Avenarius' pupil, editor of the eighth edition of Mach's *Science of Mechanics* (1921), the first to appear after Einstein's general theory of relativity), who founded the *Gesellschaft für positivistische Philosophie*, from which derived (upon the proposal of David Hilbert, one of its members), the *Gesellschaft für empirische Philosophie*: there met and worked physicians and psychologists under the supervision of Friedrich Kraus and Alexander Herzberg, together with champions of technology like August von Parseval. Later, they were joined by more philosophically minded thinkers, like Hans Reichenbach (though he was himself a student of science), who since 1926 held the chair of Philosophy of Physics at Berlin University. He was then assisted by his pupils Carl Gustav Hempel and Richard von Mises. Among the members of the Berlin School were also Kurt Grelling, Wolfgang Köhler, Kurt Lewin and Walter Dubislaw. See Barone (1953, 1986) and Vidoni (1993), pp. 147–157.

8 See Woleński (1989) and (1999) and Szaniawski (ed.) (1989). The new orientation inspired also English philosophers like Alfred J. Ayer, L. Susan Stebbing, Gilbert Ryle, John O. Wisdom and Richard B. Braithwaite, as well as Scandinavian philosophers like Jørgen Jørgensen, Eino Kaila, Arne Naess and Åke Petzäll. Vienna Circle members were also in touch with the Uppsala "empiricist" School, with the group of the Dutch philosopher and mathematician Gerrit Mannoury (1867–1956), who studied meaning and was the central figure in the Signific Circle (a Dutch counterpart of the Vienna Circle: see de Swart (ed.) (1988); after World War II, Popper also was in touch with the Significs: see his (1974a, 1976), p. 127), and with the school of logic Heinrich Scholz had established in Munich.

9 See the movement's manifesto: Hahn, Neurath, Carnap (1929). This pamphlet does not give an author's name on the title page. Indeed, it is the product of teamwork: Neurath did the writing, while Hahn and Carnap edited the text with him; other members of the Circle were asked for comments and contributions (see Feigl (1969) and Neider (1973), p. 49). See also Schlick (1930), (1934) and (1936). Beginning in 1930, scholars from Vienna and Berlin jointly published the Circle's official journal, *Erkenntnis* (whose previous name was *Annalen der Philosophie*), edited by Reichenbach and Carnap, which will be the movement's main propagation organ; in 1939 the journal changed its name into *The Journal of Unified Science*, and closed down in 1940. Also, Frank and Schlick edited a series called *Schriften zur wissenschaftlichen Weltauffassung*, in which appeared ten volumes between 1929 and 1937, mostly comprising works by Vienna Circle's members. In 1933 another series saw the light, edited by Neurath, Carnap, Frank and Hahn (whose place was taken by Jørgen

logical analysis of the assertions of the sciences (developed by Giuseppe Peano, Gottlob Frege, Alfred N. Whitehead and Bertrand Russell): such a logical analysis was the only one which was allegedly able to provide a real unification of the various sciences by showing their common logical-linguistic foundation. The outcome of the application of this method should have been twofold: on the one hand, it should have taken to the clarification and precise determination of concepts and theories of empirical sciences, besides the ultimate definition of the logical foundations of mathematics and logic;[10] on the other, to the elimination of metaphysics through the proof of the meaninglessness of its propositions and (alleged) problems.[11] It aimed not only at producing an autonomous philosophy of science, but an overall scientifically-based worldview,[12] in sharp contrast with the previous ones, which were theologically or metaphysically-based.[13]

As Peter Achinstein and Stephen Barker remarked in a volume devoted to its legacy, Logical Positivism "was a revolutionary force in philosophy, for it stigmatized metaphysical, theological and ethical pronouncements as devoid of cognitive meaning and advocated a radical reconstruction of philosophical thinking which should give pride of place to the methods of physical science and mathematical logic".[14] By combining the results of different traditions such as empiricism and formal logic, the neopositivists transformed philosophy of science into logic of science – that, not dealing any more with particular scientific theories or with their contents, is immune from the vicissitudes which trouble the scientific enterprise, and devotes itself only to defining the requirements which any scientific theory must meet. In so doing, the knowledge that looked shaky and wavering from the point of view of nineteenth-century positivistic canons, was secured to the twofold warrant of empirical verification and formal logic.[15]

Jørgensen in 1934, after his death), and from 1938 on also by Charles Morris, in which appeared six monographs; in the late 1930s it was replaced by the *Library of Unified Science*, in which only a few volumes appeared (among them, Richard von Mises' *Kleines Lehrbuch des Positivismus*, 1939). Important moments for the propagation of the new ideas were the International Congresses of Scientific Philosophy, held in Prague (1929 and 1934), Königsberg (1930), Paris (1935 and 1937), Copenhagen (1936), Cambridge (1938), Harvard University (1939) and Chicago (1941), though in a decidedly lesser tone. For a history of the movement, see Jørgensen (1951) and Stadler (1997).

10 It aimed at showing *das Gegebene*, that is, the immediately observable content.

11 In both directions the Vienna Circle carried on Mach's philosophical perspective. However, through the application of logical analysis, which characterizes the new empiricist (or positivistic) approach, it aimed at a completeness and precision which were utterly unknown to the old forms of empiricist (or positivistic) movements. See Carnap (1928a) and (1928b), together with Hahn, Neurath, Carnap (1929).

12 A real *Weltanschauung*, even if they preferred to call it *Weltauffassung* due to the metaphysical connotations that the former term had taken in the nineteenth century.

13 However, neopositivists actually developed mainly researches on scientific methodology.

14 Achinstein, Barker (eds) (1969), p. v.

15 For a more detailed account of the genesis and development of the Vienna Circle and Logical Positivism see Barone (1953, 1986), Friedman (1999), Hacohen (2000), Jørgensen (1951), Kraft (1950), Menger (1994), Neurath (1935b), and Stadler (1997).

Karl Popper, "Boundary" Philosopher Between Neopositivists and New Philosophers of Science

Having never been invited by Schlick to take part in their meetings, Karl R. Popper never became a member of the Vienna Circle. Nevertheless, he was a pupil of, got to know and had long exchanges with a number of its members.[16] Critical dialogue with Logical Positivism propelled Popper's revolution from the beginning: he used to work in virtual isolation, withdrawing into seclusion for lengthy periods, then reappeared to confront the Circle with new ideas. Circle members were, at intervals, a source of critical feedback that led him to crucial developments.[17] However, although they were a crucial context for his philosophy, Popper's differences with the Circle were significant: for not only did Popper work out innovative solutions to the problems dealt with by the Circle members, but his critique sprang from a marginal Kantian perspective foreign to Logical Positivism.[18]

As Malachi Hacohen has shown, it was Julius Kraft who introduced Popper to the unorthodox Kantian philosophy of Jacob F. Fries and Leonard Nelson:[19] their philosophies provided the background for Popper's solution of the foundation

---

[16] See Bartley (1969), (1970), (1974) and (1989); Hacohen (2000), ch. 5; Kraft (1974) and Popper (1974b), pp. 963–976. See also Popper (1974a, 1976), sections 16–17.

[17] See Hacohen (2000), chs. 4–6.

[18] For sure, logical positivists were also strongly influenced by Kant (see, for example, Coffa (1991) and Friedman (1999) and (2001)). But I am here referring to the Fries–Nelson tradition, transmitted to Popper primarily through Julius Kraft. The *Problemstellung* for Popper's epistemological revolution is set within the framework of this particular tradition. See Popper (1935, 1959), ch. V ("The Problem of the Empirical Basis"), especially section 29, and particularly p. 105, n. 3: "It seems to me that the view here upheld is closer to that of the 'critical' (Kantian) school of philosophy (perhaps in the form represented by Fries) than to positivism". On this issue, see Wettersten (1985) and (2005), Hacohen (2000), chs. 3 and 6, especially pp. 117–127, 220, 224–231 and 265, and Gattei (2005a), (2007), ch. II, and (forthcoming).

[19] In the following reconstruction I am following Hacohen (2000), ch. 3. Julius Kraft (1898–1960) came to Vienna in 1924, after completing a dissertation in Göttingen (under Nelson (1882–1927)) on the method of legal theory in Kant and Fries (1775–1843). The early discussions Popper had with him focused on the critique of Marxism and Social Democratic policies, and on Kant's epistemology, especially Fries' psychological critique of it (see Fries (1828–1831)). While they promptly reached agreement on politics, on Kant's epistemology and Fries' psychological procedure they disagreed. Nelson's influence on Popper was profound and he would refer to Fries and Nelson in all his early works: see Popper (2006) and (1979, 1994). Their philosophy was a point of departure to which he continuously returned to check his own developing views, first on the psychology of learning, then on the logic of science. In *Die beiden Grundprobleme der Erkenntnistheorie* (written in 1930–1933), Popper devoted the longest chapter to Fries' critique of Kant (see Popper (1979, 1994), ch. V). The radically shortened discussion of the empirical basis of science in *Logik der Forschung* still evinced Nelson's and Fries' role in the long-winded passage to the new philosophy (see Popper (1935), pp. 51–52). "Critical philosophy" set the problem-situation that enabled Popper to make his radical theoretical move, reformulate the question of the validity of knowledge and achieve his great breakthrough in the philosophy of science.

problem. But while Kraft accepted Fries' psychological critique of Kant and his alternative foundation for knowledge, Popper dismissed Fries' proposal as psychologistic and, by the early 1930s, disposed of foundationalism altogether.[20]

Fries, said Popper, was the first to notice the confusion of psychology and epistemology in Kant. Kant's transcendental proof showed synthetic propositions to be necessary *a priori*, but did not prove them valid: "He demonstrates the basic metaphysical statements of natural science through the possibility of experience. But this does not constitute ontological justification of a natural law. Rather, [it is] psychological justification of [...] human reason's need to presuppose laws' truth in order to regard appearances as unified in experience. The entire observation is correctly understood as psychic-anthropological".[21] The transcendental proof, Fries argued, had to be psychologically grounded, or it would be caught in a circular argument. Epistemology independent of psychology, he concluded, was impossible.[22] Fries' proposal grounded epistemology in psychology. In so doing, Nelson argued, he dispelled Kant's agnosticism concerning the "thing in itself" (*Ding-an-sich*), renewing the self-confidence of reason: immediate knowledge provided epistemology's foundation.

Popper regarded Fries' and Nelson's demonstration of endless regress in epistemology as impeccable and took it as the point of departure for epistemology. However, he thought that Fries and Nelson were wrong to assume that the task of epistemology was grounding knowledge: it was foundationalism, rather than endless regress, that made epistemology impossible. Epistemology, for Popper, was nothing but general scientific methodology: it did not justify statements, but offered rules, investigated methods and criticized procedures, pointing out contradictions and misapplications. As its subject matter was scientific practice, it required no foundation: it sought to clarify, criticize and improve practice. In Popper's eyes, Carnap and Neurath committed Fries' very same mistake: their protocols were

---

20 As a student, Nelson discovered the nearly forgotten Kantian philosopher Fries, who considered himself Kant's true successor. Fries formed a critique of Kant's transcendental proofs in epistemology, ethics, and religion: Kant held that certain propositions had an *a priori* validity because no conception of reality or morality was possible without them; Fries thought that these synthetic *a priori* propositions left too much of the world closed to the human mind, and, at the same time, ran the risk of subjectivism. He developed a methodological procedure for grounding knowledge in a universal human psychology, thereby eliminating much of Kant's agnosticism and "subjectivism". In his dissertation (1904), Nelson defended Fries against contemporary Neo-Kantians. His voluminous work in epistemology, ethics, and jurisprudence carried the imprint of Fries' "Kantianism with a greater confidence of reason" (Nelson spoke of the "Grundsatz des Selbstvertrauens der Vernunft", the "principle of the self-confidence of reason"): Popper rejected precisely this "confidence". He shared Fries' and Nelson's critique of Kant but declined their solution and offered his own: ever uncertain knowledge. Popper's arguments with Kraft over Fries and Nelson set the context for his epistemological revolution (see Popper (2006), (1979, 1994), ch. V, and (1962b)). On the early development of Popper's thought see Wettersten (1985), (1992) and (2005), Hacohen (2000), and Gattei (2004), (2005a), (2007), ch. II, and (forthcoming).

21 Fries (1828–1831), vol. I, p. xvii.

22 Fries (1828–1831), vol. I, pp. 21–30.

psychological reports in physicalist disguise.[23] Experience, or experiential language, can not directly exercise empirical control over science: science has no absolute empirical basis.

Therefore, Popper recast the basic problem of epistemology. All epistemologists, he said, confronted "Fries' trilemma". They could reconcile themselves to dogmatism, i.e. accept basic propositions without justification. Or they could admit infinite regress, whereby no statement would ever reach conclusive validation. Or else, they could opt for psychologism, justifying statements by appealing to "experience". With the exception of conventionalists, epistemologists had historically chosen, like Fries, psychologism: in order to avoid dogmatism, they called on experience, perceptions or immediate knowledge to justify statements. Popper dissented: observation and experiential reports are scientifically admissible only if they can be intersubjectively checked. Scientists' personal convictions have no epistemological significance: they could contribute to the discovery of a theory or explain subjective preference for it – but they cannot justify it. To retain objectivity, epistemology has to exclude psychologism.

Earlier, Popper had believed that infinite regress ended with verification or falsification of specific prognoses. In *Logik der Forschung* he recognized that even these prognoses – or, as he began calling them, singular (or basic) statements[24] – were theories of a lower degree of universality, testable hypotheses of their own. No foundation is required: dogmatism (i.e. the tentative and temporary acceptance of scientific statements), psychologism (i.e. scientists' subjective convictions, contributing to the consensus that ends the testing process) and endless regress all play a role in scientific work – but none constitutes a real threat to epistemology given science's hypothetical, falsifiable character.[25]

> The Kant–Fries critique reshaped Popper's epistemology, introducing issues and concepts that gained permanent hold on his philosophy: Fries' trilemma; exclusion of psychologism; theory as a system of statements; basic (singular) statement; empirical basis; methodological decision. He negotiated convention, experience, and logic, forming a unique synthesis of conventionalism and empiricism. He showed that convention and experience modified, rather than determined each other. Experience always remained problematic, but one could learn from it all the same.[26]

---

[23] Popper matured his views during the so-called "protocol sentences debate". The 1934 summer and fall issues of *Erkenntnis* carried a sharp exchange: see, in particular, Schlick (1934), Neurath (1934) and Hempel (1935), as well as the 1934 March–June correspondence between Carnap and Schlick, in The Rudolf Carnap Collection at the Archives for Scientific Philosophy of the University of Pittsburgh. Popper was initially oblivious to the debate. Neurath eventually directed him to the exchange, which provided him with new ammunition for his attack on positivist psychologism and subjectivism.

[24] The terminological change – which reflected a new view at the theoretical level – took place only in his (1979, 1994), ch. V, a section added to the original manuscript of the book when Popper, in all likelihood, was already working on *Logik der Forschung*: see Wettersten (1985) and (1992), ch. 8.

[25] See Popper (1979, 1994), pp. 107–136.

[26] Hacohen (2000), p. 233.

Science is perpetually in flux, change is its core characteristic. It progresses not by discovering unshakable truths, but by eliminating errors. The sole guarantor of progress is intersubjectivity, Kant's substitute for objectivity. Basic statements, a temporary end-point to testing, constitute science's relative, transitional, conventional "foundation" – the only foundation science needs.

> The empirical basis of objective science is *nothing absolute.* Science does not rest on a bedrock. Its towering edifice, an amazingly bold structure of theories, rises over a swamp. The foundations are piers going down into the swamp from above. They do not reach a natural base, but go only as deep as is necessary to carry the structure. One does not stop driving them down because one reached firm ground. Rather, one resolves to be satisfied with their firmness, hoping they will carry the structure.[27] (If the structure proves too heavy, and begins tottering, it sometimes does not help to drive the piers further down. It may be necessary to have a new building, which must be constructed on the ruins of the collapsed structure's piers). [...] *The objectivity of science can be bought only at the cost of relativity.* (He who seeks the absolute must seek it in the subjective).[28]

In contrast with other epistemologists and particularly the logical positivists, Popper gave up the idea that justification is a necessary condition for scientific knowledge: our scientific knowledge cannot and need not be justified.[29] Quite differently from preceding philosophies, his critical rationalism emphasized the role of trials, as to how science grows, and criticism, as to the way in which its assertions are tested. Popper himself described this process by saying that knowledge evolves through a succession of conjectures and refutations, of attempts to solve problems together with careful and thorough tests.

There is no method of discovering true theories (a recurrent illusion in Western philosophy: Plato, Aristotle, Francis Bacon, René Descartes and John Stuart Mill, to mention but a few), nor – a weakened version of this illusion – can we ascertain the truth of a scientific hypothesis: we can never verify it. Nor (a still weaker version) can we ascertain whether a hypothesis is probable, or probably true.[30] Nevertheless,

---

27 See also Popper (1935, 1959), p. 111.

28 Popper (1979, 1994), p. 136.

29 For Popper, Logical Positivism grounded science in perceptions and experiences: Schlick (1934) was a perfect example. Placing a premium on the scientist's feeling of certainty, Schlick recapitulated Fries' "immediate knowledge". Such "knowledge" was irrelevant to science. "We must distinguish between, on the one hand, *our subjective experiences or our feelings of conviction*, which can never justify any statement (though they can be made the subject of psychological investigation) and, on the other hand, the *objective logical relations* subsisting among the various systems of scientific statements, and within each of them" (Popper (1935, 1959), p. 44). Carnap and Neurath made an unsuccessful attempt to overcome the gap between psychology and logic by translating psychological behavior into physicalist language: whether phenomenalist or physicalist, their protocols were logical construction of experience, "perception statements", records of sense data, translation of observations into formal speech. They gained nothing by changing mode of expression. They remained attached to the psychological basis (see Popper (1979, 1994), pp. 429–432 and 438–439, and (1935, 1959), pp. 95–97.

30 See Popper (1983), p. 6; see also his (1935, 1959), p. 32.

our knowledge is in a way "objective", since it can provide proofs of the theory's falsity and means to learn from our errors. Growth of knowledge and criticism are closely interconnected: according to Popper, we should prefer the theoretical system that, at any given state of critical discussion, accomplishes a growth of the possibly corroborated (that is, which has survived sincere attempts of refutation) empirical content. There is no inductive process through which theories can be confirmed: within Popper's philosophy of science there is no place for any theory of justification, as logical positivists thought. Therefore, Popper's anti-inductivism exposes, in the first place, the myth of foundationalism and of the first (or ultimate) elements on the basis of which we can allegedly construct, or reconstruct, the world. On the other hand, pride of place is given to metaphysics that Popper, contrary to logical positivists, refuses to reject as meaningless and rehabilitates as part and parcel of scientific research. It does not matter whether metaphysics is not empirically testable: it must be taken into consideration as far as a theory can be rationally criticized. In other words, we have to look for its fruitfulness, its ability to solve problems, to shed new light upon them and to set new ones.[31]

The overthrow of the empiricist position is complete: Popper replaces the view (developed from Bacon onwards) of mind as a *tabula rasa*, a sort of empty bucket to be filled with the contents of experience, with the theory (of Kantian origin) of mind as a searchlight that sheds its beams (hypotheses, theories, expectations) in the attempt to grasp reality more and more clearly. The primacy logical positivists attributed to observation data Popper confers to theory.

After Popper, the so-called "new philosophy of science" – particularly Kuhn and Feyerabend – would take this view to its extreme consequences, transforming the primacy of theory into the domination of theory.[32] In this sense, critical rationalism lays itself open to a dangerous and self-destroying virus. And it is exactly at this point that critical rationalism displays its complex nature of boundary epistemology. Falsifiability needs facts: a scientific theory is tested by exhibiting a fact that clashes with one of its logical consequences. Elsewhere, critical rationalism presupposes the theoretical character of observations, and this rules out the independence and

---

[31] See also below, ch. 2, n. 154.

[32] In his *magnum opus* Popper underlines how observation is actually "theory laden" and notices that it is impossible to refute an empirical theory in any final and unquestionable way: "Theory dominates the experimental work from its initial planning up to the finishing touches in the laboratory" (Popper (1935, 1959), p. 107). And in a footnote added to the English edition, he continues: "observations, and even more so observation statements and statements of experimental results, are always *interpretations* of the facts observed; [...] they are *interpretations in the light of theories*" (*ibidem*, p. 107, n*3: this is the reason why, Popper continues, it is always "deceptively easy" to find verifications of a theory). Moreover, "our ordinary language is full of theories; [...] observation is always *observation in the light of* theories; and [...] it is only the inductivist prejudice which leads people to think that there could be a phenomenal language, free of theories, and distinguishable from a 'theoretical language' [...]" (*ibidem*, p. 59, n*1). See also pp. 42, 50 and n*1, 81–87, 106–107, 280, 412–413 and 423 of the same book, together with Popper (1945, 1966), vol. II, pp. 213–214 and 260–261, (1963a, 1989), p. 387 (where Popper speaks of facts as "soaked in theory").

autonomy of facts: a scientific fact is always laden with some theory.[33] Therefore, on the one hand, critical rationalism requires the distinction between observation language and theoretical language while, on the other, it rules it out.

Disagreements within the Popperian group were stirred up also by philosophers outside it. Thus grew a dissent that led, in the late 1950s and early 1960s, to the second revolution in twentieth-century philosophy of science. In the eyes of many, Popper was therefore a "transitional figure"[34] between Logical Positivism in the 1920s and the "new philosophy of science" in the 1960s, since he shared something with both of them.[35] Moreover, even those who disagree with his views have to concede that one of the hallmarks of the fruitfulness of Popper's conceptual framework resides in the fact that many philosophers of science who contrasted the "Standard View" either grew within it or affirmed themselves against it.[36] Thus Lakatos writes: "I think meeting Karl [Popper] has been a critical turning-point in my life and probably the best one. I suppose I learnt from him more than from anybody else in my life, in fact infinitely more than from anybody else. The problems which I inherited from him or whose solution he inspired will give me work for a lifetime".[37] And Feyerabend himself, the methodological anarchist and Popper's most recalcitrant pupil (as he liked to portray himself), more than once acknowledged that his own philosophy, or at least its premises, is rooted in Popper's own: "I for one am not aware of having produced a single idea that is not already contained in the realistic tradition and especially in Professor Popper's account of it".[38]

---

[33] Indeed, for his (1935) Popper chooses a motto from the fifth *Dialogue* (1798) of Friedrich von Hardenberg, alias Novalis: "Theories [from 1992 onwards, Hypotheses] are nets: only he who casts will catch" (in Novalis (1960), p. 668). It is interesting to read also the rest of the passage Popper is quoting from, which continues: "Hasn't America been discovered with a hypothesis? / Long live hypothesis – only she remains / eternally new, though it often defeats itself" (*ibidem*).

[34] Brown (1977), p. 67.

[35] See Pera (1981), p. 3.

[36] Kuhn is no exception. Despite the fact that his name is only occasionally mentioned, Kuhn wrote *The Structure of Scientific Revolutions* having Popper's model in mind: indeed, the philosophical conclusions Kuhn draws from his historical account clearly, albeit implicitly, refer to Popper, whose William James Lectures Kuhn attended at Harvard, in 1950. See Kuhn (1962a), p. 77.

[37] Letter to Victor Kraft, 16 June 1964, in Lakatos (12.4), item 94. And a decade later, at the beginning of his critical contribution to the Popper volumes in the Library of Living Philosophers, he writes: "Popper's ideas represent the most important development in the philosophy of twentieth century; an achievement in the tradition – and on the level – of Hume, Kant, or Whewell. Personally, my debt to him is immeasurable: more than anyone else, he changed my life. His philosophy [...] provided me with an immensely fertile range of problems, indeed, with a veritable research programme" (Lakatos (1974), p. 241). See also Lakatos (1978c), p. 222.

[38] Feyerabend (1965c), p. 251, n. 1; see also his (1961a). Feyerabend acknowledges also another debt to Popper, who already in 1949 had highlighted the unavoidable connection between observations and expectation horizons (in Popper (1949)). Even before him, Michael Polanyi had contemplated the "theory-ladenness" thesis in the context of the socio-psychological process of discovery, something which Popper always refused to take into

The American Adventure of Logical Positivism

The "analytic adventure" gave rise to a true "epistemological fracture in the body of American philosophy, a clean break that divided its history into two parts":[39] for at least two decades the Vienna Circle's approach to philosophical problems became the standard approach, shaping American philosophy. Willard Van Orman Quine played an important role in this process: he arrived in Vienna in 1932, shortly after receiving his Ph.D. at Harvard University, where he had studied under Clarence Irving Lewis and Alfred North Whitehead.[40] Quine reached Vienna when the continental adventure of the Vienna Circle, ten years after it was born (from the discussion between Schlick and Reichenbach on the philosophical meaning of Einstein's theory of relativity), was perhaps at its peak. In 1936, the year of Schlick's death,[41] there

consideration (see especially Polanyi (1946); see also Jacobs (2002) and (2003)). Feyerabend will remain very close to Popper till the end of the 1960s (see, for example, his (1965d) and their estrangement became irreversible only at the beginning of the 1970s. It is otherwise Lakatos himself who realizes that Feyerabend "contributed probably more than anybody else to the spread of Popper's ideas" (Lakatos (1970), p. 115, n. 3). And in his (1962a), which is generally regarded a sharp dissociation from Popper's philosophy, Feyerabend explicitly acknowledges how his harsh attack on the empiricism of Ernest Nagel, Carl Gustav Hempel and Paul Oppenheim took as its starting point Popper's ideas, and particularly Popper (1949), a lecture Feyerabend had attended in Alpbach in 1948, on the occasion of their first meeting (see Feyerabend (1962a), p. 91, n. 95). A small anecdote: years later, Popper republished the 1948 lecture as an appendix to his *Objective Knowledge*; this book, however, already comprised a revised version of that very lecture as ch. 5 (Popper (1957a)): there were no apparent reasons to include it, then. However, in a note he records that the lecture was delivered in Alpbach 1948 (see Popper (1949, 1972a), p. 341), a coded way of saying that Feyerabend attended it; and in the "Bibliographical Note" appended to the revised edition of that lecture (see (1957a, 1972a), pp. 204–205) he highlights Feyerabend's debt, adding that "Feyerabend's acknowledgement seems to have been overlooked by the authors of various papers on related subjects (*ibidem*, p. 205). To this remark, Feyerabend would harshly reply in 1980, when the wound that divided them had become incurable, upon the republication of his "Explanation, Reduction, and Empiricism": see Feyerabend (1962a, 1981a), p. 47, n. 6 (where Lakatos is also criticized for "his usual propagandistic flair"). See also Watkins (2000), p. 48.

[39] Borradori (1991/1994), p. 5.

[40] The American philosophical environment within which Quine had been trained was quite different from the European one, that had been transformed by the impact of the logical empiricist movement. American philosophy was still moving in the wake of Charles Sanders Peirce's and William James' pragmatism, later developed by John Dewey in the context of New Deal. At any rate, there began to surface a new interest for logic, particularly for its foundational role in mathematics (following the trend inaugurated in Germany by Gottlob Frege and Georg Cantor, a mathematician of Russian origin, which was eventually systematized in *Principia Mathematica* by Bertrand Russell and Alfred N. Whitehead, who had moved to England in 1924). In fact, Quine was totally devoted to mathematical logic, following the interest of his own teacher, Lewis, a leading exponent of the logicistic wing within pragmatism that contributed to the diffusion of *Principia Mathematica* and created the conditions for a fruitful interchange with European Logical Positivism.

[41] Schlick was murdered by his former pupil Johann Nelböck, a Nazi student whose thesis (in ethics) Schlick had examined. Upon shooting his former teacher on the flight of

was the diaspora of the last members of the Vienna Circle to remain in Vienna: as a consequence of political and racial persecutions effected by Nazis in Europe, in the mid-1930s a sizeable fraction of middle-European philosophy emigrated to the United States, establishing itself in the major overseas universities.[42]

---

steps leading to the Department of Philosophy at the University of Vienna, Nelböck was sentenced to life imprisonment, but was released when the Nazi troops occupied Vienna, in 1938 (*Anschluß*). In 1941 he applied successfully for a full acquittal, claiming to have done society a good service by killing a Jewish professor (actually, Schlick was not a Jew, but a descendant from Prussian aristocracy). He died in 1954. See Haller (1995).

[42] Several members of the Vienna Circle were Jews (or of Jewish origin), or anyway supported liberal and socialist ideas that would have been hardly tolerated in the new regime. Moreover, the Circle's "scientific conception of the world" constituted a threat for the highly pseudoscientific racial theories at the heart of Nazi propaganda. Therefore, simultaneous to the progressive spreading and establishment of Logical Empiricism at an international level, there gradually took place also the dissolution of the Vienna Circle as a result of the expatriation of most of its members due to the impending Nazi persecutions (see Stadler (1995) and Dahms (1995)). However, what was a tragic loss for the Austrian culture and civilization, and more generally for the German-speaking world, turned out to be at the same time a great breakthrough for the English-speaking world. Philosophers, scientists and mathematicians in exile established themselves in England, the Commonwealth and the United States of America, where they exerted a huge influence on the later developments of the philosophy of science (see Feigl (1969) and Hull (1995)). Such a process was aided and fostered by a number of English and American philosophers who, after studying in Austria in close contact with the Vienna Circle, had then gone back home. This was the itinerary followed by Quine, who helped many members of the Circle to reach Harvard University, as did Alonzo Church at Princeton University: these two colleges, together with the younger ones of Berkeley, in California, and Pittsburgh, in Pennsylvania, are still the major propulsive centres of analytic philosophy. On the opposite side of the ocean, Alfred J. Ayer spread the Vienna Circle's ideas in England with his (1936), that gained a tremendous success (worth noting are also the criticisms and further elaborations to which Ayer subjected his earlier ideas in the introduction to the second edition of the book, published in 1946). Fundamental were also the contributions of two Cambridge philosophers, Ludwig Wittgenstein and Frank P. Ramsey. Several members of the Vienna Circle (especially Waismann and Schlick: see Waismann (1967) and Wittgenstein, Waismann (2003)) regarded Wittgenstein's philosophy as a crucial turning point and the *Tractatus Logico-Philosophicus* was carefully read and commented during the Circle's meetings. Indeed, Wittgenstein was named as one of the three "leading representatives of the scientific world-conception" in the movement's manifesto (see Hahn, Neurath, Carnap (1929/1973), p. 318); the other two were Albert Einstein and Bertrand Russell (another Cambridge philosopher who was very close to Logical Positivism, at least in its early stages, and also took an active part in the 1935 Paris Congress). Frank P. Ramsey – "a meteor in the philosophical sky", as Jérôme Dokic and Pascal Engel portray him (see their (2001/2002), p. 1) – was one of the most remarkable productive minds of his generation. Despite his young age (he died suddenly when he was not yet 27 years old) his production ranged over most of the domains which were at the centre of intellectual interest in Cambridge at that time: mathematics, logic, ethics, economics and philosophy. His thought had a profound influence on Wittgenstein (contributing to the birth of his *Philosophical Investigations*: see Wittgenstein (1953), p. x) and he was one of the main interlocutors also of Russell, Moore and Keynes. Most importantly, his reflections proved fundamental for what was not yet known as analytic philosophy, significantly contributing to its rise during the first half of the twentieth

From this first journey of Quine's to Europe the history of the Vienna Circle has never ceased to intersect with that of American philosophy. Indeed, such a journey gave birth to the tradition of analytic philosophy, a term with which scholars usually refer to that particular area of philosophy, characterized by logical and semantic approaches, grown in the shadow of this migratory wave. Quine, in other words, played a major role in transplanting the new, logically-empirically oriented philosophy of the Vienna Circle from the then jeopardized European continent into the fertile American soil.

In England the situation was much different: "when I began lecturing at Oxford during 1950", Stephen Toulmin recalls, "there was nothing one could call an orthodoxy in the philosophy of science in Britain".[43] Not so in the United States: here philosophy of science underwent the sudden and decisive influence of Logical Positivism.[44] It was, in particular, the influence of Carl Gustav Hempel and Herbert Feigl, Rudolf Carnap and Hans Reichenbach, Johann von Neumann and Philipp Frank – all of whom were scholars who had greatly admired Ludwig Wittgenstein's *Tractatus Logico-Philosophicus*, but were dissatisfied by Wittgenstein's later works. They developed a general method in philosophy which linked the logical techniques developed by Russell and Whitehead in *Principia Mathematica* to an empiricist epistemology of Machian origin. The exponents of this line of thought located in the formal rigour and empirical foundations of natural science (the "deductive validity" of its inferences and the "verifiability", typically inductive, of its assertions) the touchstone to assess the adequacy of intellectual activity in any other area as well. Their image of natural science reflected two factors: on the one hand, they came to the philosophy of science mainly from the philosophy of mathematics and formal logic; on the other, theirs was a long-term programme that aimed at the construction of a single exhaustive axiomatic system, at whose heart was Russell's mathematical analysis, which could represent (at least in principle, and maybe with the help of some additional axiom) the totality of our scientific knowledge. This attitude gave birth to the "Unified Science" movement of the 1930s.[45]

---

century: see Mellor (ed.) (1980) and Sahlin (1990) and (1997), together with Ramsey (1931), (1978), (1990). See also Galison (1993), (1995) and (1996), together with Stadler (1997) and Stadler, Webel (eds) (1995).

43 Toulmin (1977), p. 145. "The 1960s" – writes Toulmin at the opening of this essay – "were a time of striking changes throughout the intellectual and artistic worlds. None of those who grew up and entered academic life during the preceding thirty years could live through that decade without feeling that the landmarks of his mental world were being eroded, shifted, or even swept away" (*ibidem*, p. 143). The academic world in which Toulmin himself was educated was actually marked by a rigid division of disciplines and university departments that left no room for an interdisciplinary and cross-boundary work among different intellectual fields.

44 "In 1960 logical empiricism was *the* Anglo-American philosophy of science": Giere (1988), p. 22.

45 "Unified science", that is, as Charles Morris defines it, "the scientific study of the scientific enterprise in its totality" (Morris (1938), p. 74). Such a movement later developed an organizational form in a series of annual international congresses, programmed during the first of them, held in Prague in 1934. In particular, the third of these congresses (held in

The meeting with American pragmatism: scientific empiricism

Philosophers who emigrated from the German-speaking world found their natural allies among the young American pragmatists who, like Ernest Nagel, had a particularly formalistic mind. For both groups of scholars each authentic problem in the philosophy of science had to be confronted according to its logical structure, rather than according to the psychology of scientific discovery, or the historical evolution of scientific concepts. "Indeed, the one point of most general agreement among them all", as Toulmin again notes, "was probably the acceptance of Hans Reichenbach's distinction between the 'context of discovery' and the 'context of justification'".[46] The process of scientific discovery, comprising both the psychological vicissitudes of single scientists and the collective history of scientific groups, had to be the subject of social and behavioural sciences, having nothing to do with the philosophy of science: "that could begin only when questions of justification arose, that is, when scientific work provided material to formulate an explicit *argument*, whose validity, evidential force, and cogency could be exposed to logical scrutiny".[47] The central interests were consequently directed towards the calculus of probability, confirmation (or validation) theory and other technical procedures concerning the analysis of formal relations between scientific statements,[48] while scientific concepts, in conformity of which these very statements were structured, were taken for granted. Another major characteristic of this tradition was the general prohibition to mistake formal or logical conclusions for empirical data.[49]

To sum up, then, there were three basic beliefs: in the first place, a careful analytical examination of arguments emerging from scientific justification highlights how natural science works following a characteristic method. Second, the fundamental procedures of this method can be expressed by formal algorithms that map empirical observations to theoretical propositions, in the terms of which the former have to be explained. Third and last, the rationality of the natural sciences resides in their conforming to this set of formal procedures. All this accomplished, on the one hand, the construction of a real *organon* in the Baconian sense and, on the other, the exclusion of those elements which affected the rational structure of science, that is, the *idola* of history, psychology and sociology. These two components united in

---

Paris, in 1937), took the shape of a conference devoted to the project of the *International Encyclopedia of Unified Science*.

[46] Toulmin (1977), p. 146.

[47] Toulmin (1977), p. 146.

[48] Classical instances of this approach to metascience are Hempel (1952) and (1965a), Carnap (1950a) and Nagel (1961). The two volumes edited by Neurath, Carnap and Morris in 1955 and 1970 include all the essays of the original *Encyclopedia of Unified Science*. These works, together with others, discuss themes and perspectives later developed in the subsequent three decades – that is, the theoretical basis of the so-called "Standard View" (or "Received View") of the philosophy of science in the English-speaking world effectively epitomized in Nagel (1961): see, for example, Suppe (1974) and (1977).

[49] Toulmin reads in this rigid distinction the prohibition to mistake "rational" conclusions for "non-rational" or "irrational" material.

the ambition to prove the essential rationality of scientific method and provide an analytic description of this rationality in algorithmic terms.

The project for the International Encyclopedia of Unified Science

The meeting between logical-empiricist themes and pragmatism's various currents, differently represented by Charles S. Peirce, William James, Percy W. Bridgman, John Dewey, George H. Mead and Charles Morris, proved extremely fruitful. Due particularly to Morris, it gave rise to a new philosophical trend, "American scientific empiricism" (the expression, coined by Morris himself,[50] simply designates the movement arisen out of the fusion of European logical neo-empiricism and American pragmatism), and to a new expression of the unity of the sciences, namely, the *International Encyclopedia of Unified Science*, set up by Carnap, Frank, Neurath, Morris, Jørgensen and Rougier in Chicago, in 1938.[51]

American scientific empiricism, which saw in Morris' semiotics its first (and perhaps also single) programmatic and originally synthetic wording, and in the *Encyclopedia* the boldest attempt at its radical realization,[52] presents itself both as the last product of the movement of ideas and researches that had problematically grown in Europe through the experiences of the Vienna Circle and the Berlin School, and as the new expression of the autonomous American pragmatist tradition.

---

50 In his (1937).

51 With a decidedly programmatic purpose, Neurath opened the *Encyclopedia* with these words: "Unified science became historically the subject of this *Encyclopedia* as a result of the efforts of the unity of science movement, which includes scientists and persons interested in science who are conscious of the importance of a universal scientific attitude" (Neurath (1938), p. 1); on Neurath's project, see his (1935c), (1936a), (1936b), (1937a), (1937b), (1938) and (1946). See also Reisch (1994) and (1995).

52 "The resulting comprehensive point of view, embracing at once radical empiricism, methodological rationalism, and critical pragmatism, may appropriately be called scientific empiricism. […] It is an empiricism genuinely oriented around the methods and the results of science and not dependent upon some questionable psychological theory as to the 'mental' nature of experience. It is an empiricism which, because of this orientation and the use of powerful tools of logical analysis, has become positive in temper and co-operative in attitude and is no longer condemned to the negative skeptical task of showing defects in the methods and results of its opponents. Such a point of view, characteristic in the main of this *Encyclopedia* […] signalizes the widest possible generalization of scientific method. The field of application of this point of view is science itself" (Morris (1938), pp. 68–69).

Consisting in a series of monographs, each "devoted to a particular group of problems"[53] and supplied with a "highly analytical index",[54] the *Encyclopedia* was not meant to construct *the* system of science – rather, it aimed at integrating methods and contents of particular sciences. "The *International Encyclopedia of Unified Science*", writes Neurath in the very first monograph, significantly titled "Unified Science as Encyclopedic Integration", "aims to show how various scientific activities such as observation, experimentation, and reasoning can be synthesized, and how all these together help to evolve unified science. These efforts to synthesize and systematize wherever possible are not directed at creating *the* system of science; this *Encyclopedia* continues the work of the famous French *Encyclopédie* in this and other respects".[55] The idea is just to create the basis for an international cooperation among philosophers and scientists: "The maximum of co-operation – that is the program!".[56]

---

[53] Neurath (1938), p. 24. "The collaborators and organizers of this work are concerned with the analysis and interrelation of central scientific ideas, with all problems dealing with the analysis of sciences, and with the sense in which science forms a unified encyclopedical whole. The new *Encyclopedia* so aims to integrate the scientific disciplines, so to unify them, so to dovetail them together, that advances in one will bring about advances in the other. The *Encyclopedia* is to be constructed like an onion. The heart of this onion is formed by twenty pamphlets which constitute two introductory volumes. These volumes, entitled *Foundations of the Unity of Science*, are completed in themselves but also serve as the introduction to what will follow. The first 'layer' of the onion which will inclose this 'heart', consisting of the first two volumes, is planned as a series of volumes which deal with the problem of systematization in special sciences and in unified science – including logic, mathematics, theory of signs, linguistics, history and sociology of science, classification of sciences, and educational implications of the scientific attitude. [...] The following 'layers' may deal with more specialized problems; the interests of the reader and the collaborators in the particular problems will lead the members of the Committee of the Organization and the Advisory Committee to consider various possible lines of development" (*ibidem*, pp. 24–25).

[54] Neurath (1938), p. 24.

[55] Neurath (1938), p. 2. Morris continues: "The *Encyclopedia* presents a contemporary version of the ancient encyclopedic ideal of Aristotle, the Scholastics, Leibniz, the Encyclopedists, and Comte" (Morris (1938), p. 75). Neurath explicitly refers to Jean Baptiste d'Alembert's "Discours préliminaire": he intended the new *Encyclopedia* to have the same historical impact as Diderot's and d'Alembert's – however, all the two projects shared were the difficulties in the process of realization.

[56] Neurath (1938), p. 24. In a letter of 1935 (see Morris (1960)) Neurath wrote he was at work on the project of the *Encyclopedia* as least as early as 1920, and that he first talked it over with Einstein, Hahn, Carnap and Frank. The idea of "Encyclopedism based on logical empiricism" (Neurath (1938), p. 24) was set up by Neurath in "An International Encyclopedia of Unified Science", a paper read at the First International Congress for the Unity of Science (Paris, 1935). It was then supported and fostered by the members of the Encyclopedia Committee of Organization (Carnap, Frank, Jørgensen, Morris, Rougier and Neurath himself), who on the same occasion spoke about the problems, the importance and the logical basis of the project. In 1936 Neurath established *The Institute for the Unity of Science*, as a section of the *Mundaneum Institut* he had founded in The Hague in 1934; it was also set up an Organization Committee of the International Congresses for the Unity of

As years went by, the influence of Logical Positivism on the American philosophical environment gradually dissolved: subject to a slow but unremitting criticism from within, due to drastic revisions of its original theses by its very advocates, beginning from the 1950s it was heavily tackled also by philosophers outside the movement, whose criticism affected the very heart of their programme, the philosophy of science.

In his 1967 "Logical Positivism" entry to *The Encyclopedia of Philosophy* John Passmore announced the "death" of the movement, at least in its original form.[57] Among the most significant symptoms of the crisis was the end of publications (in 1969) of the series of monographs of the *International Encyclopedia of Unified Science*, which Neurath, together with Carnap and Morris, had started in 1938.[58] However, already in the mid-1940s the Second World War and the sudden death of Neurath (1945) led to the relinquishment of the original project, that contemplated some twenty-six volumes, each comprising ten monographs plus a pictorial companion in the *Visual* (or *Picture* or *Isotype*) *Thesaurus*. Neurath had contemplated English, German and French editions of the work; conceiving it as genuinely international in scope, he thought of contributors from Asiatic countries as well as from the West. The first volume (in two books) of the first introductory part saw the light only in 1955; the second one, whose publication was continuously postponed, appeared in 1970. In the "Preface" to the two-volume 1970 edition Carnap and Morris conclude that "There are no plans at this time to proceed further with the *International Encyclopedia of Unified Science*, to which these monographs were intended to be the introduction".[59]

Science, composed of the same person plus L. Susan Stebbing. See Barone (1953, 1986), pp. 350–354, Jørgensen (1951), pp. 890–891, and Morris (1960).

[57] "Logical Positivism, considered as the doctrine of a sect, has disintegrated. In various ways it has been absorbed into the international movement of contemporary empiricism, within which the disputes which divided it are still being fought out. [...] Even among those philosophers who would still wish to make the contrasts on which the logical positivists insisted, few would believe that they can be made with the sharpness or the ease which the logical positivists at first suggested. Logical positivism, then, is dead, or as dead as a philosophical movement ever becomes" (Passmore (1967)), p. 56. Frederick Suppe titled one of the sections of his (1977) "Swan Song for Positivism" (pp. 619–632).

[58] See Neurath, Bohr, Dewey, Russell, Carnap, Morris (1938).

[59] In Neurath, Carnap, Morris (eds) (1955, 1970), p. vii. In a 1960 survey article Morris writes: "The *Encyclopedia of Unified Science*, though now only a fragment of what had been planned, has had historical significance. The monographs are still very much alive. The movement of which the *Encyclopedia* was a part continues to develop vigorously in its own way. The Institute for the Unity of Science continues its activity in the United States under the leadership of Philipp Frank. Whether the larger plans for the *Encyclopedia* are ever to be resumed is a problem for another generation" (Morris (1960), p. 521). The Institute for the Unity of Science undertook the editorship of the *Encyclopedia* in 1949, with Philipp Frank; the last editor was Herbert Feigl. The possessions and estate of the Institute (together with its hopes) were transferred to the Philosophy of Science Association.

The Structure of Scientific Revolutions appears as the last monograph but one;[60] and if we consider the second edition of Kuhn's book (1970), containing an important postscript, it is actually the *Encyclopedia*'s last substantial entry. The presence of Kuhn's seminal work in the most ambitious project of Logical Positivism can be explained by the fact that Kuhn was actually a physicist who lent himself to history and philosophy of science, something which was very much in tune with logical positivists' scientific-scientistic views.

The Revolt against Empiricism

While the protagonists of the first revolution in the philosophy of science of the twentieth century constituted, as we have seen, a compact and close-knit group, the protagonists of the second revolution do not form a unitary school of thought, either in terms of education or interests. Thomas Kuhn, Norwood Russell Hanson, Michael Polanyi, Stephen Toulmin and Paul Feyerabend make up a quite heterogeneous group. Nevertheless, however pronounced the differences, it is possible to identify a shared attitude – and however different their biographies, and therefore the stimuli they received and the influences upon them, we can see how they all shared a more or less extreme form of the idea of the primacy of theory over observation, held by Popper, with the thesis of the gestaltic nature of vision, held by Wittgenstein.

In particular, Popper influenced Feyerabend's early views. The thesis of the theoretical character of observations, that constitutes one of the basic premises of the "new philosophy of science", was advanced by Popper already in the early 1930s. In the same line, it is possible to trace back to Popper the thesis of the theoretical character of meanings, that already appears in *Logik der Forschung* (1934) and is later reasserted in the 1959 English edition of the book; not to mention the idea of contextual theoretical character of the laws for the acceptance of basic assertions, another key feature of the "new philosophy of science".[61] As to the gestaltic nature of vision, Hanson often followed or drew on Wittgenstein's *Philosophical Investigations*.[62] Reference to Wittgenstein was explicit in Toulmin, who attended his lectures in Cambridge in 1941 and 1946–1947.[63] Feyerabend carefully studied Wittgenstein's *Philosophical Investigations*,[64] and at the same time he often referred

60 The last one is Gerhard Tintner, "Methodology of Mathematical Economics and Econometrics", published in 1968. This closes the second of the two introductory volumes of the projected series. Volumes 3–9, which should have given substance to the *Encyclopedia*, were never completed.

61 On these points, see Pera (1981), ch. 9, and (1982a), ch. 5.

62 See, in particular, Hanson (1958), ch. 1, which refers to Wittgenstein (1953), Part II, section XI.

63 See Toulmin (1953). The "Preface" and "Introduction" to this booklet anticipate, in Wittgensteinian form, a typical point of the new philosophy of science, namely, "the adoption of a new theory involves a *language-shift*" (*ibidem*, p. 13).

64 See Feyerabend (1955) and (1978c), pp. 114–116: here Feyerabend claims he actually rewrote Wittgenstein's book: "While in London I read Wittgenstein's *Philosophical*

to Kuhn as well. And the very notion of "paradigm", in one of its main meanings, is of Wittgensteinian origin.[65] Crucial were then the new studies in the history of science, the decline of Logical Positivism and Quine's attack on the distinction between analytic and synthetic.[66] "By the mid-fifties, it was becoming clear to some younger philosophers of science that certain crucial questions on the subject could not be tackled with any hope of success, unless they set aside all formal or 'logical' issues, and paid attention instead to the processes of historical change out of which

---

Investigations in detail. Being of a rather pedantic turn of mind I rewrote the book so that it looked more like a treatise with a continuous argument" (pp. 115–116).

[65] For a history of the term, see Cedarbaum (1983), I. Bernard Cohen (1985), p. 480, Hoyningen-Huene (1989a/1993), pp. 132–133, and Toulmin (1972), pp. 106–107. Toulmin refers to Wittgenstein's lectures in Cambridge (1938–1947), to a book by one of Wittgenstein's pupils, William H. Watson (see his (1938)), and to his own (1961). The term "paradigm" appears also in Wittgenstein (1953), Part I, §§50, 55, 57, 300 and 385. If already Aristotle (the study of which proved so important for the development of Kuhn's views of science) used "paradigm" in the sense of exemplar (see his *Posterior Analytics*, II, 24, 68b38), according to Cedarbaum and Toulmin it was Georg Christoph Lichtenberg, a mathematician and physicist of the eighteenth century, who introduced the notion of paradigm into contemporary debates: indeed, he developed a pattern of scientific change based on "paradigms". However, Lichtenberg (unlike Kuhn, who was not aware of his works) saw his own paradigms as a grammatical analogue to the sciences and even wrote about "*paradigmata*" according to which the various sciences are to be "declined": he had in mind a variety of formulas of procedure embodied in sensible form and relatable from one science to another, and across natural sciences to philosophy. Joseph P. Stern sees Lichtenberg's paradigms as "archetypical configurations or Goethean 'Urphänomene'; issuing somewhere between facts and laws, these 'paradigms' would in themselves be actual parts of natural science (as grammatical paradigms are parts of natural language)" (Stern (1959), p. 103). However, Lichtenberg never thoroughly developed such a concept, confining himself to arguing for primitive or self-explicative elements of scientific knowledge that could be considered the analogue of the grammatical standards, and to which more complex phenomena could be referred (see also Toulmin (1972), p. 106). Neither Toulmin nor Cedarbaum, however, mention that already in 1935, reviewing Popper (1935), Otto Neurath frequently employed the word "paradigm" in the sense of "ideal model" (Neurath (1935a), pp. 353, 357 and 361). In turn, Moritz Schlick employed it to refer, roughly, to an "exemplary case" (see Schlick (1918/1985), p. 32, and (1986/1987), p. 45). Lichtenberg was well known to Schlick, who quotes him in ch. 20 of his (1918/1985). Schlick's usage of the term might have been mediated by Cassirer (see Cassirer (1910), p. 243), who employed "paradigms" in order to refer to exemplary illustrations of certain principles and theorems of pure mathematics; Cassirer is also quoted by Schlick in his (1918, 1925), ch. 40 (see Hoyningen-Huene (1989a/1993), pp. 133–134, n. 7). Later in the century, "paradigm" is employed by Wittgenstein, Watson, Hanson and Toulmin, but none of these authors interprets paradigm changes as essentially discontinuous changes. With Kuhn the term established itself in its epistemological meaning, and the success met by Kuhn's works will take it much beyond that. Finally, it must be noted that Kuhn himself refers to a work whose title contains the word "paradigm": Bruner, Postman (1949) (see Kuhn (1962a), p. 63, n. 12).

[66] See Quine (1951).

the basic concepts, theories, and methods of science have emerged, and to which they are continually subject".[67]

In other words, the rationality of science cannot depend exclusively on the formal validity of the inferences drawn within the scientific theories of a given historical period.[68] On the contrary: we can recognize and understand the origins of their explanatory power only when we come to recognize and understand what is *implicit* (or, to use Polanyi's expression, "tacit") in the processes of conceptual transformations.[69] At a certain moment things changed and the self-limitations historians and philosophers of science inflicted upon themselves were tacitly removed: the focus of attention quite rapidly shifted from the "structure" of scientific theories to the "dynamics" of scientific transformations, and recourse to history became utterly natural, almost automatic.

As Marcello Pera highlighted,[70] three classes of problems play a central role in this transition. First, the very applicability to concrete scientific practice of the impressive formal and abstract apparatus of inductive logic worked out by logical positivists is questioned: philosophers no longer ask whether formal algorithms are practically applicable in the form in which they are proposed (few logicians had ever demanded that), but whether there exists a possibility, at least in principle, to restructure those very algorithms so that they become compatible with conceptual conclusions with which scientists come to terms in their everyday research activity. This becomes the central question of philosophical reflection on science, and it becomes at once clear that it may be resolved only by appealing to history.[71]

---

[67] Toulmin (1977), pp. 147–148. The rationality of science, Toulmin argues, cannot depend solely on the formal validity of the inferences drawn within the scientific theories of any given time: "We can recognize the source of science's explanatory power only if we come to understand also what is involved in the process of conceptual change: [...] a formalist approach is *necessarily* insufficient, and the architectural metaphors of formal logic – 'structure' and the rest – must be set aside in favor of some other analysis of scientific work" (*ibidem*, p. 148).

[68] As to this point, the distinction between the "context of discovery" and the "context of justification" is emblematic.

[69] As Toulmin once again remarks, this progressive awareness was considerably due to the diffusion, in the 1950s, of a "new" discipline such as the history of science (see, for example, Sarton (1952)). Before the 1950s, in the States, historians and philosophers of science walked along parallel paths, refraining from any collaboration. If the task of philosophy was that of establishing, as I said, the formal *organon* of science, the task of scientific historiography consisted in setting up "rational constructions" of past scientific achievements. Its limited task was that of tracing and arranging the intellectual stages through which those who came before us managed to build, brick by brick, the edifice of Knowledge we are inhabiting today. The key text to understand the change of approach is Agassi (1963), which criticizes what was then the received view and advances a new approach; on Agassi's book see Kuhn (1966) and Munz (1985), chs. 1–2.

[70] See, for example, Pera (1982a) and (1982b).

[71] The new trend became dominant in the 1970s, when debates over key episodes of the history of science developed: see for example Lakatos (1971a) and (1971b), together with Lakatos, Zahar (1976).

Secondly, it becomes more and more evident that the approach of orthodox empiricism was founded on assumptions that are now being questioned: the possibility of isolating pure "observation facts" quite independently from any theoretical consideration (Vienna Circle's rigorous *Protokollsätze*), or the distinction between the "context of discovery" and the "context of justification".

Third and last, the logical positivists' concern for the predictive power of sciences is integrated with an equal care for their explanatory power, that is set at the very basis of our understanding of the scientific enterprise.

In sharp contrast with Logical Positivism, then, the most relevant characteristics of the new perspective are the rejection of formal logic as the primary tool for scientific analysis and the recourse, in its stead, to a detailed study of the history of science, with references, at times, also to sociology.

> Most scientific research consists, in this view, of a continuing attempt to interpret nature in term of a presupposed theoretical framework. This framework plays a fundamental role in determining what problems must be solved and what are to count as solutions to these problems; the most important events in the history of science are revolutions which change the framework. Rather than observations providing the independent data against which we test our theories, fundamental theories play a crucial role in determining what is observed, and the significance of observational data is changed when a scientific revolution takes place: Perhaps the most important theme of the new philosophy of science is its emphasis on continuing research, rather than accepted results, as the core of science. As a result, analysis of the logical structure of completed theories is of much less interest than attempting to understand the rational basis of scientific discovery and theory change.[72]

Once a historical and dynamical perspective was acquired, the next step consisted in highlighting the fact that no scientific research is carried out without assumptions. Scientific research is always conducted within conceptual frameworks, systems of categories, sets of beliefs or general world views. Stephen Toulmin speaks of "models

[72] Brown (1977), p. 10. As to this change of perspective, William Newton-Smith notices that "Much scientific activity consists in accounting for or explaining change. The shifting of allegiances from theory to theory which will be referred to as *scientific change* is itself a type of change that requires explanation" (Newton-Smith (1981), p. 3).

and ideals, principles of regularity and explanatory paradigms",[73] Michael Polanyi of "conceptual frames",[74] and Thomas Kuhn chooses the term "paradigm".[75]

Moreover, the assumptions on the basis of which scientific research operates are all-pervasive. Feyerabend talks about assumptions and "ways of looking at the world", ascribing their discovery to Kant;[76] Kuhn about a "strong network of commitments – conceptual, theoretical, instrumental, and methodological"[77] – that tells us what the world and its science are like; along the same line, Toulmin speaks of "principles of regularity, conceptions of natural order, paradigms, ideals, or what-you-will: intellectual patterns which define the range of things we can accept (in Copernicus' phrase) as 'sufficiently absolute and pleasing to the mind'".[78]

Once the all-pervasiveness of assumptions is taken on, the circle closed by stating that scientific theories are assumptions in this very sense. It is the "all-pervasive character of theoretical assumptions"[79] Feyerabend speaks of, arguing that "scientific theories are ways of looking at the world; and their adoption affects our general beliefs and expectations, and thereby also our experiences and our conception of reality".[80] He also assimilates theories to "natural languages".[81] Kuhn refers to "*Gestalt* switches" that would take place suddenly and transport the scientist onto "another planet": "after a revolution", he says, "scientists are responding to a different world".[82] Polanyi speaks of a "new world" that opens in front of the eyes of the student;[83] according to him, members of different schools of thought "think differently, speak a different language, live in a different world, and at least one of the two schools is excluded to this extent for the time being (whether rightly or wrongly) from the community of science".[84] Hanson talks about "patterns" or "conceptual structures" which allow us

---

73 Toulmin (1961), pp. 42–43. They "are not always recognized for what they are; differences of opinion about them give rise to some of the profoundest scientific disputes, and changes in them to some of the most important transformations of scientific theory" (*ibidem*, p. 43). This idea is scattered throughout Toulmin's book: see especially ch. 3 ("Ideals of Natural Order (I)"), particularly pp. 44–46 and 53–60. "For, though Nature must of course be left to answer to our interrogations for herself, it is always *we* who frame the questions. And the questions we ask inevitably depend on prior theoretical considerations. We are here concerned, not with prejudiced belief, but rather with preformed concepts; and, to understand the logic of science, we must recognize that 'preconceptions' of this kind are both inevitable and proper – if suitably tentative and subject to reshaping in the light of our experience" (*ibidem*, p. 101).

74 See, for instance, Polanyi (1958), pp. 59–60.

75 The term is first employed in Kuhn (1959a), p. 165. The concept, however, is already present in his (1957), expressed by words like "model", "theory" or "conceptual scheme" (pp. 40–41). See also Kuhn (1962a), p. viii, and (1977a), p. xix.

76 Feyerabend (1962a), p. 29.

77 Kuhn (1962a), p. 42.

78 Toulmin (1961), p. 81; see also pp. 42–43 and 115.

79 Feyerabend (1962a), p. 29.

80 Feyerabend (1962a), p. 29.

81 Feyerabend (1975), p. 225.

82 Kuhn (1962a), p. 111.

83 Polanyi (1958), p. 101.

84 Polanyi (1958), p. 151.

to understand the sense data, "determine [...] how the facts hang together"[85] and set what the problems are. Similarly, Toulmin says that different ideals make us see the world differently.[86]

Thus, beginning in the 1950s, a number of philosophers from different philosophical backgrounds launched a compact attack on the methods and conclusions achieved by Logical Positivism.[87] Though not bound to any single person or work, it seems plausible to date the "second revolution" in the philosophy of science of the twentieth century in the five years spanning from 1958 to 1962. Heralded by a few previous ferments, in those crucial years the major works of the main actors of the revolution appeared in press. Preceded by Kuhn's *The Copernican Revolution* (1957), in which many features of the new philosophical trend are already at work, in 1958 Hanson published *Patterns of Discovery* and Polanyi *Personal Knowledge* (second edition, 1962).[88] Then, in 1959,[89] we have Kuhn's important article "The Essential Tension"; two years later, Toulmin's *Foresight and Understanding* (1961). Finally, 1962 saw the publication of both Kuhn's seminal work, *The Structure of Scientific Revolutions*, and Feyerabend's important article, "Explanation, Reduction, and Empiricism", both regarded as a sort of manifesto of the new philosophical approach.[90] The Bastille of Logical Positivism seemed to have been taken once and for all.

---

[85] Hanson (1958), p. 118.

[86] See Toulmin (1961), pp. 55–59.

[87] The collection Feigl, Maxwell (eds) (1961) offers a good idea of state of the philosophy of science shortly before the appearance of Kuhn's *The Structure of Scientific Revolutions*.

[88] Some of Polanyi's views appeared also in his (1946) and (1951), that anticipate some ideas concerning the theory-ladenness of observation statements.

[89] In this very year appeared the English translation of Popper's *magnum opus*, *The Logic of Scientific Discovery* – too late, perhaps: on the one hand, its revolutionary impact had considerably soothed by then, and some of Popper's radical new ideas are now credited to his pupils and critics; on the other, Popper's misconstrued image as a minor logical positivist had been spread by the heirs of the Vienna Circle into the English-speaking world, and would be reinforced a few years later by Lakatos' stereotypical picture of Popper as a naïve falsificationist (see Lakatos (1969) and (1970)). After being ignored or overlooked for three decades, in the 1960s Popper's ideas turned out to be common knowledge, shared by all, if not totally trivial.

[90] Respectively, Kuhn (1957), Hanson (1958), Kuhn (1959a), Toulmin (1961), Kuhn (1962a) and Feyerabend (1962a).

Chapter 2

Kuhn and the "New Philosophy of Science"

> GALILEO: "[...] we shall question everything all over again. And we shall go forward not in seven-league boots but at a snail's pace. And what we discover today we shall wipe it off the slate tomorrow and only write it up again once we have again discovered it".
>
> *Bertolt Brecht*

In the 1960s and 1970s the protagonists of what has later become known as the "new philosophy of science" gave rise to a wide and rich debate that involved several philosophers from different countries. Since the present work aims at offering a critique of one of the characteristic theses of the post-Popperian philosophy of science, and particularly Kuhn's interpretation of it, I need first to analyse in some detail the positions of the other participants in the debate and how they influenced Kuhn's own position. I will do that by paying particular attention to a key moment: the International Colloquium in the Philosophy of Science held in London in 1965, which saw the meeting (and the clash) of some of the most important philosophers of science of the twentieth century.

The Early Phase of the Debate

Michael Polanyi

Some philosophers, like Michael Polanyi, thought that Popper's epistemology was not so much to be revised – as Lakatos attempted to do two decades later – as to be rejected almost entirely, since the neopositivistic assumptions it was founded upon could hardly be shared.

In particular, Polanyi deemed unacceptable Popper's illusion that an objective and impersonal knowledge is possible. Against such a "myth" Polanyi argued for a conception that highlights the tacit and unspoken dimension of our knowledge, both in its scientific expression and in its more general and common manifestations. It is a "personal knowledge", a "post-critical philosophy"[1] that does not believe in objective criteria of falsification or testability any more, rejects "the ideal of scientific detachment" and aims at establishing "an alternative ideal of knowledge, quite generally".[2] The words in the title of Polanyi's major philosophical work –

1 See Polanyi (1958, 1962). See also his (1962) and (1966).

2 Polanyi (1958, 1962), p. vii.

Personal Knowledge – may seem to contradict each other, since knowledge is usually deemed impersonal, universally established and objective. However, such alleged contradiction "is resolved by modifying the conception of knowing".[3] It becomes "an active comprehension of the things known, an action that requires skill. Skilful knowing and doing is performed by subordinating a set of particulars, as clues or tools, to the shaping of a skilful achievement, whether practical or theoretical".[4] In this way

> Acts of comprehension are to this extent irreversible, and also non-critical. For we cannot possess any fixed framework within which the re-shaping of our hitherto fixed framework could be critically tested. Such is the *personal participation* of the knower in all acts of understanding. But this does not make our understanding *subjective*. Comprehension is neither an arbitrary act nor a passive experience, but a responsible act claiming universal validity. Such knowing is indeed *objective* in the sense of establishing contact with a hidden reality; a contact that is defined as the condition for anticipating an indeterminate range of yet unknown (and perhaps yet inconceivable) true implications. It seems reasonable to describe this fusion of the personal and the objective as Personal Knowledge.[5]

Being an intellectual commitment, Polanyi continues, personal knowledge is "inherently hazardous".[6] Ultimately, according to Polanyi, "into every act of knowing there enters a passionate contribution of the person knowing what is being known, and [...] this coefficient is no mere imperfection but a vital component of his knowledge".[7] It is a theme dear to most "new philosophers of science", and particularly to Kuhn: underlining the key relevance of those aspects of the cognitive act that, as for Kant's "light dove", were regarded as nothing more than annoying obstacles.[8]

Therefore, on the one hand, Polanyi takes issue with the alleged neutral and objective character of knowledge (and of scientific knowledge in particular),

---

3 Polanyi (1958, 1962), p. vii.

4 Polanyi (1958, 1962), p. vii.

5 Polanyi (1958, 1962), pp. vii–viii; on Polanyi and Feyerabend see Preston (1997b).

6 Polanyi (1958, 1962), p. viii. "Hypothesizing is a dangerous game", wrote Novalis in the very same dialogue from which Popper took the motto for his (1935) (see Novalis (1960), p. 668).

7 Polanyi (1958, 1962), p. viii.

8 In the "Introduction" to the *Critique of Pure Reason* Kant exposed the illusion of Platonic idealism: just as the "light dove", feeling the resistance of air, might imagine that its flight would be much easier in empty space, so Plato left the world of the senses and the obstacles it poses to the intellect, and "ventured out beyond it on the wings of ideas, in the empty space of pure understanding" (Kant (1781), A5). In an analogous way Wittgenstein, in his *Philosophical Investigations*, unveils the illusion upon which he had built the *Tractatus Logico-Philosophicus*: a univocal relationship between the logical essence of language and the *a priori* order of the world. The presupposition according to which language corresponds to the crystalline purity of logic sets us on a slippery ice surface "where there is no friction and so in a certain sense the conditions are ideal, but also, just because of that, we are unable to walk. We want to walk: so we need *friction*. Back to the rough ground!" (Wittgenstein (1953), Part I, §107). It is not an unavoidable and annoying friction, but the very thing that makes the walking or flying (in plain terms: motion, optics, physics) possible.

highlighting how subjective elements are intertwined with objective ones. On the other hand, by insisting on the tacit component, which rational discourse can not account for but nonetheless trusts, he opens up an opportunity for irrationalism (even if he thinks that such an outcome can be avoided through the inter-subjective tests that scientists practise by observing one another, reciprocally criticizing and encouraging one another).[9]

Polanyi portrays science as the "art of knowing", showing how knowledge is produced or gained not by abiding to rigorous logical or experimental procedures, but by a personal effort and engagement in which all methodological elements work only as tools that scientists freely employ according to an "art" guided by personal commitment that, after all, has an ethical scope. In the final analysis, such a commitment is the actual force, rule and guide of the process from which scientific doctrines are born, develop and get into trouble. It is a commitment that does not come out of the blue, rather, it is the outcome of structured organic dispositions and, above all, of a system of background knowledge acquired by learning a language, thus constituting the premises and the presuppositions of any further knowledge. Writes Polanyi:

> Scientific discovery reveals new knowledge, but the new vision which accompanies it is not knowledge. It is *less* than knowledge, for it is a guess; but it is *more* than knowledge, for it is a foreknowledge of things yet unknown and at present perhaps inconceivable. Our vision of the general nature of things is our guide for the interpretation of all future experience. Such guidance is indispensable. Theories of the scientific method which try to explain the establishment of scientific truth by any purely objective formal procedure are doomed to failure. Any process of enquiry unguided by intellectual passions would inevitably spread out into a desert of trivialities. Our vision of reality, to which our sense of scientific beauty responds, must suggest to us the kind of questions that it should be reasonable and interesting to explore. It should recommend the kind of conceptions and empirical relations that are intrinsically plausible and which should therefore be upheld, even when some evidence seems to contradict them, and tell us also, on the other hand, what empirical connections to reject as specious, even though there is evidence for them – evidence that we may as yet be unable to account for on any other assumption.[10]

In his critical remarks on Kuhn's "The Function of Dogma in Scientific Research" Polanyi stresses the strong coincidence of his own ideas with Kuhn's and highlights how "the dependence of research upon a deep commitment to established beliefs receives the very minimum of attention today".[11] Moreover, if on the one hand Polanyi underlines the "essential tension" felt by the scientist, on the other he poses a fundamental problem, that of the distinction between an essential anomaly and a

---

[9] See Polanyi (1966), p. 72. Once again, these elements are also in Popper (and Kuhn), with his idea of "world 3", a world whose inhabitants are objects produced by human minds: they are objective because they are criticizable, inter-subjectively testable – and they give rise, once they are created, to consequences that their creators neither intended nor foresaw.

[10] Polanyi (1958, 1962), p. 135. The similarities with Kuhn's own views are striking: compare the last sentences of this passage with Kuhn (1962a), chs. 4–5, *passim*; see also Preston (1997b) and Gattei (2000b), pp. 299–302.

[11] Polanyi (1963), p. 375.

mere personal failure of the individual researcher who has not yet learned how to apply certain rules and certain techniques properly:

> It is clear that the doctrine of universal doubt[12] makes no sense within the context of modern science. A scientist who would query and try to test all the information transmitted to him by scientific journals and textbooks would be condemning himself to total sterility.[13] But on the other hand scientific originality includes a capacity to doubt accepted beliefs. [...] We have to face then both rejections of authority that are futile and other rejections of authority to which science owes its greatest advances. Is there any rule for distinguishing between the two? Or for that matter, any rule by which scientists can distinguish their own failure to apply the current framework of scientific beliefs from the presence of an essential anomaly incompatible with these beliefs [...]? There are of course no such rules.[14]

Thus Polanyi accepts Kuhn's "excellent paper" to the 1961 Oxford Symposium "only as a fragment of an intended revision of the theory of scientific knowledge. Otherwise", he concludes, "it would not only fail to answer the questions it raises, but appear altogether to ignore them".[15]

"I doubt", replies Kuhn,

> that Mr. Polanyi is well pleased with my notion of paradigm [...]. It therefore seems worth emphasizing that, though I have only recently recognized it as such, Mr. Polanyi himself has provided the most extensive and developed discussion I know of the aspect of science which led me to my apparently strange usage. In his perceptive and challenging book, *Personal Knowledge*, Mr. Polanyi repeatedly emphasizes the indispensable role played in research by what he calls the 'tacit component' of scientific knowledge. This, if I understand him correctly, is the inarticulate and perhaps inarticulable part of what the scientist brings to his research problem: it is the part learned not by precept but principally by example and by practice.[16]

---

12 Deriving from Descartes, such a doctrine was inherited by the founders of the Royal Society when, more than three centuries ago, they chose "*nullius in verba*" (that we might translate as "we accept no authority") as their motto.

13 On the issue of trust in knowledge, see Hardwig (1991), in which it is argued that knowledge rests not on evidence, but on trust, in its various forms: "Modern knowers cannot be independent and self-reliant, not even in their own fields of specialization. In most disciplines, those who do not trust cannot know; those who do not trust cannot have the best evidence for their beliefs. In an important sense, then, trust is often epistemologically even more basic than empirical data or logical arguments: the data and argument are available only through trust" (pp. 693–694). The alternative to trust, Hardwig argues, is often ignorance: an untrusting, suspicious attitude would impede the growth of knowledge, perhaps even without substantially reducing the risk of unreliable testimony. Provided it is not abused, trust is a positive value for any community of finite minds.

14 Polanyi (1963), pp. 379–380.

15 Polanyi (1963), p. 380. See also Agassi (1966a), that closes in the same line.

16 Kuhn (1963b), p. 392. After his exchange with Polanyi in Oxford in 1961, Kuhn adds a reference to him in *The Structure of Scientific Revolutions*, then in proofs: see Kuhn (1962a), p. 44, n. 1. See also Polanyi (1963).

Therefore, Kuhn shifts the problem from the individual to the community: "I was concerned", he says, "not to find a methodological rule for individual scientists (e.g. Mr Popper's principles of falsificationism) but rather to characterize the state of the scientific community within which a new theory is invented and accepted".[17]

Indeed, as Kuhn indicates at the very beginning of the published version of his paper,[18] this is actually a fragment (or an extreme synthesis) of a larger project, namely *The Structure of Scientific Revolutions* – but this fragment contains all its core features, and Polanyi and Kuhn would not agree as to the shape of the puzzle into which the fragment must be fitted.

Norwood R. Hanson

In the very same year in which Polanyi published *Personal Knowledge* Norwood Russell Hanson saw through the press a book dealing with the philosophical aspects of microphysics, *Patterns of Discovery*. Among several interesting things, this book argues that empirical observation is not theoretically neutral, but it is conditioned by the observer's beliefs and attitudes. It is the thesis of the theory-ladenness of observation statements, according to which every observation is, by its very nature, laden with theory and therefore it makes no sense distinguishing between theoretical language and observation language. In so doing, Hanson outlines a conception of knowledge that rejects one of the fundamental dogmas of Neopositivism, the one referring to the neutral character of observations. More than that: according to Feyerabend he actually demonstrates "the chaotic character of science and thereby shows the tremendous abyss that exists between a certain philosophical picture of science and the real thing".[19]

Logical Positivists did not hesitate to substitute their "rational reconstructions" for actual scientific practice and the history of science, by sharply distinguishing, with Hans Reichenbach, the "context of discovery" from the "context of justification".[20] An analogous attitude we may find in Popper: "I shall distinguish sharply between the process of conceiving a new idea, and the methods and results of examining it logically".[21] Indeed, he sets himself the task to give "a logical skeleton of the procedure of testing".[22]

For Hanson such a separation is untenable:

> The slogan contrast between 'the context of justification' and 'the context of discovery' is often advanced to stifle queries that are fundamentally *conceptual* in character. Too many explorations into the *concept* of discovery have been dismissed by contemporary analysts

[17] Kuhn (1963b), p. 392.

[18] See Kuhn (1963a), p. 347, n. 1.

[19] Feyerabend (1970c), p. 277.

[20] See Reichenbach (1938); however, the distinction goes back at least to Herschel (1830).

[21] Popper (1935, 1959), p. 31.

[22] Popper (1935, 1959), p. 32.

as turning on issues of psychology and history, when it is our very *ideas* of discovery, of creativity, and of innovation which are at issue in such inquiries.[23]

A logic of discovery, he says, should deal with exactly with that.

It makes no sense talking about physical theories in which interpretation plays no role: phenomenalism failed to provide an adequate answer to questions like whether observations can do without interpretations or how an observation free of all interpretations should be. It was Wittgenstein's merit, Hanson stresses, to highlight, through "his *analysis* of complex concepts such as *seeing*, *seeing as*, and *seeing that* [...] the crude, bipartite philosophy of sense datum versus interpretation as being the technical legislation it really is. By means of *philosophy* he destroyed the dogma of the immaculate perception".[24] With a remarkable metaphor Hanson concludes by saying that "immaculate observation" is only a dogma:[25] if we had to empty out geometrical optics of its content, we would be left not with the theory of geometrical optics, but with geometry *simpliciter*. To deprive geometrical optics of interpretation is like depriving optics of optics. The theory of geometrical optics, in the positivistic

---

[23] Hanson (1969), p. 74.

[24] Hanson (1969), p. 74; see also p. 75. See the references in n. 32, below.

[25] See Hanson (1969), p. 74. Also Karl Popper, by drawing on Parmenides (see Popper (1992) and (1998a), chs. 3–6), observed that the theoretical/observation polarity needs to be taken seriously. However, his aim was different: he meant to underline, against the positivists' dogma, the primacy of theory upon observation. If every good empiricist rightly insists on saying that without any comparison with experience physics would reduce to a solitary's monologue, the critical rationalist reminds us that observation without any prejudice or expectation is a plain illusion. Long before Galileo (but also Plato and Democritus) Parmenides showed that in order to make phenomena observed intelligible we must refer to hidden structures (the whole of modern physics, from Galileo to Newton, not to mention contemporary physics, systematically appeals to theoretical entities that are not reducible to observables – indeed, in so doing it makes perceptual data comprehensible and scientifically employable). Popper translates Parmenides and praises the goddess, who exhorts those who listen to her not to "let experience / Much tried habit, constrain [him]" (Popper (1992), p. 14). The critique of any form of empiricism that confines the researcher to the limits of what he thinks he can dominate with his own senses is the premise both of the growth of knowledge and of intellectual emancipation. As Galileo saw (in the third day of his *Dialogue Concerning the Two Chief World Systems*): "there is no limit to my astonishment when I reflect that Aristarchus and Copernicus were able to *make reason do such violence to sense* that, in defiance of the latter, the former became mistress of their belief" (Galilei (1632/1953), p. 328, my emphasis; all the following sentences are taken from pp. 327–328). If "sensible experience" was plainly contrasting with "the diurnal rotation of the earth" and its "annual movement", yet the "brightness of intellect" of innovative scientists consists precisely in "[doing] such violence to their own senses" to devise, in a typically speculative way, an explanation of what seems to be "plainly contrary" to the theory. In Galileo's times the phases of moon were no longer at stake, but the alleged refutation of the movements of the earth (around itself and around the sun). Yet Aristarchus and Copernicus – not to mention Galileo himself – in "[preferring] what reason told them" are heirs of Parmenides, regardless of the correctness of the latter's opinions, and even of the correctness of their own opinions. Through Galileo's trial vicissitudes science earned itself not only the possibility of expression but, most importantly, the right to make mistakes. See Gattei (1995).

sense, is not optic, but geometry; in the same way, the result of "extracting" physical interpretation from physical theory is not physics, but sheer algebra.

Physical theory is therefore an "*indissolubly* complex concept", but realizing that should not prevent us from trying to study it: "Complexity never constitutes a good argument for reluctance to undertake analysis. [...] at the frontiers of research into the unknown not only are the observations 'theory-laden', but even the most provisional theories *already* have their interpretations, their applications, their observations, built firmly into the system itself".[26]

If Kuhn, in 1962, is able to argue without too much ado for the incommensurability of the "two chief world systems", the Ptolemaic and the Copernican, it is because he refers to Hanson's analyses. And from Hanson Kuhn explicitly takes the "*Gestalt* switch" image for describing the shift from one paradigm to another.[27]

Theories are for Hanson "a conceptual *Gestalt*";[28] they are not the product of the accumulation of scattered observation data, rather, they are "what makes it possible to observe phenomena as being of a certain sort, and as related to other phenomena. Theories put phenomena into systems".[29] In this sense theories are "patterns" that make experience intelligible.[30]

A long tradition, that counts among his foremost exponents a positivist like Auguste Comte or a conventionalist like Pierre Duhem, had already highlighted that our observations are "theory-laden". However, according to Hanson, it was Ludwig Wittgenstein and the *Gestalt* psychology theorists who showed that our perceptions are the outcome of a process of perceiving in accordance with specific conditions, as becomes clear from the famous *Gestalt* pictures. In his *Philosophical Investigation*s

[26] Hanson (1969), p. 77.

[27] If, on the one hand, Kuhn's position seems to be more radical than Hanson's, it is certainly true that, on the other, their views do not significantly differ. As opposed to the case of *Gestalt* pictures (that of the duck-rabbit, for example, Necker's cube, the bird-antelope, or the surprising frog-horse – not to mention many of Maurits C. Escher's drawings), for which the subject realizes that his perception has swayed and can always learn how to guide it, scientists cannot do that. Moreover, contrary to what usually happens with *Gestalt* pictures, for which there is a perfect symmetry between the two readings and the shift from one to another is quite easy, in actual scientific practice the shift is irreversible: "scientists do not see something *as* something else; instead, they simply see it. [...] In addition, the scientist does not preserve the gestalt subject's freedom to switch back and forth between ways of seeing" (Kuhn (1962a), p. 85). In order to account for this latter aspect psychological considerations about individuals' perceptions are no longer sufficient: we need to consider also the sociological phenomenon of research tradition.

[28] "Just as a *Gestalt* organizes the unrelated perception elements into meaningful wholes, so a theory arranges its own group of facts; and just as different *Gestalten* give rise to different perceptions, so different scientific theories give rise to observations of different facts. Therefore, not only observations do presuppose theories, in the sense that the latter give meaning and importance to the former, following the original request of criticism and the one typical of Popper's critical rationalism, but observation data are entirely produced by theories": Pera (1982a), p. 110. On *Gestalt* psychology and *Gestalt* pictures see Mach (1900), especially chs. VI–VII and X–XI, Koffka (1935), Katz (1944) and Musatti (1965).

[29] Hanson (1958), p. 90.

[30] It is a point later developed by René Thom in his (1990).

Wittgenstein stressed how in such cases there is not the slightest resemblance between a head seen from one side or the same head seen from another.[31]

In his turn, Hanson underlines that in such cases "the concept of seeing [...] does not designate two diaphanous components, one optical, the other interpretative",[32] shifting these remarks to scientific practice: "Let us consider Johannes Kepler: imagine him on a hill watching the dawn. With him is Tycho Brahe. Kepler regarded the sun as fixed: it was the earth that moved. But Tycho followed Ptolemy and Aristotle in this much at least: the earth was fixed and all other celestial bodies moved around it. *Do Kepler and Tycho see the same thing in the east at dawn?*".[33] Of course, "The same configuration is etched on Kepler's retina as on Tycho's",[34] but the vision of the Sun is not the vision of its retinical image: vision is an *experience*. The elements of Kepler's and Tycho's experiences "are identical; but their conceptual organization is vastly different".[35] In fact, Tycho sees a mobile Sun, while Kepler sees a static Sun.[36]

Pure observation does not exist, and therefore a neutral observation language does not exist either: this is something already known to conventionalists like Duhem and Poincaré, and to sophisticated positivists like Neurath. Popper too has always held that theory precedes observation: we can not see anything if we do not have any expectation about what we will observe.[37] For all these reasons Hanson attacks both the dichotomy between the "context of discovery" and the "context of justification", and that between theoretical and observation language (dealing with allegedly "pure" sense data). Every statement is essentially theory-laden and authentic scientific discoveries are not the outcome of induction or deduction, but of *retroduction*, that is, of the individuation of a new pattern of concepts, hypotheses and formulas within which we can frame phenomena of a different nature. This conception bears a very important consequence for our assessment of conceptual

---

[31] Wittgenstein suggests to see in the flashing of a shape the arrangement of perceptions under the aegis of similarity, to move then to the thesis that "seeing as" is seeing according to a rule, that is, following an interpretation. "Seeing as" is not a mere form of vision, nor an interpretation that adds to the perception *a posteriori*, without getting a grip on it; but it is the echo of a thought in the perception, its giving life to it from within, changing its shape and determining its substance: thought does not add to perception, but is the rule that allows us to apply a certain image in a certain way. And language games are the dimension in which we should look for the relation between thought and perception. See Wittgenstein (1953), Part I, §§ 74, 139–141, and Part II, section XI; see also Wittgenstein (1921), 5.5423, and (1980), vol. I, §§868–869.

[32] Hanson (1958), p. 9. Seeing does *not* have two separate components: "Kepler and Tycho just see the sun. That is all" (*ibidem*).

[33] Hanson (1958), p. 5.

[34] Hanson (1958), p. 6.

[35] Hanson (1958), p. 18.

[36] For a criticism of Hanson's position see Kordig (1971). Kordig's considerations are then more precisely stated in Petroni (1990), particularly on pp. 29–35.

[37] Even if Popper himself would dislike and reject taking this intuition to an extreme. On his personal copy of Lakatos, Musgrave (eds) (1970), he wrote: "Very bad: a good idea of mine taken to an impermissible limit. All this is just rubbish".

change in science: to go back to the above-mentioned example, between Tycho and Kepler will not stand up the judgement of "pure" science, but the complex weaving of different views. A scientific revolution will consist, for Kuhn, in the shift from one view to another.

Stephen E. Toulmin

Stephen Edelston Toulmin moves in the same direction as Hanson: he holds that our perceptions are not free from intellectual pre-comprehension, since we see the world through the spectacles of some basic concepts, which in the final analysis constitute the fundamental concepts elaborated by science in the course of its history. These "paradigms" or "ideals of natural order", as Toulmin calls them, are the explicative paradigms which we refer to in our efforts to make nature intelligible, that is, to give it an order or frame it into a system. A "continual interaction of theory with fact"[38] is required to that end, that forbids us to exclude any traces that have not been thoroughly explored: in other words, Toulmin's aim is to weaken the distinction between theoretical and observation statements, advancing a holistic conception of theories. Toulmin also anticipates a (radical) incommensurability thesis:

> Men who accept different ideals and paradigms have really no common theoretical terms, in which to discuss their problems fruitfully. They will not even *have* the same problem: events which are 'phenomena' in one man's eyes will be passed over by the other as 'perfectly natural'. These ideals have something 'absolute' about them [...].[39]

The value of *Foresight and Understanding* (1961), that reproduces Toulmin's Mahlon Powell Lectures, given at Indiana University in 1960, does not lie only in a sort of anticipation of some of Kuhn's theses, published shortly after this book, in 1962.[40] Indeed, he does not limit himself to inquiring into the ways in which

---

[38] Toulmin (1961), p. 95. This interaction, Toulmin continues, expresses "the way in which theories are built on facts, while at the same time giving significance to them and even determining what are 'facts' for us at all", (*ibidem*).

[39] Toulmin (1961), p. 57.

[40] It is interesting to notice that Toulmin was also one of Kuhn's interlocutors at the Symposium on the History of Science held at Oxford University on 9–15 July 1961, whose proceedings were edited by Crombie and published in 1963: see Toulmin (1963). This was the first, decisive public test of Kuhn's novel ideas, that had been previously discussed only in private form (see, for example, Feyerabend (1995a) and (2006)). Kuhn (1963a) was received with a barrage of criticism by A. Rupert Hall, Michael Polanyi, Bentley Glass, Stephen E. Toulmin and Edward F. Caldin (see Crombie (ed.) (1963), pp. 370–386), to which Kuhn replied in his (1963b). Kuhn's paper is also extremely telling for the impact it had on the audience attending the conference, which comprised numerous protagonists of the history and philosophy of science debates of the twentieth century: among them, in addition to the already-mentioned people, were historians like Maurice Bowra, Herbert Butterfield, I. Bernard Cohen, Alistair C. Crombie, Henry Guerlac, Alexandre Koyré, Vasco Ronchi and Shmuel Sambursky; scientists like David Bohm; and philosophers like Isaiah Berlin, Geoffrey J. Warnock and Gerald J. Whitrow. Protagonists of the epistemological debates of the following years also attended: Norwood R. Hanson, Rom Harré, Mary B. Hesse, William C. Kneale and

a scientific theory is constructed, but also analyses the problem of the predictive ability of scientific statements. This turns, at a first level, into a characterization of the aims of scientific activity, and at a second level into the individuation of "models" or "ideals of natural order", understood as general schemes – in a sense that will be amply theorized by Kuhn – that allow a scientific theory to come into being. At a first level, then, Toulmin's interest is addressed to the horizon of expectations that comes to the scientist from the very boundaries of common sense, while at a second level it moves towards the way in which a theory succeeds (or fails) to correspond to a general scheme of "ideal objects".

Toulmin tries "to extend and to reapply Wittgenstein's analysis of 'language games' and 'methods of representation' to the life and work of natural science",[41] showing how the actual "language games" operating in that life are in fact played and so indicating how and why the formal inductive logic approach adopted by English logicians ever since John Stuart Mill "missed the serious philosophical points that arise in the course of scientific work".[42] Scientific theories and scientific explanations employ, according to Toulmin, "'representations' of many different kinds, and they can be usefully analyzed as deductive or axiomatic systems only in special cases and on special conditions".[43] In other words, rather than allowing the axiomatic method of analysis any monopoly, he feels the necessity of "a functional taxonomy of explanatory procedures and techniques, which would relate these procedures to the problematics of different kinds of scientific inquiry".[44]

Particularly interesting is Toulmin's evolutionary model: indeed, the application of such a model to science has the advantage, he says, of explaining both the stability and the mutability of concepts; the process of variation and selection entails a balance between two kinds of factors: an innovative one, that accounts for the appearance of new variants, and a selective one, that modifies the conceptual population through the transmission of the winning variants. Innovation and transmission are processes oriented towards the adaptation to the ecological needs and necessities of the very many problematic situations that the conceptual population is designed to manage and exploit. Instead of "form" and "validity", the key terms to understand science become "adaptation" and "mutation".[45]

The influence of Benjamin Lee Whorf and Ludwik Fleck

As we saw at the end of the previous section, Hanson, Toulmin, Kuhn and Feyerabend, together with Polanyi's contribution, do not constitute a single school

---

Imre Lakatos. Feyerabend reviewed the volume of proceedings, raising very interesting points about, among others, Kuhn's paper: see his (1964).

41 Toulmin (1977), p. 145.

42 Toulmin (1977), p. 145. Toulmin does not refer to him here, but I believe Carnap (for instance) was working along the same lines as well.

43 Toulmin (1977), p. 145.

44 Toulmin (1977), p. 146: this is the implication of Toulmin's, first book, his (1953).

45 This perspective will prove central for Kuhn, both in *The Structure of Scientific Revolutions* (see Kuhn (1962a), pp. 171–173) and in his later writings (see especially his (1989a), (1991a), (1992) and (1993a)).

of thought. However, they may very well be seen as indicating a precise trend. A common denominator is, on the one hand, the influence of the later Wittgenstein; on the other, the rejection of formal logic as the primary tool to analyse the scientific enterprise, and the appeal in its stead, to history of science. In other words, they all privilege the external to the internal approach: science is a product of human activity and as such it is not indifferent to the social, cultural and psychological context in which the individual researchers find themselves immersed. Finally, for all of them (and especially for Feyerabend and Kuhn) a constant point of reference is the critical rationalism of Karl Popper and his workshop.

Particularly influential on Kuhn are also, as he himself recognizes, Benjamin Lee Whorf and Ludwik Fleck. Whorf (1897–1941), an American linguist, pupil of Edward Sapir at Yale University, was encouraged by his teacher to study the language of Hopi Indian tribes in the south of California. The realization that the structure of this language was very far from that of European languages led Whorf to uphold a radical linguistic relativism, according to which linguistic categories determine the world conceptions, if not the very thought structure, of those who employ them.[46] Writes Whorf:

> The phenomena of language are background phenomena, of which the talkers are unaware or, at the most, very dimly aware [...]. These automatic, involuntary patterns of language are not the same for all men but are specific for each language and constitute the formalized side of the language, or its 'grammar' – a term that includes much more than the grammar we learned in the textbooks of our school days.
>
> From this fact proceeds what I have called the 'linguistic relativity principle', which means, in informal terms, that users of markedly different grammars are pointed by their grammars toward different types of observations and different evaluations of externally similar acts of observation, and hence are not equivalent as observers but must arrive at somewhat different views of the world.[47]

Ludwik Fleck (1896–1961) was a Polish microbiologist who survived the extermination camps of Auschwitz and Buchenwald. In the post-war years, he taught for a period at the Institute of Microbiology of the University of Lublin. In 1957 the communist authorities gave him the permission to emigrate to Israel, where he taught at the Medical Faculty of the University of Jerusalem. In the "Preface" to *The Structure of Scientific Revolutions* Kuhn says he encountered Fleck's "almost unknown monograph"[48] (*Enstehung und Entwicklung einer wissenschaftlichen Tatsache*, 1935) and acknowledges how it is "an essay that anticipates many of my own ideas".[49] The reference was ignored for a long time, until some critics, reading Fleck's book (which spread especially after it was translated into English, with a

[46] This idea, which had been already advanced by Whorf's teacher, was labelled the "Sapir–Whorf hypothesis" in the 1950s. Whorf's most important works are collected in his (1956).

[47] Whorf (1956), p. 221. See Feyerabend (1975), ch. 17, and Kuhn (1964), p. 258.

[48] Kuhn (1962a), p. viii.

[49] Kuhn (1962a), p. ix.

foreword by Kuhn himself[50]), realized its relevance. Fleck's monograph[51] played a major role in the fundamental process that led Kuhn to frame the ideas he drew from history of science, *Gestalt* psychology and language theory in the sociology of the scientific community.

However, Fleck's ideas go well beyond this scope. The "historical fact" analysed in the 1935 book is syphilis, and the author aims at showing that a scientific fact is not something given – rather, it is the construction of a community of experts. Indeed, that of syphilis is a very special case, but it allows Fleck to formulate more general theses. For example, he criticizes Logical Empiricism, which ignores the essential historicity of scientific knowledge and its being socially conditioned.[52] Moreover, he contrasts the positivists' "pure" data with the thesis according to which there is an interaction between facts and theories, and indeed briefly outlines Wittgenstein's view that pure visual perception is supplemented (at a later stage and not necessarily) with an interpretation or a meaning: every visual perception is always a kind of "meaningful seeing" (*Sinn-Sehen*). It makes no sense separating the datum from the historical context in which it appears: "it is all but impossible to make any protocol statements [*Protokollsätze*] based on direct observation and from which the results should follow as logical conclusions. [...] Every statement about 'First Observations' is an assumption. If we do not want to make any assumption, and only jot down a question mark, even this is an assumption of questionability, which places the matter in the class of scientific problems. This is also a thought-stylized assumption".[53]

Furthermore, scientific terms possess a meaning only within a certain context, or "style of thought" (*Denkstill*), that from several points of view reminds us of a Kuhnian paradigm. It fundamentally conditions the activity of a scientific community, or "collective of thought" (*Denkkollektiv*), possesses a "tendency towards self-preservation" (*Beharrungstendenz*) and to make impossible the ascertainment of facts that may contradict it. On its basis scientists are "trained" and learn certain experimental procedures or ways to interpret observations (Fleck amply underlines the role of textbooks on which scientists form themselves). Every style of thought is incommensurable with others:[54] they may not have anything in common, and one can easily regard as physical reality what for other styles of thought does not even exist. In the clash between different styles of thought it is often possible to resort only to "demagogy", since each and every argument appears to be nothing but a *petitio principii*. Finally, the introduction of a style of thought is often equivalent to a *Gestalt* switch.[55]

---

[50] See Kuhn (1979c).

[51] Fleck (1935), to which could be also added a selection of his papers, Fleck (1983). See also Cohen, Schnelle (eds) (1986).

[52] Fleck (1935/1979), p. 21: "At least three-quarters if not the entire content of science is conditioned by the history of ideas, psychology and the sociology of ideas and is thus explicable in these terms".

[53] Fleck (1935/1979), p. 89.

[54] Fleck (1935/1979), p. 62.

[55] See Fleck (1935/1979), p. 92. As we can see, affinities with Kuhn's own model are quite remarkable. See also Baldamus (1972), (1977) and (1979), Buzzoni (1986), pp. 51–66,

London 1965: Kuhn Versus Popper

"[...] limitations of space have drastically affected my treatment of the philosophical implications of this essay's historically oriented view of science. Clearly, there are such implications, and I have tried both to point out and to document the main ones. But in doing so I have usually refrained from detailed discussion of the various positions taken by contemporary philosophers on the corresponding issues".[56] Thus wrote Kuhn in February 1962, seeing *The Structure of Scientific Revolutions* to the press.

However, Kuhn's book is ground-breaking not only for the new ideas that it purports to introduce but also because these ideas appear to clash with the dominant tradition in the philosophy of science.

In his review, Joseph Agassi underlines, among other aspects, how Kuhn regrettably fails to confront Popper, and actually "entirely ignores [him]".[57] A few years later, however, when the debates over his book (started already in 1961[58]) are more heated than ever, Kuhn and his critics are offered the opportunity for a close and detailed confrontation. The occasion is the International Colloquium in the Philosophy of Science, held at Bedford College in Regent's Park, London. Exactly four years after facing a particularly challenging audience in Oxford,[59] Kuhn faced a new and perhaps even more experienced one: up to that moment, his views were the most serious challenge to the critical rationalism of Popper and his workshop.[60]

Campelli (1999), Cohen, Schnelle (eds) (1986), Rossi (1981) and (1983), Schäfer, Schnelle (1980), Schnelle (1982), Stock (1980) and Wittich (1978) and (1981).

[56] Kuhn (1962a), p. x. And a few pages before: "Space limits of the *Encyclopedia* made it necessary [...] to present my views in an extremely condensed and schematic form" (*ibidem*, p. viii).

[57] Agassi (1966a), p. 121. Indeed, in *The Structure of Scientific Revolutions* we can find only a handful of passages that refer to Popper and to his model for the growth of scientific knowledge, and most of them are indirect. The most important of them is at the beginning of ch. 8, devoted to "The Response to Crisis": "No process yet disclosed by the historical study of scientific development at all resembles the methodological stereotype of falsification by direct comparison with nature" (Kuhn (1962a), p. 77). In his own copy of the book, Popper highlights (with a cross in the margin, as he used to do) only this sentence, that certainly constitutes a heavy personal lunge. Moreover, it is to be noted how rare are references to *philosophers* of science (Hanson is among the very few exceptions); much more frequent are references to works by *historians* of science, and to classics of the *history* of science.

[58] During the Symposium on the History of Science, held in Oxford in 1961.

[59] The Bedford Colloquium was held 11–17 July 1965; the Oxford Symposium was held 9–15 July 1961.

[60] Together with his pupils, Popper formed a group that was later described as a "circle", or a "school". However, as John Watkins later observed, "there *ought* not have been [a school]" (Watkins (1987), p. 213): indeed, in his (1959), Popper contrasted the *school* of Pythagoras, where the master's doctrine was to be preserved and handed down to later generations unchanged, with the *tradition of critical discussion* inaugurated by Thales and marvellously continued by his pupils and successors. Rather than a school, Popper tried to run a "workshop", as Joseph Agassi describes it in his (1993a). On the difference between "school" and "workshop" see Agassi (1981b) and Agassi, Jarvie (1979), p. 440.

Towards the Bedford Colloquium

Jointly organized by the British Society for the Philosophy of Science and the London School of Economics and Political Science, under the auspices of the Division of Logic, Methodology and Philosophy of Science of the International Union of History and Philosophy of Science, the International Colloquium for the Philosophy of Science was held at Bedford College, Regent's Park, from 11 to 17 July 1965. Thanks to the huge efforts of the honorary secretary, Imre Lakatos (with the help of the honorary joint secretary, John Watkins), it gathered in London the most influential protagonists of the then ongoing debates about the philosophy of science and remained the talk of the philosophical community for quite some time.

The "myth", as Agassi labels it,[61] reports three significant events. First, the alleged reconciliation between Popper and Carnap, the two old antagonists. Both were hoping to see the other accept their own view of philosophy and of the difference between them: Carnap wanting to have Popper admit that there was a misunderstanding rooted in Popper's exaggeration of the difference between them, and Popper to see Carnap admit that there was a genuine disagreement. No one yielded: on induction they were as far apart as ever, and their confrontation looks somehow artificial.[62] More significant for Popper were his opening address, "Rationality and the Search for Invariants"[63] and the beginning of a distressing quarrel with William W. Bartley III,[64] who had been Popper's favourite pupil.[65] Finally, the clash between Kuhn, on the one side, and Popper and his disciples on the other.

The proceedings of the conference, bound in four volumes published between 1967 and 1970,[66] are among the most influential books in the debates in the philosophy of science of the second half of the past century. In particular, the fourth

---

[61] See Agassi (1986), a review of Radnitzky, Andersson (eds) (1978).

[62] See Lakatos (ed.) (1968), pp. 258–314, and Lakatos' own assessment of the confrontation, his (1968): Lakatos wanted the meeting between them to have the opportunity to have his word in the debate and offer his own reading of it.

[63] Later published as Popper (1998b). In it Popper returns to the question of metaphysical research programmes in physics, tracing the history and the dominance of the programme (that originated with the speculations of Parmenides) of explaining change by concentrating on what does not change.

[64] See Lakatos, Musgrave (eds) (1968), pp. 40–119; Bartley's papers are his (1968a) and (1968b), while Popper's reply is in Popper (1968a). For a discussion and reconstruction of the quarrel and of the reasons behind it, see Gattei (2002a).

[65] "In this instance the suspicion persists that the sharp tone of Bartley's paper, which Popper felt as an aggressive personal affront, was encouraged by a third party determined to make mischief" (Miller (1997), p. 395). With much tricky manoeuvring by Lakatos ("a first-rate master of intrigue": Agassi (1993a), p. 156) both before the meeting (siding with Bartley against Popper) and after it (siding with Popper against Bartley), the two fell out publicly and embarrassingly. This led to a big rift, a wound that never fully healed, though a friendship and interaction were resumed some twelve years later, when Bartley undertook the task of seeing Popper's *Postscript to The Logic of Scientific Discovery* through the press. Things, however, were not as free and easy as they once were: they abstained from reference to each other's views.

[66] Lakatos (ed.) (1967) and (1968), Lakatos, Musgrave (eds) (1968) and (1970).

volume "is a rational reconstruction and expansion rather than a faithful report of the actual discussion. The whole volume arises from one symposium, the one held on 13 July on *Criticism and the Growth of Knowledge*".[67]

Though in the English-speaking world the volume is known as "Kuhn-centred", the items are clearly two: Kuhn and Lakatos. Those who attended the 1965 session of the colloquium actually witnessed the debate between Kuhn, Popper and his disciples, but the 1970 readers of the book had in their hands something more than that: in fact, Lakatos took the occasion to develop, in one of his most famous essays, his methodology of scientific research programmes.

In order to better understand the climate and background of the clash between Kuhn and Popper (and Feyerabend), I offer a reconstruction of the various steps that led to the confrontation.[68] The animated steps which led to the Colloquium and to the session of 13 July, in particular, are revealing of the character (and possibly the future projects) of Lakatos. It becomes quite clear that he wanted somehow to take over Popper's leadership both within the London School of Economics and the wider international philosophical community.[69]

In a notice directed to the members of the British Society for the Philosophy of Science,[70] on behalf of the organizing committee[71] Watkins announces the

---

[67] Lakatos, Musgrave (eds) (1970), p. vii. Although Agassi's harsh comment on the volume ("Jokes and profusion of scholarly nonsense aside, I do not know what to do with this wretched volume. [...] I do not even know what problems all these papers are facing, and I have many interesting details to quote from the various papers which may or may not give an image similar to Sterne's detailed and chaotic but rather charming picture of inept provincial life": Agassi (1971), p. 323), is motivated on the theoretical level, since the contributors to the volume often get involved in abstract disputes on the possible interpretations of their respective theories, from a historical point of view the relevance of this book is undeniable. Popperians and anti-Popperians set up their respective positions, though with some excess of analysis.

[68] What follows is based on my researches in the archives and personal libraries of both Popper and Lakatos, and also on several discussions with some of the protagonists.

[69] In a letter to Kuhn of 24 April 1973 he proudly describes the birth of "Lakatosian *school*" of philosophy of science within the LSE: "Here at LSE, since 1965, a new young generation grew up, for whom naive falsificationism and Karl [Popper] are mere historical curiosities; they set out, however, to try out your and my philosophical ideas on the testing ground on the *history of physics*, and, to some extent, also on the *history of economic thought*. (A still younger generation works on the history of chemistry and biology.) They are competent, and, indeed, brilliant young men, and I am rather proud of them" (Lakatos Archive (13.512)). The letter is marked "private and confidential", and mentions Elie Zahar, John Worrall, Peter Clark, Peter Urbach, and "also" Paul Feyerabend. After his death, John Worrall confirmed that: "Fortunately he [...] leaves behind him (and it was of this achievement that he was most proud) a thriving research programme manned, at the London School of Economics and elsewhere, by young scholars engaged in developing and criticising his stimulating ideas and applying them in new areas" (Worrall (1976), p. 7).

[70] Undated, but it is likely to have been written a few months before the Colloquium. See Popper Archive (80.1).

[71] Consisting of William C. Kneale (chairman), Stephan Körner, Karl R. Popper, Heinrich R. Post, John O. Wisdom, Imre Lakatos (honorary secretary) and John Watkins (honorary

International Colloquium in the Philosophy of Science; among other things, he says that Lakatos would be abroad till the end of May: he was in the United States for a series of lectures and seminars.[72] The "provisional program" of the Colloquium, attached to Watkins' notice, describes the session as follows:

July 13, Tuesday	*Criticism and the Growth of Knowledge I*
Chairman:	Sir Karl R. Popper
9:15–10 T.S.	Kuhn: *Dogma versus Criticism*
10:15–11 P.	Feyerabend: *Criticism versus Dogma*
11:15–12:45	Discussion

Fundamental here are the contrasting words "criticism" and "dogma", chosen in order to emphasize the differences and characterize the two opposing positions – two diametrically opposed positions. Not only was the echo of Kuhn's incisive paper read in Oxford in 1961 still very strong,[73] but on the role and function of "dogma" in science hinged the very contrast with Popper: as Kuhn clearly said in his paper, "it is precisely the abandonment of critical discourse that marks the transition to a science".[74]

Furthermore, it should have been Feyerabend who replied to Kuhn's challenge,[75] arguing for Popper's critical rationalism.[76] However, for health reasons, Feyerabend decided not to attend the Colloquium[77] and his place was taken by Watkins, who

joint secretary).

[72] See Kuhn's harsh *post scriptum* in a letter to Lakatos dated June 23, 1965, in Lakatos Archive (13.512); this *post scriptum* is not present in the copy of the same letter Kuhn sent to Popper (Popper Archive (80.9)).

[73] It is reasonable to think that Lakatos had the idea to set up the Bedford Colloquium (something Popper at first strongly opposed) after hearing Kuhn's paper in Oxford.

[74] Kuhn (1970a), p. 6. "I suggest that Sir Karl has characterized the entire scientific enterprise in terms that apply only to its occasional revolutionary parts. His emphasis is natural and common [...]. Nevertheless, neither science nor the development of knowledge is likely to be understood if research is viewed exclusively through the revolutions it occasionally produces" (*ibidem*). See also Lakatos (1978c), p. 207.

[75] "I had read earlier drafts of Kuhn's book and had discussed their contents with Kuhn" (Feyerabend (1970a), p. 219). Traces of these discussions remain in a fragment of their correspondence: see Feyerabend (1995a) and (2006).

[76] Actually, Lakatos first asked Jagdish Hattiangadi to prepare a rejoinder, but Kuhn objected: Hattiangadi (a pupil of Popper's) was going to move to Princeton to study with him, and his critical remarks would have put him in a difficult position. Hattiangadi's remarks were therefore limited to a few comments from the audience.

[77] In a letter to Lakatos from Berkeley, dated 9 June 1965, Feyerabend writes that he "fell ill again" and that he will "presumably spend the next two weeks in a hospital in the Mid-West (hence, do not try to get in touch with me for the next two or three weeks – I shall be out of circulation)" (in Lakatos Archive (13.272); Feyerabend often suffered from strong pains in his back, due to a wound he received during the Second World War, and he often went to hospital for brief periods: see his (1995b)). He also proposes to write a brief essay which Lakatos may read or have otherwise presented at the conference: "the content of my essay is very clear in my mind, though whether this clarity will be preserved when it is put to paper I do not know" (in Lakatos Archive (13.272)). Something similar happened a few years

received drafts of Kuhn's paper rushed across the Atlantic as they left Kuhn's typewriter.[78]

Lakatos as well was supposed to offer a response to Kuhn:[79] in a letter to Watkins dated March 2, 1965, Lakatos proposes the following draft programme:[80]

Tuesday, July 13	*Criticism and the Growth of Science*
(Popper in the Chair)	
10–12:45	Lakatos, Kuhn, Feyerabend
3–5:30	Panel: Grover Maxwell, Hanson, Hesse, Quine, Medawar, Koestler

However, the burdensome organizing commitments prevented Lakatos from completing his intervention, which was not presented.[81]

Until the very last days Kuhn's participation was suspended, since Lakatos, back from the States, only a few weeks before the Colloquium, changed the programme once again:[82]

July 13, Tuesday	*Philosophy of Science I*
Chairman:	R. Hall
3.00–4.30	S. Kuhn and J.W.N. Watkins: *Dogma and Revolution in the History of Science*
5.00–6.30	Discussion

later, shortly after Lakatos' death, for the coming conference on methodology in physics and economics which he and Spiro Latsis were organizing in Nauplion, Greece. Everyone wanted Feyerabend to take Lakatos' place, but Watkins received "a letter from him enclosing a tape and saying that if, as seemed rather likely, he did not turn up in person at Nauplion we could play this recording of his lecture instead. I angrily returned it to him" (Watkins (2000), p. 50). However, on that occasion Watkins' angry outburst worked and Feyerabend showed up.

78 See Watkins (1970), p. 25.

79 "Feyerabend and Lakatos were to have given the other papers; but the first could not come and the second found that, in arranging this colloquium, he had brought into existence a many-headed monster attending to whose multiplying demands would keep him busy twenty-four hours a day" (Watkins (1970), p. 25).

80 Lakatos Archive (14.8). In a hand-written note Lakatos also asks whether to invite Toulmin and Ayer for the discussion.

81 He contributed to the Colloquium with his (1968), on Carnap and Popper. His original reply to Kuhn is Lakatos (1969), an earlier version of which is the unpublished 1967 typescript "Demarcation Criterion and Scientific Research Programmes" (in Lakatos Archive (6.6)). This latter paper is particularly telling, since in it Lakatos aims to outline his view of the Popper–Kuhn confrontation, arguing that Popper's and Kuhn's views are "perfectly compatible" (p. 1). There he sketches Popper's demarcation criterion and a slightly modified formulation of Kuhn's idea of normal science, Popper's and Agassi's idea of metaphysical research programmes and Kuhn's paradigms; he often refers to Popper's *Postscript*. Lakatos (1970) is a considerably revised and expanded version of both this work and Lakatos (1969).

82 See the draft dated 15 June 1965 in Lakatos Archive (14.1), and Lakatos' letter to Kuhn on 18 June, in Lakatos Archive (13.512).

The unexpected and surreptitious change made Watkins "very happy. [...] For Kuhn, however, the programme change was not so agreeable. He had expected that Feyerabend and Lakatos would write independent papers".[83] He became furious with Lakatos, for he was left not only without Feyerabend,[84] but also without the chairman, Popper, his main interlocutor.[85] In two letters, to Popper and Lakatos, Kuhn stated his decision not to attend the Colloquium,[86] postponing his confrontation with Popper to the volumes of The Library of Living Philosophers, for which his paper was originally prepared.[87] Only a phone call by Popper made him change his mind and come to London, where he was met by Popper and his wife.[88]

The final programme of the Tuesday afternoon session, as it appears in the booklet with the official programme of the Bedford Colloquium, is the following:[89]

July 13, Tuesday	*Philosophy of Science* I
Chairman:	Sir Karl Popper
p.m. 3.00–4.30	T. S. Kuhn and J. W. N. Watkins:
	Criticism and the Growth of Knowledge
5.00–6.30	Discussion[90]

[83] Watkins (1970), p. 25.

[84] With whom Kuhn had very close interactions and intellectual exchanges when they were both members of the Philosophy Department of the University of California at Berkeley: see Feyerabend (1970a), pp. 197–198, and also the live confrontation documented in Feyerabend (1995a). Kuhn acknowledged Feyerabend's criticism in the "Preface" of the *Structure*: Kuhn (1962a), p. xii.

[85] The official motivation for Popper's withdrawal was the committee's decision that no name should occur twice on the programme, and Popper was supposed to give the opening address, "Rationality and the Search for Invariants", later published as Popper (1998b); moreover, it would have been unfair to Kuhn not only to have to face an opponent (Watkins), but also to have a chairman allied with him. See the letter from Popper to Kuhn on 7 July 1965, in Popper Archive (317.17).

[86] Kuhn's letter to Popper (in Lakatos Archive (13.512) and Popper Archive (80.9)) and to Lakatos (in Lakatos Archive (13.512)), both dated 23 June 1965, where Kuhn laments also the change in the title of the session, which had been decided more than a year before.

[87] In the following years, it took long discussions with the editor and the advisory board of The Library of Living Philosophers to have Kuhn (1970a) first published in Lakatos, Musgrave (eds) (1970).

[88] See Kuhn's letter to Popper on 30 June 1965 (in Popper Archive (317.17). Here Kuhn once again insists that the title of the session be changed back to "Criticism and the Growth of Knowledge", and that Popper be officially associated with the programme either as chairman or as speaker.

[89] Lakatos Archive (14.1), p. 3. A slightly revised version appears in the first volume of the proceedings: Lakatos (ed.) (1967), p. vii.

[90] See also Gattei (2000a).

Kuhn's lunge

Kuhn begins his paper by highlighting the close resemblance between his own position and that of the session's chairman, Karl Popper:[91] both deal with the very same data, but from them get different *Gestalten*.[92] In order to clarify the differences, Kuhn draws attention to four sets of typically Popperian phrases that he would have never used in the same places.

First, Popper says that scientists presuppose their theories and then test them, hence the idea that knowledge grows through a continuous overthrowing of ideas.[93] On the contrary, according to Kuhn, scientists first assume a "constellation"[94] of theories shared by the scientific community, and then put to test not that constellation, but their very ability and ingenuity to solve the puzzles they face during their research activity. Therefore, a possible failure to do so reflects on themselves, not on the theory:[95] "in no usual sense [...] are tests directed to current theory. On the contrary, when engaged with a normal research problem, the scientist must *premise* current theory as the rules of the game. His object is to solve a puzzle, preferably one at which others have failed, and current theory is required to define that puzzle and to guarantee that, given sufficient brilliance, it can be solved".[96] Only occasionally a

---

91 According to Kuhn, they both reject the idea that science progresses by accumulation; both emphasize the revolutionary process by which an older theory is overthrown and replaced by a new one (Kuhn adds that the two are "incompatible" (Kuhn (1970a), p. 2), while Popper does not stress this feature and often speaks of theories that survive as special cases of new ones, of which they constitute a good approximation: this conviction of Popper's, says Feyerabend, "only betrays his inability to distinguish formal from content-related (semantic) issues": Feyerabend (1978b), p. 179, n. 56); both, in particular, emphasize "the intimate and inevitable entanglement of scientific observation with scientific theory" and "are correspondingly sceptical of efforts to produce any neutral observation language" (Kuhn (1970a), p. 2); finally (even if this list is not exhaustive), both insist that scientists aim at providing explanations of observed phenomena and do so in terms of real objects (though it will become clear that they disagree about the meaning of this latter term).

92 The original version of Kuhn's paper (a copy of which is preserved in Popper (80.9)) was actually titled "Logic of Discovery or Psychology of Research: a *Gestalt* switch?": see Kuhn (1970a), pp. 2–3, and below, n. 142.

93 "Revolution in perpetuity", as Kuhn stresses once again in his last interview, "is a contradiction in terms" ((I-1997), p. 177).

94 Kuhn (1970c), p. 175.

95 "[...] in the final analysis it is the individual scientist rather than current theory which is tested" (Kuhn (1970a), p. 5).

96 Kuhn (1970a), pp. 4–5. That is why Kuhn chooses the term "puzzle" instead of "problem": "Puzzles are, in the entirely standard meaning here employed, the special category of problems that can serve to test ingenuity or skill in solution. [...] the really pressing problems, e.g. cure for cancer or the design of a lasting peace, are often not puzzles at all, largely because they may not have any solution. [...] Though intrinsic value is no criterion for a puzzle, the assured existence of a solution is" (Kuhn (1962a), pp. 36–37). Kuhn rejects Popper's choice of words as too harsh (Kuhn (1970b), pp. 233–234): Popper calls failed predictions "refutations", while he prefers "anomalies" (he borrowed the term from Reichenbach (1944)). There is not much to a name, surely – but by any name, refutations of successful theories are

repeated failure to solve puzzles within the context of the paradigmatic theory leads to casting doubt on the validity of the theory itself.

---

genuine discoveries: the value of a theory makes its refutation valuable too. Popper decidedly prefers "problem" to "puzzle". Already in *The Logic of Scientific Discovery* he had remarked the positivist-cum-Wittgensteinian flavour of this term: "The positivist dislikes the idea that there should be meaningful problems outside the field of 'positive' empirical science – problems to be dealt with by a genuine philosophical theory. [...] He wishes to see in the alleged philosophical problems mere 'pseudo-problems' or 'puzzles'. Now this wish of his – which, by the way, he does not express as a wish or a proposal but rather as a statement of fact – can always be gratified. For nothing is easier than to unmask a problem as 'meaningless' or 'pseudo'. [...] The dogma of meaning, once enthroned, is elevated forever above the battle. It can no longer be attacked. It has become (in Wittgenstein's own words) 'unassailable and definitive'" (Popper (1935, 1959), p. 51). In his reply to Kuhn he further develops this point: "The choice of this term [puzzles] seems to indicate that Kuhn wishes to stress that it is not a really fundamental problem which the 'normal' scientist is prepared to tackle: it is, rather, a routine problem, a problem of applying what one has learned" (Popper (1970), p. 53). And he continues: "I do not know whether Kuhn's use of the term 'puzzle' has anything to do with Wittgenstein's use. Wittgenstein, of course, used it in connection with his thesis that there are *no genuine problems* in philosophy – only puzzles, that is to say, pseudo-problems connected with the improper use of language. However this may be, the use of the term 'puzzle' instead of 'problem' is certainly indicative of a wish to show that the problems so described are not very serious or very deep" (*ibidem*, n. 1); see also Popper (1974a, 1976), p. 122. This discussion has also an interesting pedagogical aspect that lies at the root of Popper's reaction to Kuhn and is very telling of the different metaphysical approach behind their respective positions. It is not simply a matter of a *Gestalt*-shift, as Kuhn regards it (see below, n. 142) – rather, what distinguishes them is their very theory of rationality: in Popper's view Kuhn's "normal scientist" is "a person one ought to be sorry for. [...] The normal scientist [...] has been taught badly. I believe, and so do many others, that all teaching on the University level (and if possible below) should be training and encouragement in critical thinking. The 'normal' scientist, as described by Kuhn, has been badly taught. He has been taught in a dogmatic spirit: he is a victim of indoctrination. He has learned a technique which can be applied without asking the reason why [...] He is, as Kuhn puts it, content to solve 'puzzles'" (Popper (1970), pp. 52–53). Feyerabend will raise the very same point: see the title of his own reply to Kuhn, Feyerabend (1970a). Once again (see above, ch. 1, pp. 16–19) Popper might have taken inspiration from Julius Kraft and Leonard Nelson: a scholar of Socrates, Plato and Kant, Nelson developed a method for the teaching of philosophy that he derived from philosophy itself. He highlighted the intellectual activity related to the process of learning a subject, rather than the specific details of that subject, claiming that a dispute over an issue leaves in the pupil more profound traces than the traditional teaching of it: see Nelson (1949). In the 1920s Popper (as well as Wittgenstein, who left academia for elementary school teaching in remote Austrian villages) was deeply involved in the Austrian school-reform movement led by Otto Glöckel and supported, among others, by Karl Bühler (Popper's teacher at the University of Vienna and at the Pedagogic Institute), also publishing a few papers and reviews in *Schulreform* and *Die Quelle*, journals that regularly discussed theoretical and practical issues of school reform (see Popper (1925), (1927), (1931) and (1932)). On this, see Bartley (1969), (1970) and (1974), together with Hacohen (2000), ch. 3, especially pp. 107–116. On the pedagogical aspects of Popper's discourse and their consequences, see Wettersten (1987a) and (1987b), Agassi (1987), Long (1987), Zecha (ed.) (1999) and Segre (2002). See also Agassi (1984).

Revolutions, as Kuhn understands them, are very rare episodes in the history of science.[97] That is why Popper, as Kuhn sees him, "characterized the entire scientific enterprise in terms that apply only to its occasional revolutionary parts".[98] As many others before him, he overlooks the function of normal science, and "neither science nor the development of knowledge is likely to be understood if research is viewed exclusively through the revolutions it occasionally produces".[99] In fact, "a careful look at the scientific enterprise suggests that it is normal science, in which [Popper]'s sort of testing does not occur, rather than extraordinary science which most nearly distinguishes science from other enterprises".[100]

According to a second set of phrases, very often used by Popper, science is a particular case of the process through which we learn from our errors. True, says Kuhn – but learning from our errors makes sense only against the background of a set of accepted rules and procedures, that can be employed to identify a single failure in applying them. Therefore, according to Kuhn, learning from our errors takes place only during the periods of normal science. The Popperian use of the term "error", referring to theoretical systems (or frameworks) that dominated science in the past, such as Newton's mechanics, is therefore mistaken.

Kuhn moves then to the most Popperian of terms, "falsification", and notices that however Popper repeatedly and explicitly acknowledged that conclusive falsifications cannot exist, he actually behaves as if it could.[101] "Having barred conclusive

---

[97] At least, this is what Kuhn thinks at the time of writing *The Structure of Scientific Revolutions*, and certainly until the mid-1970s. In his later writings, however, he seems to change his views. For he no longer refers to "big" changes, such as the Copernican revolution, involving major paradigm-shifts and changes of entire world-views. He rather considers "smaller" events, relatively minor revolutions taking place within a given specialty or subspecialty. Indeed, in his last published papers Kuhn describes the progress of the sciences in terms of the progressive speciation of ever more isolated scientific disciplines, not communicating with each other. In this context, revolutions become much more frequent but are confined to a considerably lesser scale, that of the discipline or sub-discipline within which they occur. This has two significant consequences. First, it turns Kuhn's model much closer to Popper's view of science as "revolution in permanence". Second, it considerably scales down the scope and relevance of the incommensurability involved in a revolution: whereas, on a larger (world-view) scale, there may be significant changes in the conceptual and terminological network of a theory, on a smaller (intra-discipline) scale such changes give rise at most to difficulties that may easily overcome and dealt with.

[98] Kuhn (1970a), p. 6.

[99] Kuhn (1970a), p. 6.

[100] Kuhn (1970a), p. 6.

[101] "What is falsification if it is not conclusive disproof?", asks Kuhn (in his (1970a), p. 15). Kuhn claims, with Popper's books in hands, that locutions like "falsification" and "refutation" are antonyms of "proof". But in the "Index of Subjects" of the very book of Popper he claims to have read, he could have found "Disproof, no conclusive disproof can be produced, 42, 50, 81–87" (Popper (1935, 1959), p. 471). No doubt, he read this book having in mind his own (pre-formed) picture, or paradigm. Kuhn (1970a) is actually a detailed criticism of a non-existent philosopher, the legendary naïve falsificationist Karl R. Popper, also described by Lakatos (in his (1969)) as "$Popper_1$". At times Lakatos felt the necessity to operate with numbered Poppers: "$Popper_0$" was "the dogmatic falsificationist who never published a

disproof", says Kuhn, "he has provided no substitute for it, and the relation he does employ remains that of logical falsification. Though he is not a naïve falsificationist, [Popper] may, I suggest, legitimately be treated as one".[102] There is no exclusively logical criterion, Kuhn claims,[103] that can dictate the conclusions scientists must draw when facing an anomaly.[104] The criteria Kuhn has in mind are not logical ones, but criteria that allow us to understand the values that make scientists react as they actually do: in other words, we have to devote ourselves to a sociological analysis of the scientific community. Rather than employing terms such as "refutation", Kuhn speaks of a paradigm that is no longer able to sustain a puzzle-solving tradition. When a sufficiently high number of scientists become convinced of this inability, they decide to transfer their commitment to another paradigm (if any) that is able to keep the promises the old one proved unable to keep.

"Almost everything said so far rings changes on a single theme. The criteria with which scientists determine the validity of an articulation or an application of existing theory are not by themselves sufficient to determine the choice between competing theories".[105] Popper erred by transferring specific characteristics of everyday research to the (occasional) revolutionary episodes and therefore ignored the everyday enterprise entirely. "In particular, he has sought to solve the problem of theory choice during revolutions by logical criteria that are applicable in full only when a theory can already be presupposed",[106] that is, during periods guided by a paradigm.

---

word" (Lakatos (1970), p. 181), first introduced and allegedly criticized by Ayer in his (1936), p. 38, and then by many others, including Béla Juhos and Ernst Nagel (according to Lakatos himself, Popper_0 is the outcome of a prejudiced misreading: "his *Logik der Forschung* is the strongest ever criticism of dogmatic falsificationism" (Lakatos (1970), p. 181, n. 2); "Popper_1" was the naïve methodological falsificationist (another non-existent Popper by Lakatos' own admission, whom he also at times confused with Popper_0); and Popper_2 was the sophisticated methodological falsificationist – all this to agonize over which was the "real" Popper. See also Watkins (2002), p. 5, and Lakatos' unpublished essay "On the so-called 'deductive' model of explanation", in Lakatos Archive (5.4), referred to in Motterlini (2002), pp. 32–33.

[102] Kuhn (1970a), p. 14. In his (1983), p. xxxiv, Popper explains this "really astonishing" passage by hypothesizing that "Kuhn, early in his career, formed a theory of my views which became his paradigm of Popper: Popper was the man who replaced verificationism with ('naïve') falsificationism. Kuhn formed this paradigm (according to his own indications) before he ever read any of my writings. When at last he read *The Logic of Scientific Discovery*, he read it in the light of this paradigm. Many passages of this book (one on the page immediately after my introduction of the idea of falsification) showed that I did not conform to his paradigm. But, as we have learnt from Kuhn, paradigms are not given up so easily".

[103] In a letter to Popper dated 30 June 1965, summarizing the main points of his paper, Kuhn claims that logical formulations of a theory are necessarily incomplete; this is the reason why he thinks that the notion of paradigm better approximates to what is the actual form of scientific research; the letter is in Popper Archive (317.17).

[104] See Kuhn (1970a), p. 19. "Rather than a logic, [Popper] has provided an ideology; rather than methodological rules, he has supplied procedural maxims" (*ibidem*, p. 15). Compare these words with those Feyerabend addresses to Kuhn in their exchange on the draft of *The Structure of Scientific Revolutions*: see Feyerabend (1995a), pp. 355, 360 and 367.

[105] Kuhn (1970a), p. 19.

[106] Kuhn (1970a), p. 19.

Finally, after affirming the impossibility of defining a satisfying notion of verisimilitude, and therefore of speaking of progress in terms of ever better approximations to the truth,[107] Kuhn underlines that *explanation* in science must, "in the final analysis, be psychological or sociological. It must, that is, be a description of a value system, an ideology, together with an analysis of the institutions through which that system is transmitted and enforced".[108] Popper opposes a psychological or sociological approach to science – and yet, remarks Kuhn, he advances and supports an ideology and a system of values for science, therefore working in this scope: Popper rejects the "psychology of knowledge", but it is "a long step from the rejection of the psychological idiosyncrasies of an individual to the rejection of the common elements induced by nurture and training in the psychological make-up of the licensed membership of a *scientific group*".[109]

Kuhn and Popper often hold close positions, as becomes clear from the former's discussion of the latter's term "falsification": I think Popper never believed in a purely logical criterion of demarcation, since already in his *Logik der Forschung* he had underscored the relevance of *methodological decisions* within a set of rules of scientific method.[110] On the other hand, a decidedly conspicuous disagreement marks their views of criticism. Indeed, if they could agree in principle – for Popper theories must be falsifiable, while for Kuhn what distinguishes science from non-science is a puzzle-solving tradition – their underlying approach is in sharp contrast. For Popper the testing process undergone by a theory is a special case of the unceasing critical discussion of foundations which alone can warrant the rational character of science without bringing it down into dogmatism, while for Kuhn history suggests

[107] In Kuhn's view, scientific progress is not a progressive path, a series of better and better approximations to the "Truth". Indeed, he distinguishes between the world in itself and the world of phenomena (even if he occasionally claims that we can do also without the concept of world-in-itself): for him the reality which is usually addressed in everyday or scientific contexts is *a* world of phenomena, not *the* (single) world of phenomena, and certainly not the world in itself. In the web of similarity and dissimilarity relations that constitute a given world of phenomena, we find a blend of objective and genetically subjective elements (not at the individual level, but at the social one: if we want to find an idealistic element in Kuhn's idea of reality, this has a *social*, not an individual nature). When examining a web of this kind we cannot separate those two moments. As a consequence, it is not possible to "purify" the world of phenomena from its subjective components, in order to achieve a "pure" picture of the objective elements, absolute reality or the world in itself. On the contrary, the concrete properties of the world in itself are inaccessible to us: even if we feel the resistance that world offers against our epistemic attempts, we are not in the position to grasp this very resistance in itself.

[108] Kuhn (1970a), p. 21. In a letter to Popper dated 30 June 1965 (in Popper Archive (317.17)) Kuhn claims that the line Popper seems sometimes to be drawing between history on the one hand and psychology, on the other, appears to be arbitrary: Kuhn believes that we can (and ultimately must) understand the nature of the growth of scientific knowledge through the understanding of the nature of the community responsible for its creation and protection.

[109] Kuhn (1970a), p. 22. From this and the above quoted passage it is clear that Kuhn holds a view of science as an *institution*: see Watkins (1970), p. 26, where Kuhn's picture of the scientific community is likened to Popper's picture of a closed society.

[110] See Popper (1935, 1959), ch. II, especially pp. 49–50. See also Gattei (2002a).

that "it is precisely the abandonment of critical discourse that marks the transition to a science".[111]

The clash involves the very basic assumptions of falsificationism.[112] Popper insists on the rational nature of science, marked by the openness of mind of its practitioners, that allows it to grow and progress. Extreme flexibility of thought and creative boldness are balanced by a relentless demand of refutability of our hypotheses. Scientists, according to Popper, should make an effort to refute their own theories rather than seek confirmations of them. The hallmark of intellectual honesty is in stating in advance under what conditions we would be ready to give up our theories. Without considering irrelevant questions of meaning, Popper is firmly convinced that scientific theories progress towards an ever better correspondence with reality.

By contrast, Kuhn seems to be drawing a picture of a scientific community like a closed society, formed by closed-minded people, bounded by and committed to certain procedural models – paradigms – that guide their theoretical and experimental activity.[113] Practitioners of a certain discipline attempt to frame nature in the bounds given by the paradigm. Revolutions are rather occasional events, usually the outcome of the scientists' inability to assimilate and analyse facts in the way the paradigm suggested. They are processes akin to religious conversions and commit the members of a scientific community to a new system of theories, practices and methods. Radical, but more often subtle meaning changes of key theoretical terms see to it that scientists bound to the new paradigm manage only partially to communicate with those supporting the old one. Although some of these changes can lead to actual improvements in the level of understanding of nature, Kuhn does not speak of approximation to the truth. He seriously threatens the rational image of science Popperians had carefully depicted. The harsh reactions to Kuhn's paper,[114]

---

[111] Kuhn (1970a), p. 6. Severity of tests and a problem-solving tradition: *both* characterize science, according to Popper. That is why Popper's and Kuhn's lines of demarcation so often coincide – but such a coincidence, Kuhn hastens to point out, is "only in their *outcome*; the process of applying them is very different, and it isolates distinct aspects of the activity about which the decision – science or non-science – is to be made" (*ibidem*, p. 7). The example is astrology: Popper excludes it from sciences for the way in which its practitioners explained their failures – Kuhn because though astrologers "had rules to apply, they had no puzzles to solve and therefore no science to practise" (*ibidem*, p. 9). What was lacking, in other words, was a puzzle-solving (or research) tradition, that is, the kind of activity that "normally" characterizes all sciences acknowledged as such: "To rely on testing as the mark of a science is to miss what scientists mostly do and, with it, the most characteristic feature of their own enterprise" (*ibidem*, p. 10).

[112] At the opening of his impromptu rejoinder, Popper says he is "in the strange position of being at the same time the chairman and the bone of contention" of the discussion (Popper Archive (75.5), p. 1).

[113] John Watkins suggested this very parallel in 1961 after reading the manuscript of *The Structure of Scientific Revolutions*: see Watkins (1970), p. 26.

[114] In his rejoinder to Kuhn, Watkins criticizes the secondary role played by tests in Kuhn's conception of the scientific enterprise and tackles the idea of normal science, which he identifies with "periods of theoretical stagnation" (Watkins (1970), p. 32). He even suggests that "Kuhn sees the scientific community on the analogy of a religious community and sees

the result also of a strong emotional involvement in the questions at issue, make use of terms such as "irrationalism" or "mob psychology", and make no mystery of the negative and dangerous consequences of Kuhnian ideas.

Popper's rejoinder

"Professor Kuhn's criticism of my views about science is the most interesting one I have so far come across".[115] He acknowledges Kuhn's merit for having highlighted an aspect of the scientific enterprise he had "completely overlooked",[116] namely, the existence of periods of normal science.[117] But, continues Popper, this is "the activity

science as the scientist's religion" (*ibidem*, p. 33). Normal science is also the critical target of Toulmin's paper, according to whom the distinction between normal and revolutionary science simply does not hold water. In accordance with the author of *The Structure of Scientific Revolutions*, he expresses the hope that in epistemology logic is joined and supported by sociology and psychology in epistemology. But he invites Kuhn (resuming Toulmin (1963), that commented on Kuhn (1963a)) both to give up his suggestion that science necessarily involves a certain form of dogmatism, and to give more details on the precise ways in which paradigms are employed (see Toulmin (1970)). Margaret Masterman's apparently friendly rejoinder draws attention to the ambiguity of Kuhn's term "paradigm" (highlighting its use "in no less than twenty-one different senses" (Masteman (1970), p. 61), inconsistent with one another (see *ibidem*, pp. 61–65) in *The Structure of Scientific Revolutions*), but stresses the relevance of normal science and the central role played by paradigms in the concrete practice of science (in particular, she emphasizes the paradigm's feature of existing *prior* to a theory). Interestingly, Masterman spots an important aspect missed by the other commentators: a paradigm is something scientists use *in the absence* of a theory (see *ibidem*, pp. 66–68 and 73–76).

[115] Popper (1970), p. 51. In some notes taken for a possible second edition of *Criticism and the Growth of Knowledge* Popper softens his generous remark, describing Kuhn's criticism as "*one* of the most interesting". In 1963 *The Structure of Scientific Revolutions* was widely discussed during Popper's seminar at LSE, during which Jagdish Hattiangadi presented a paper on it, later developed into a master's thesis, Hattiangadi (1965) (see Watkins (1970), p. 25). Popper replies to Kuhn's criticism in his (1970), (1974c), (1974d), (1975), (1976) and (1983), pp. xxxi–xxxv. However, in order to reconstruct his position I made ample use also of the private correspondence and unpublished material I found in Popper's archives. I am particularly referring to the papers collected in Popper Archive (75.5–10) and to an unfinished essay especially devoted to Kuhn, "Revolution and Continuity in Science", dated 6 February 1972 (Popper Archive (120.11)). The latter was probably a draft of Popper's reply to Kuhn for the Schilpp volumes, which, together with the other "Replies", he was writing in those years (they were eventually published as his (1974b)).

[116] Popper (1970), p. 52. The same he says in his second reply to Kuhn, Popper (1974c), and in his (1974d), his reply to Wisdom (1974a).

[117] Kuhn, however, emphasized its relevance too much, as Popper underscores during in the discussion (see Popper Archive (75.5)). In Popper Archive (80.9) are kept two rather different drafts of Kuhn's paper (which he had sent to Popper in view of their confrontation). On the second and more complete one Popper notes that normal science is "Kuhn's main discovery […]. Very important and very new: I certainly did not see it". Indeed, his rejoinder hinges on it. But Popper immediately asks himself whether Kuhn had not emphasized it too much, and contrasts it with what he calls "The heroic age of science".

of the non-revolutionary, or more precisely, the not-too-critical professional: of the science student who accepts the ruling dogma of the day; who does not wish to challenge it; and who accepts a new revolutionary theory only if almost everybody else is ready to accept it – if it becomes fashionable by a kind of bandwagon effect".[118] Such an attitude is regarded by Popper as "a danger to science and, indeed, to our civilization".[119]

For Popper science is an essentially critical enterprise, and therefore it is revolutionary in permanence.[120] Science is always revolutionary because it is thought in evolution, that is, critical thought. For him[121] the discovery of something is always a discovery *against* something else, because, as in the case of Christopher Columbus, it collides with a constellation of established prejudices.[122] The creative scientist does not seek an easy consensus, but gives rise to a frank dissent, even if difficult to handle, since "only in the change of a system [...] it is clearly shown the character of a science that draws teachings from reality, from experience".[123] Thus Copernicus went beyond the tradition affirmed by Ptolemy, Newton went beyond Galileo and Kepler, and Einstein beyond Newton. The overthrows of established ideas (the so-called "revolutions of ideas") are not exceptional episodes, but constitute the usual condition of scientific activity: science grows as a *revolution in permanence*.

On his part, Popper always stressed "the need for some dogmatism: the dogmatic scientist has an important role to play. If we give in to criticism too easily, we shall

---

[118] Popper (1970), p. 52.

[119] Popper (1970), p. 53.

[120] See Popper's "Revolution and Continuity in Science" (Popper Archive (120.11)) and Popper (1974c). From the evolutionary point of view developed by Popper, the routine seems to be characterizing the way in which animals learn, or the way in which they adapt themselves to the environment. Man, on the contrary, by means of the invention of language – that has, among others, descriptive and argumentative functions – "has begun to replace *routine* more and more by *critical approach*" (Popper (1974c), p. 1146), and science is the most advanced application of the critical approach to the growth of knowledge. Popper sees in science, taken in an evolutionary context, "the conscious and critical form of an adaptive method of trial and error" (*ibidem*, p. 1147): for this reason we can learn from our errors, in a permanently revolutionary process, constantly characterized by revolutions at various levels. See also Agassi (1966b) and (1973).

[121] See Popper (1972a).

[122] Popper is a firm supporter of dissent: "I am not an admirer of philosophical discipline" (Popper (1983), p. 7, where he also tells the story of the soldier who found that his whole battalion – except himself, of course – was out of step: "I constantly find myself in this entertaining position. And [...] I am content as long as enough members of the battalion are sufficiently out of step with one another"). He actually thinks it is possible to spot the secret of the flourishing of Greek philosophy, that at every new generation produced a new cosmology of surprising originality and profundity (see Popper (1959)), exactly in the *tradition of critical discussion*. The possibility of fighting with words instead of swords is, for Popper, the very basis of our civilization, and particularly of its legal and parliamentary institutions, as well as the hallmark of scientific reason.

[123] Popper (1979, 1994), p. 136.

never find out where the real power of our theories lies".[124] But this does not seem to be the kind of dogmatism Kuhn advocates: "He believes in the domination of a ruling dogma over considerable periods; and he does not believe that the method of science is, normally, that of bold conjectures and criticism".[125]

Already in 1934 Popper had spoken of falsifiability in terms of ability to lay oneself open to criticism,[126] and affirmed that no theory is falsified in a conclusive way. And even if he does not mention him explicitly, in his 1965 impromptu reply to Kuhn Popper is addressing Quine, who sits in the audience:[127] our blaming a particular hypothesis (or theory) in our theoretical system for clashing with reality is itself a conjecture, a new hypothesis that needs testing. It transforms our theoretical system into a new one, in its turn to be tested. In any case, it is always the whole system that is in question.[128]

---

[124] Popper (1970), p. 55; see also his (1940), p. 312, n. 1, (1963a), p. 247, and (1974b), p. 984. "That it is desirable that a theory should be defended with a certain dogmatism, so that it is not knocked out too quickly before its resources have been explored, Popper has never denied; but such dogmatism is healthy only as long as there are other people who are not inhibited from criticizing and testing a tenaciously defended theory. If *everyone* were [...] to preserve the current theories of science against awkward results, then those theories would, according to Popper, lose their scientific status and degenerate into something like metaphysical doctrines" (Watkins (1970), p. 28). See also Toulmin (1961), p. 81: "One cannot even label a false trail as such without exploring it some way first".

[125] Popper (1970), p. 55.

[126] See Popper (1935, 1959), ch. VI, "Degrees of Testability".

[127] This is a part of Popper's reply that was not included in the printed version, Popper (1970); traces of it remain in the notes Popper took during Kuhn's paper in view of his rejoinder, which are kept in Popper Archive (75.5).

[128] See Popper Archive (75.5), p. 2. Popper's views on the Duhem–Quine thesis are also expressed in his (1935, 1959) pp. 49–50, (1963b), pp. 238–239, and (1974b), p. 982. The so-called Duhem–Quine thesis is taken to be a sound criticism of Popper's falsificationism, according to which if an observation instance is not consistent with the prediction drawn from a hypothesis, the hypothesis is falsified. I do not think it is. The thesis claims that it is never possible to deduce any observable statement from a single hypothesis alone: hypotheses have always to be conjoined with other assumptions about background conditions, the reliability of measurements, the initial conditions and so on – in Duhem's words, "an experiment can never condemn an isolated hypothesis, but only a whole theoretical system" (Duhem (1906, 1914/1954), p. 183; or again: "the physicist can never subject an isolated hypothesis to experimental test, but only a whole group of hypotheses; when the experiment is in disagreement with his predictions, what he learns is that at least one of the hypotheses constituting this group is unacceptable and ought to be modified; but the experiment does not designate which one should be changed" (*ibidem*, p. 187)); or, to use Quine's, "our statements about the external world face the tribunal of sense-experience not individually but as a corporate body" (Quine (1951), p. 41). Therefore, the standard view goes, the falsification of a theory by an observation is not as straightforward as Popper's (quite naïve) schema suggests. The point raised is logical and from the logical point of view we have to accept it. However, let us first note a difference in attitude: while the Duhem–Quine thesis aims at rescuing a theory from criticism, Popper invites and encourages criticism. No conclusive proof (or disproof) can ever be produced, for it is always possible to say that experimental results are not reliable, or that the discrepancies that are asserted to exist between the experimental results and the

Moreover, Kuhn's arguments are logical ones: "Kuhn suggests that the rationality of science presupposes the acceptance of a common framework. He suggests that rationality *depends* upon something like a common language and a common set of assumptions. He suggests that rational discussion, and rational criticism, is only possible when we have agreed on fundamentals".[129] This is for Popper the thesis of *relativism*, and it is a logical one: "the myth of the framework", as he labels it, "a logical and philosophical mistake".[130] Admittedly, in every moment we are prisoners caught in the framework of our language, theories, past experiences and expectations – but we are prisoners "in a Pickwickian sense: if we try, we can break out of our framework at any time.[131] Admittedly, we shall find ourselves again in a framework, but it will be a better and roomier one; and we can at any moment break out of it again".[132] Critical discussion, in other words, is always possible, and the contrary thesis (i.e. the incommensurability thesis, the idea that different frameworks are like mutually untranslatable languages) is a dangerous *dogma* – "the central bulwark

---

theory are only apparent and that they will disappear with the advance of our understanding. By insisting on strict proof (or disproof) in the empirical sciences we will never benefit from experience, we will never learn from it how wrong we are (see Popper (1935, 1959), p. 50). Secondly, as a matter of fact, when a theory is falsified the whole system of which it is a part gets falsified with it. Newton's theory is a system: if we falsify it, we falsify the whole system. We may perhaps put the blame on one of its laws or another, but this means only that we conjecture that a certain change in the system will free it from falsification – or, in other words, we conjecture that a certain alternative system will be an improvement. Attributing the blame for a falsification to a certain subsystem is a hypothesis, a conjecture like any other, though perhaps hardly more than a vague suspicion, if no definite alternative suggestion is being made. And the same applies the other way round: the decision that a certain subsystem is not to be blamed for the falsification is likewise a typical conjecture. The attribution or non-attribution of responsibility for failure is conjectural, like everything else in science. What matters is the proposal of a new alternative and competing conjectural system that is able to pass the falsifying test. On the distinction between "the Duhem thesis", "the Quine thesis" and "the Duhem–Quine thesis" see Gillies (1993), Part II, especially ch. 5; see also Lakatos (1970), pp. 184–189, and the papers collected in Harding (ed.) (1976).

[129] Popper (1970), p. 56.

[130] Popper (1970), p. 56; see also Popper (1976) and (1994a). Popper's criticism of relativism is particularly effective in his (1962a) and (1963b).

[131] Kuhn correctly highlights the existence, in science, of a community of professionals whose training has been mostly by indoctrination. We live within communities and are a product of a century-old tradition (if each time we had to start from the beginning, as the positivists wanted, it is reasonable to think that we would reach more or less the point Adam reached: our progress beyond him is due to the existence of a tradition); we grew within a cultural framework and we are in need of it. Popper is fully aware of this. What he (together with Watkins) dislikes, under the rubric of "normal science" is mental rigidity and, contrary to Kuhn, he wishes to fight it. This is the meaning of the expression (somewhat *à la* Leon Trotzky) "revolution in permanence": although we are prisoners caught in the framework the tradition provided us with, we can always try to pull down the walls of the prison and escape. All we need is the will to do that: see the notes in Popper Archive (75.5).

[132] Popper (1970), p. 56. Popper spots this as the central point of disagreement with Kuhn and repeats it also a few days before the Colloquium, in a letter to Kuhn dated July 7 (Popper Archive (317.17).

of irrationalism".[133] The "myth of the framework" exaggerates a difficulty into an impossibility: however difficult, there is nothing more fruitful than the clash between different cultures of ideas. Denying this possibility is a mistake, since authentic progress springs from it. Incommensurability, in other words, however often taken for granted as a problem, reveals itself rather as a *solution*, an all too easy way out of problems: instead of confronting them, we deem them insurmountable, label them incommensurable and set them aside.

Just like Kuhn, Popper highlights some points of agreement,[134] but several are also the mutual misunderstandings. Kuhn, for instance, keeps regarding Popper as a naïve falsificationist,[135] even if in *The Copernican Revolution* he seems to accept Popper's ideas on the revolutionary character of the evolution of science, breaking away from them only for his "fideism": "a scientist *must believe* in his system before he will trust it as a guide to fruitful investigations of the *unknown*".[136] "But the scientist", Kuhn continues in the passage quoted by Popper, "pays a price for his commitment [...]. A single observation incompatible with his theory [may demonstrate] that he has been employing the wrong theory all along. His conceptual scheme must then be abandoned and replaced".[137] This is perfect falsificationism,

---

133 Popper (1970), p. 56; see also Popper (75.5), (1972b), pp. 215–216, and (1974d). For a detailed analysis of the "myth of the framework", or "myth of the paradigm", see Pera (1981), pp. 205–219. The incommensurability thesis is rooted in the very ideas of the founder of critical rationalism: Popper's insistence on the theoretical dependence of basic statements and on the conventional nature of their choice turns out to provide the implicit premises for the incommensurability thesis the myth of the paradigm draws its nourishment from. In fact, Popper's philosophy does contain the germs (but also the antidote, I think) of its own destruction. Incisively writes Pera: "The overall image we get from this situation is that of a theoretical construction in unstable balance, as a consequence of forces pushing in opposed directions: towards the empirical basis and the method of experience, on the one hand; and towards the lack of foundations and the science without experience, on the other. The image, in short, of a science erected on piles, and of boundary a philosophy of science" (Pera (1981), p. 219; the image of a science erected on piles is Popper's: see his (1935, 1959), p. 111).

134 As to falsifiability and the impossibility to provide conclusive falsifications, and the role both of them play in the history of science and in scientific revolutions, there is no difference, according to Popper, between their respective positions: "Kuhn's and my views coincide almost completely" (Popper (1983), p. xxxi).

135 It is "the legend of Popper", according to which he advanced a criterion of demarcation similar to that of logical positivists, simply replacing verification with falsification: see Kraft (1974), Popper (1974a, 1976), sections 16–17, and (1974b), pp. 963–976. See also Hacohen (2000), ch. 4, especially pp. 208–213.

136 Kuhn (1957), p. 75. In reporting these words in his (1983), p. xxxii, Popper italicizes the words "must believe" because fideism is the only point in this passage where Kuhn deviates from him: he would have said "may believe" or alternatively "may accept" his system only tentatively.

137 Kuhn (1957), p. 75. In reporting this passage (in his (1983), pp. xxxii–xxxiii), Popper slightly modifies it (Kuhn's actual words are: "But the scientist pays a price for this commitment to a particular alternative: he may make mistakes. A single observation incompatible with his theory demonstrates that he has been employing the wrong theory all along. His conceptual scheme must then be abandoned and replaced"). In particular, he

Popper remarks – actually, a thoroughly naïve falsificationism. It is therefore Kuhn,

---

changes "demonstrates" into "may demonstrate", and says he dislikes "conceptual scheme" and prefers "theory" instead. Apart from these differences, Popper continues, what Kuhn describes is "something like a 'methodological stereotype of falsification', to cite Kuhn's allusion to me in his later book, *The Structure of Scientific Revolutions*" (Popper (1983), p. xxxiii): the reference is to Kuhn (1962a), p. 77. Kuhn regards Popper as a naïve falsificationist, as the man who replaced verificationism with naïve falsificationism – again, this is nothing but "the *legend* or *paradigm*" (Popper (1983), p. xxxiv), a "legend" Popper relentlessnessly fought for his entire life. Indeed, if Popper is directly criticized in *The Structure of Scientific Revolutions* (albeit at times implicitly), Popper's falsificationist model is Kuhn's critical point of reference already in *The Copernican Revolution*. It is very interesting to read the pages from which Popper is quoting: there Kuhn says that the outline he provides (the one Popper quotes) "is the logical structure of a scientific revolution. A conceptual scheme, believed because it is economical, fruitful, and cosmologically satisfying, finally leads to results that are incompatible with observation; belief must then be surrendered and a new theory adopted; after this the process starts again" (Kuhn (1957), pp. 75–76). But this thoroughly Popperian reading, Kuhn immediately remarks, however useful ("because the incompatibility of theory and observation is the ultimate source of every revolution in the sciences", *ibidem*, p. 76) does not correspond to the actual practice of science: "historically the process of revolution is never, and could not possibly be, so simple as the logical outline indicates. As we have already begun to discover, observation is never *absolutely* incompatible with a conceptual scheme. [...] though scientists undoubtedly do abandon a conceptual scheme when it seems in irreconcilable conflict with observation, the emphasis on logical incompatibility disguises an essential problem. What is it that transforms an apparently temporary discrepancy into an inescapable conflict? How can a conceptual scheme that one generation admiringly describes as subtle, flexible, and complex become for a later generation merely obscure, ambiguous, and cumbersome? Why do scientists hold to theories despite discrepancies, and, having held to them, why do they give them up?" (*ibidem*). If Kuhn explicitly (albeit sketchily) criticizes Popper in *The Structure of Scientific Revolutions*, he implicitly addresses him already in *The Copernican Revolution*, while at the same time offering the historical background and also some interesting glimpses of the philosophical views that underlie his historical work. This substantiates my claim that the critical reference of Kuhn's philosophy has always been Popper's falsificationism, not Logical Positivism or Empiricism (although he grows intellectually within this philosophical tradition, he will address it only later in his life). Indeed, as Kuhn himself reports (see, for example his (1970a), p. 1, n. 3, or (I-1997a), p. 286; see also Popper (1974c), p. 1144) Popper's 1950 William James Lectures at Harvard proved to be crucial for Kuhn's own philosophical development. Indeed, *The Copernican Revolution* is the outcome of a series of lectures in the history of science held from 1951 to 1956 (that is, immediately after the meeting with Popper's philosophy) for the course of General Education at Harvard; *The Structure of Scientific Revolutions* was written shortly after it, and Popper is among the very few philosophers whose thought is explicitly addressed. Finally, Kuhn particularly valued the 1965 confrontation with Popper at the Bedford Colloquium (so much that he nearly withdrew from the programme when he realized Popper was not going to be his discussant, as previously agreed with the organizer, Lakatos: see above, p. 54). And their clash, again, proved crucial: although it had already received a number of important reviews, I think *The Structure of Scientific Revolutions* would have hardly had the impact it had (and, in a sense, still has) on the philosophy of science debates had it not been for the enormous success of the volume of proceedings grown out of that conference. Indeed, contrary to Lakatos' own philosophical-cum-political ambitions, that volume established Kuhn, not himself.

says Popper, who adopts "the methodological stereotype of falsification", and indeed "a far more simplistic stereotype of falsificationism than anything I myself ever said in my writings, my lectures, or my seminars".[138] Otherwise, Popper agrees with Kuhn when he underlines that observation is never absolutely incompatible with a theory, a consequence of the theoretical nature of observation terms.

More important, however, are the differences. More than on anything else, it is on truth that Popper's and Kuhn's views diverge.[139] The original title of Kuhn's paper – "Logic of Discovery or Psychology of research: a *Gestalt* switch?" – seems to suggest the actual incommensurability of their positions, the impossibility of switching from one to another without a full change of perspective, that is, without a different approach to science and philosophy – a different theory of rationality.[140]

---

[138] Popper (1983), p. xxxiii.

[139] "I do not doubt that this is one of the points on which we are most deeply divided": Popper (1970), 56. Kuhn, says Feyerabend, "has failed to do one important thing. He has failed to discuss the *aim* of science" (Feyerabend (1970a), p. 201).

[140] That this was Kuhn's original idea is testified in a letter of his to Lakatos (22 May 1964, in Lakatos Archive (13.512)). In it, besides thanking Lakatos warmly for the invitation to take part in the 1965 Bedford Colloquium ("I do want to be taken seriously by philosophers and I am having less success with that goal in the U.S. than makes me happy") and asking him more information about the participants and the topics to be discussed, Kuhn speaks of the invitation ("another of the nice things that have happened to me lately") to contribute to the two volumes on Popper's philosophy in The Library of Living Philosophers (Schilpp (ed.) (1974)) and describes his own effort to clarify the points that divide him from Popper: "This", he writes, "seems to me particularly essential because I am so conscious of being close to him and because I know perfectly well that he's got a full answer for every point I'll raise. Nevertheless, in some ways we're as far apart as the duck and rabbit". The same image closes Kuhn (1970a), where some passages taken from Popper's writings are read as "further evidence of the gestalt switch that still divides us deeply" (p. 22): "Though the lines are the same, the figures which emerge from them are not. That is why I call what separates us a gestalt switch rather than a disagreement […]. How am I to persuade [Popper], who knows everything I know about scientific development and who has somewhere or other said it, that what he calls a duck can be seen as a rabbit? How am I to show him what it would be like to wear my spectacles when he has already learned to look at everything I can point out to through his own?" (*ibidem*, p. 3). However, at the end of a draft of this very paper, Kuhn invites Popper just to do that: "I hope he will try my spectacles for a time" (this draft was sent to Popper in view of the Bedford Colloquium and is kept in Popper Archive (80.9)).

Popper believes in "absolute" or "objective" truth, in Tarski's sense:[141] for him scientific knowledge can be regarded as knowledge without a knowing subject.[142] He believes in scientific progress as a progress towards truth, that is, the growth of knowledge. Kuhn is sceptical on this point, and Popper calls him a "relativist":[143] it is "the deepest issue" that divides them,[144] and was not dealt with by Watkins. Even if we do not have any method of discovering scientific theories, of ascertaining the truth of a scientific hypothesis, or whether a hypothesis is "probable", or "probably true", we can improve our knowledge through a confrontation (and a clash) between different theories, or hypotheses.[145]

Popper holds a correspondence theory of truth (whose origin he traces back to Xenophanes, Democritus and Plato, and finds quite explicitly in Aristotle) and finds Kuhn's ideas, on this fundamental question, "affected by relativism; more specifically,

---

141 See Popper (1962a) and (1963b). In 1935 – shortly *after* the publication of *Logik der Forschung* (see Popper (1972c), pp. 319–324 – Alfred Tarski introduced Popper to his own correspondence theory of truth, thus solving one of the major difficulties of Popper's realism. Popper immediately accepted such theory and endorsed in his later writings, always retaining a profound sense of gratitude towards Tarski, to whom he Popper (1972a) is dedicated. Indeed, thanks to Tarski, Popper was able to speak of the correspondence between propositions and facts in one language. The access to reality remained problematic, but Tarski succeeded in relegitimating the common sense idea of truth: truth could be a regulative idea, always sought and never sure to have been obtained. The "linguistic turn" of the Vienna Circle no longer threatened Popper's metaphysical realism and his idea of objective knowledge. See Tarski (1932) and (1935); Popper (1935, 1959), p. 274, n*1, (1955), (1963b), (1972c), (1974a, 1976), pp. 98–99, and (1979, 1994), pp. xxii–xxxii. By rejecting the correspondence theory of truth and appealing to a sort of coherence theory of truth, Kuhn does not succeed in escaping from the logical positivists' "linguistic turn": see Gattei (2002a), (2002b), (2003), and ch. 5, below.

142 See his "Epistemology Without a Knowing Subject" (Popper (1968b)), and "Addendum 1" to Popper (1982a), titled "Indeterminism is not enough", pp. 113–130.

143 Popper's meaning for this word is clearly described in his (1962a).

144 These are the words Popper uses in an unpublished typescript (in Popper Archive (75.5), p. 9; in his (1970), p. 56, this remark is slightly softened). As I see it, this is the major difference between Popper and Kuhn. It is a *logical* difference, as it concerns the role played by truth in scientific research and in the appraisal and choice of different theories. Kuhn conflates the concept of truth with the criterion of truth, thus claiming that it makes no sense to speak of truth in the absence of a decisional procedure to determine it. This is a mistake, as I shall argue below, in ch. 5. But it is also a *metaphysical* difference, as it concerns their different approaches to science and philosophy, the different solution they provide to the problem of rationality. Therefore, to use an expression Popper himself used in another context (see his (1974b), p. 1193), in Kuhn's philosophy remains a "whiff" of neopositivism (in Popper's case it was of "inductivism") – which explains the presence of *The Structure of Scientific Revolutions* in the *International Encyclopedia of Unified Science*, the logical positivists' most ambitious project, and the warm welcome it received by Rudolf Carnap (see Reisch (1991), Earman (1993), Irzik, Grünberg (1995) and Irzik (2000)). However, as in the case of Popper's philosophy, such alleged "whiff of [neopositivism]" risks to turn into a "full-blown storm" (see Newton-Smith (1981), p. 68).

145 See, on this point, Popper Archive (75.5), together with Popper (1972a), (1974c) and (1983).

by some form of subjectivism and of elitism, as proposed for example by Polanyi. Kuhn seems to me also affected by Polanyi's fideism: the theory that a scientist *must* have faith in the theory he proposes […]".[146] And he goes on emphasizing the relevance of "objective rational criticism" for science.[147] Furthermore, in Popper's eyes Kuhn identifies scientific revolutions with ideological revolutions, overlooking "the many purely *scientific* revolutions that are *not* connected with *ideological* revolutions".[148]

Lakatos' proposal

Imre Lakatos accepts Kuhn's challenge and tries to oppose it by saying that the image of science he proposes is a subjectivist and a psychologistic one.[149] His attempt aims to reply to the historical criticism Kuhn raises against falsificationism.

Lakatos deems Kuhn a relativist and summarizes his views as follows: "For Kuhn scientific change – from one 'paradigm' to another – is a mystical conversion which is not and cannot be governed by rules of reason and which falls totally within the realm of the *(social) psychology of discovery*. Scientific change is a kind of religious change".[150] And he provokingly asks, then, whether science is "reason or religion":[151] he asks, in other words, whether a rational philosophy of science is at all possible, or we must content ourselves with psychological explanations of the growth of science. When he asks whether science is rational or not, he does not ask, at first, whether history of science can be rationally reconstructed but, rather, whether it is possible to defend a normative rational methodology. His discussions of falsificationism and of the methodology of scientific research programmes have therefore a methodological scope.

In the first volume of the *Postscript to The Logic of Scientific Discovery* Popper examines *metaphysical* research programmes (such as, for instance, Descartes'

---

146 Popper (1983), pp. xxxi–xxxii.

147 Popper (1983), p. xxxii.

148 Popper (1983), p. xxxii. See also Popper (1975).

149 Not without some exaggerations, sometimes Lakatos goes so far as to accuse Kuhn's position of "mob psychology", charging Kuhn himself with irrationalism (see Lakatos (1970), p. 178: "There are no super-paradigmatic standards. The change is a bandwagon effect. Thus *in Kuhn's view scientific revolution is irrational, a matter for mob psychology*") – charges that Kuhn counters by reproaching Lakatos for losing sight of the history of science. The Kuhn–Lakatos dispute shows how it is difficult, in the heated phases of a debate, to hold one's own position consistently: Lakatos and Kuhn, in the various reformulations of their respective perspectives, end up coming closer to each other more than they would have wanted. See what Kuhn writes to Lakatos in a letter dated 7 July 1969, in view of the publication of their papers in *Criticism and the Growth of Knowledge*: "My replies are occasionally pretty sharp – Karl [Popper] is the only one with whom I have tried strenuously to restrain myself. […] I think you won't misunderstand me if I say I read your paper […] as far more Kuhnian than Popperian" (in Lakatos Archive (13.512); "this might be the case", replies Lakatos in a letter of 14 July 1969, *ibidem*).

150 Lakatos (1970), p. 93.

151 Lakatos (1970), p. 91.

research programme in physics) and says that they have an important heuristic function, since they indicate the direction of our search and the kind of explanation that may satisfy us, also allowing for something like an appraisal of the depth of a theory.[152] In the same line Lakatos introduces *scientific* research programmes and stresses that they also have a heuristic function. Furthermore, he thinks that they can explain continuity in the scientific enterprise. Indeed, one of Lakatos' aims is that of finding a rational heuristic of science, in order to show that new hypotheses and new discoveries are the result of a continuous and rational development of thought.[153]

The unit of evaluation is no longer a theory, but a set of theories: Lakatos' methodology of scientific research programmes is designed to be applied not to the elaboration of a single, isolated theory, but to a series of correlated theories that form together a programme to be assessed as a whole: philosophers must not chop it up into segments or units that, taken individually, may even lose the significance they have within a given research context. Among other things, Popper required of a good theory that it "is not refuted too soon":[154] he acknowledged, that is, that a healthy

---

[152] See Popper (1983), pp. 192–193.

[153] As we saw in the previous chapter, in his (1935) Popper rehabilitated metaphysics after Logical Positivism's attempt to ban it from the philosophy of science. Later, in the three volumes of the *Postscript to The Logic of Scientific Discovery*, Popper's project goes further, ascribing metaphysics both a historical and a heuristic role and introducing the idea of *metaphysical research programmes* (see Popper (1983), pp. 189–193 and 194–216, as well as Popper (1982a), pp. 87–109 and 113–130, and (1982b), pp. 159–211). Metaphysical conjectures, according to Popper and, following him, Agassi and Watkins, are programmes for the future development of science, for they indicate the problems scientists have to deal with and the directions of their research. In the late 1950s Agassi develops a new theory of how science and metaphysics might be integrated without endangering science by equating it to metaphysics and without explaining metaphysics away by demanding that it be a science. According to him, metaphysical theories may play two important roles in science. They may serve as research programmes and they may serve to interpret a physical theory, that is, they may provide a unified picture of the objects of the world as described by scientific theories. On the one hand, then, they serve science by helping to pose problems and to generate theories. They fulfil a heuristic role, already proposed by Popper, but they do so both systematically and critically. They are not merely lucky sources of interesting ideas – rather, they are integrated as research programmes into scientific research itself. On the other hand, they serve to solve the philosophical problems of some interest for scientists and philosophers alike. The integration of the two should make each better: both metaphysics and science can be critical together, posing problems for the other as well as methods of criticism. Metaphysics may pose problems for science by pointing to the lack of physical explanations of some phenomena interpretable by some metaphysics. Science can pose problems for metaphysics by offering physical theories in need of metaphysical interpretation. Attempts to solve such problems may lead to better theories, which take us closer to the truth. Lakatos' philosophy of mathematics (as developed in his (1963–1964)) lacks a view of research programmes and of metaphysics. His later philosophy of science incorporates both Kuhn's key concept of paradigm and Agassi's view of metaphysics and metaphysical research programmes. See Agassi (1964a) and (1976). On the relationship between Agassi's views and Lakatos', see Bartley (1976), Berkson (1976) and Wettersten (1992), pp. 241–243.

[154] Popper (1963a), p. 247. "The dogmatic attitude of sticking to a theory as long as possible is of considerable significance. Without it we could never find out what is in a theory

dose of dogmatism is necessary.[155] Lakatos makes Kuhnian "tenacity" a central feature of his methodology: "Purely negative, destructive criticism, like 'refutation' or demonstration of an inconsistency does not eliminate a programme. Criticism of a programme is a long and often frustrating process and one must treat budding programmes leniently".[156]

Lakatos himself describes his "methodology of scientific research programmes" (MSRP)[157] as a refinement of Popper's own methodology, and at the same time as an attempt to account, in the world 3 of objective knowledge, for Kuhn's views. According to Lakatos the basic units of the growth of knowledge are scientific research programmes, each one of which is identified by its background metaphysics. This could be expressed in the form of heuristic rules: "some tell us what paths of research to avoid (*negative heuristic*), and others what paths to pursue (*positive heuristic*)".[158] It is the metaphysics – something which we can individuate, analyse and criticize, as Popper himself had realized and highlighted – that guides the members of a scientific community in their search, and the history of science is also the history of rival metaphysics, each at the heart ("hard core") of competing research programmes.[159] But Lakatos moves beyond Popper's position: there are

---

– we should give the theory up before we had a real opportunity of finding out its strength; and in consequence no theory would ever be able to play its role of bringing order into the world, of preparing us for future events, of drawing our attention to events we should otherwise never observe" (Popper (1940), p. 312, n. 1). Lakatos finds all this baffling and odd, and by referring to Popper (1970) notices that "these remarks cannot be regarded as anything but a reluctant admission of an undigested anomaly in the Popperian research programme" (Lakatos (1969), p. 167, n. 55. Lakatos' point of view is pursued in Zahar (1982), (1983a) and (1983b).

[155] However, Kuhn would not agree with Popper on quantity: "One need not make neither resistance nor dogma a virtue to recognize that no mature science could exist without them" (Kuhn (1963a), p. 349); "it is precisely the abandonment of critical discourse that marks the transition to a science" (Kuhn (1970a), p. 6); "even resistance to change has a use [...]. By ensuring that the paradigm will not be too easily surrendered, resistance guarantees that scientists will not be lightly distracted and that the anomalies that lead to paradigm change will penetrate existing knowledge to the core" (Kuhn (1962a), p. 65).

[156] Lakatos (1970), p. 179, emphasis suppressed.

[157] Lakatos expounds and develops it mainly in his (1970) and (1971a). For a development and improvement of the original model, and for its application to important case-studies, see Lakatos, Zahar (1976) and Zahar (1973) and (1989).

[158] Lakatos (1970), p. 132.

[159] For the positivist Auguste Comte the metaphysical phase preceded (both epistemologically and historically) the mature science one (the "positive" phase); for a new positivist like Ernst Mach metaphysics infected reliable scientific theories (like Newtonian mechanics) and constituted a dangerous threat for science; as the logical empiricists argued (at least in the early phases of the development of their *wissenschaftliche Weltauffassung*), it should be rejected and expelled offhand. According to Popper, and especially to neopopperians such as Joseph Agassi and John Watkins, metaphysics is closely linked with scientific thought: from a historical point of view, metaphysical theories (or "*haunted-universe doctrines*", as Watkins refers to them in his (1958), p. 344) are the source from which spring empirical theories; from a heuristic point of view, metaphysics provides the scientist with extremely important regulative ideas, as long as, by voicing new ways of conceiving the world (and the

no metaphysical research programmes that provide scientific theories with their frameworks from the outside. Metaphysics is not only the source or the catalyst for science – rather, it is the very core of scientific enterprise. A research programme is therefore an aggregate of theories that develops from a number of methodological decisions taken by the researchers that promote the programme. Such decisions serve to individuate those hypotheses that are to be regarded as unfalsifiable in virtue of some methodological decree.[160] They constitute the core of the research programme, which embodies the "tenacity" (or dogmatism) that characterizes "normal" scientists. In its rigidity, the core of the programme recalls Kuhn's paradigm: indeed, on the basic issues, there is ample agreement, and they are questioned only during periods of crisis. Such a core, however, is the fruit of a long historical development: "The actual hard core of a programme does not actually emerge fully armed like Athena from the head of Zeus. It develops slowly, by a long, preliminary process of trial and error".[161]

Around the core proliferates a protective belt: while the negative heuristic of the programme forbids scientists to direct the arrow of *modus tollens* at the hard core, they must use their ingenuity "to articulate or even invent 'auxiliary hypotheses',

---

knowledge we have of it), it suggests methods to explore it (see Agassi (1964a), together with Watkins (1958), (1975) and (1978); see also Antiseri (1982) and Popper (1983), pp. 189–216, (1982a), especially pp. 87–109, and "A Metaphysical Epilogue", in his (1982b), pp. 159–211). According to Watkins, it is an "influential metaphysics" which acts on science from outside, while Agassi proposes "to view some metaphysics as the foundation of science; to view it as often conflicting with existing scientific theories and as incentives to alterations which would remove the conflict" (Agassi (1964b), p. 272). Moving beyond Kuhn, who speaks of "metaphysical paradigms", or "metaphysical parts of paradigms" (see his (1970c), p. 184), Lakatos puts metaphysics in the hard-core of his scientific research programmes, that is at the very heart of scientific enterprise.

[160] From the logical point of view (Duhem–Quine thesis) it is impossible to direct the *modus tollens* against a certain part of a falsified theoretical system. But we can decide to regard some parts of the theoretical system as "unproblematic", and therefore not to change them. A research programme is made irrefutable "by the methodological decision of its protagonists" (Lakatos (1970), p. 135). Lakatos regards the test statements of naïve falsificationism as irrefutable by methodological decision and adopts the same idea applying it to the hard core of a research programme.

[161] Lakatos (1970), p. 133, n. 4. The image is Duhem's: see his (1906, 1914/1954), p. 221, but see also the whole chapter VII of the book. Lakatos' reference to Duhem is important because it highlights that (with Popper) what is to be assessed is not an individual theory but a succession of theories, or a whole theoretical system – but (unlike Popper) Lakatos also stresses that the elements constituting that succession are usually connected by a remarkable continuity, a characteristic that closely resembles Kuhn's "normal science", that Lakatos deems fundamental in order to understand the history of science. According to Lakatos, closely following Duhem, history shows that no physical theory has ever been created out of whole cloth – rather, it has proceeded by a series of retouchings which from almost formless initial sketches have gradually led the system to more finished states: "A physical theory is not the sudden product of a creation; it is the slow and progressive result of an evolution" (Duhem (1906, 1914/1954), p. 221).

which form a *protective belt* around this core":[162] to this they must direct the arrow of *modus tollens*. Such a belt, formed by auxiliary hypotheses, observation theories, initial conditions and so on, must "bear the brunt of tests and get adjusted and re-adjusted, or even completely replaced, to defend the thus-hardened core".[163] Together with the statements of the hard core of the programme, these continuous adjustments and readjustments (negative heuristic) lead to the absorption of anomalies and recalcitrant instances, explain already know facts and predict new facts (positive heuristic).

> Few theoretical scientists engaged in a research programme pay undue attention to 'refutations'. They have a long-term research policy which anticipates these refutations. This research policy, or order of research, is set out – in more or less detail – in the *positive heuristic* of the research programme. The negative heuristic specifies the 'hard core' of the programme which is 'irrefutable' by the methodological decision of its protagonists; the positive heuristic consists of a partially articulated set of suggestions or hints on how to change, develop the 'refutable variants' of the research-programme, how to modify, sophisticate, the 'refutable' protective belt.
>
> The positive heuristic of the programme saves the scientist from becoming confused by the ocean of anomalies.[164] The positive heuristic sets out a programme which lists a chain of ever more complicated *models* simulating reality: the scientist's attention is riveted on building his models following instructions which are laid down in the positive part of his programme. He ignores the *actual* counterexamples, the available '*data*'.[165]

Only when successive modifications of the protective belt are no more able to cope with anomalies and predict new facts, the research programme turns regressive.

Lakatos asks whether there are objective (as opposed to socio-psychological[166]) reasons on the basis of which it is possible to reject a research programme, that is, "to eliminate its hard core and its programme for constructing protective belts?".[167] His answer is that "such an objective reason is provided by a rival research programme which explains the previous success of its rival and supersedes it by a further display of *heuristic power*".[168]

---

162 Lakatos (1970), p. 133.

163 Lakatos (1970), p. 133.

164 Every theory, Lakatos remarks, floats in an ocean of anomalies: it is refuted at the very moment of its birth. Positive heuristic – very much akin to Kuhn's "puzzle-solving tradition" or the "exemplars" guiding research – gives guidance for the articulation of specific theories only within the context of a given research programme.

165 Lakatos (1970), p. 135. And in n. 1 he continues: "If a scientist (or mathematician) has a positive heuristic, he refuses to be drawn into observation. [...] Occasionally, of course, he will ask Nature a shrewd question: he will then be encouraged by Nature's *YES*, but not discouraged by its *NO*".

166 The target is Kuhn, of course, at least in Lakatos' reading.

167 Lakatos (1970), p. 155.

168 Lakatos (1970), p. 155. As he explains in a footnote, "heuristic power" is here employed "as a technical term to characterize the power of a research programme to anticipate theoretically novel facts in its growth" (*ibidem*, n. 3). In other words, it is the programme's explanatory power.

But how is it possible to determine the heuristic power of a research programme? How can we compare it with that of another programme? Moreover, if we can (now) assess whether a research programme is able to explain the previous success of a rival research programme, how is it possible for us to do so in the future, without knowing *a priori* its heuristic power? Lakatos offers no answer to these questions (indeed, in his later writings he tends to apply his methodology only retroactively), and this constitutes the Achilles' heel of his proposal. Feyerabend will tackle him on this very point.

However vitiated from the start by the impossibility of combining Popper's view with Kuhn's – this very impossibility is the topic of the present work – Lakatos' model is a very subtle attempt to provide a synthesis of Popper's and Kuhn's ideas of science.[169] On the one hand, Kuhn's paradigm is transformed in the ideas of a core and a set of rules of the game (positive heuristic) that apply in the protective belt. Moreover, in Lakatos' model there is a strong dynamical element that Kuhn might approve of. On the other, we find also Popper's notions of conjecture, test, corroboration and falsification (Lakatos' distinction between progressive and regressive programmes embodies Popper's criteria of the increase of empirical content and of corroborated empirical content). Lakatos picks up Popper's interest in the *ad hoc* character of various theories and for the notion of empirical content. Nevertheless, he concedes a certain propensity towards conventionalism in science. Furthermore, Lakatos' approach is certainly not inductive. His model has the special advantage, as opposed to Popper's, of explaining why particular hypotheses (such as Newton's idea of punctiform planets moving along elliptical orbits around a punctiform Sun) can be accepted in the early phases of a research programme, even if it is then known that they are in conflict with clear experimental results. The position of the intransigent falsificationist is untenable for, on the one hand, he assures the falsifiability of all scientific theories, but on the other he dogmatically demands that we assign a neutral and infallible character to the empirical basis. Therefore, Lakatos offers a new criterion of demarcation, that excludes falsifiability and limits itself to say simply that "a theory is 'scientific' (or 'acceptable') if it has an empirical basis'".[170]

---

[169] Donald Gillies (a pupil both of Popper's and Lakatos' at LSE) suggests that for an analysis of revolutions in science, and particularly in mathematics, we use both the Kuhnian concept of paradigm and a modified version of Lakatosian research programme (that, however different and nonetheless often confused, are both necessary to provide an adequate account of the growth of mathematical knowledge). His idea is that in a revolution new research programmes are introduced that, although they may initially be supported only by a small group of people (or, as it happens more often in mathematics, by a single person) eventually lead to the emergence of a new paradigm, that gets accepted by the whole community: this is the case of the revolutionary programmes of Gottlob Frege and Giuseppe Peano, for instance. See Gillies (1992); on Peano's programme see Segre (1994).

[170] Lakatos (1970), p. 109, emphasis suppressed. For a criticism of Lakatos' model that accepts some of Kuhn's proposals (like the acknowledgement of the relevance of the scientific community, that eventually prefers a research programme over another), see Musgrave (1976a).

Feyerabend's position

"In the years 1960 and 1961 when Kuhn was a member of the philosophy department at the University of California in Berkeley", writes Feyerabend at the opening of his contribution,[171] "I had the good fortune of being able to discuss with him various aspects of science".[172] However, although he recognizes Kuhn's problems and tries to account for some aspects of science Kuhn has drawn attention to (the omnipresence of anomalies, for example), he declares himself "unable to agree with the *theory of science*"[173] Kuhn proposes. Least of all is he prepared to accept the ideology which he thinks forms the background of Kuhn's thinking:[174] such an ideology, as Feyerabend sees it, can "only give comfort to the most narrowminded and the most conceited kind of specialism. It would tend to inhibit the advancement of knowledge. And it is bound to increase the anti-humanitarian tendencies which are such a disquieting feature of much post-Newtonian science".[175] Even more troubling for Feyerabend, perhaps, is the second feature of Kuhn's philosophical proposal: "are we here presented with *methodological prescriptions* which tell the scientist how to proceed; or are we given a *description*, void of any evaluative element, of those activities which are generally called 'scientific'?".[176] Feyerabend does not see in any of Kuhn's writings any straightforward answer: they are ambiguous and lend support to both interpretations.[177] He sees such ambiguity as intended and charges Kuhn of being willing "to fully exploit his propagandistic potentialities".[178]

Then, assuming that Kuhn's aim is indeed to offer but a description, he goes into the details of Kuhn's paper. The existence of a puzzle-solving tradition, the characteristic of normal scientific activity and therefore the discriminating feature of science, is not able "to exclude, say, Oxford philosophy, or, to take an even more extreme example, *organized crime* from our consideration. For organized crime, so it would seem, is certainly puzzle-solving *par excellence*. Every statement which Kuhn makes about normal science remains true when we replace 'normal science'

[171] As I said, Feyerabend did not attend the Bedford Colloquium, but contributed to the volume of proceedings.

[172] Feyerabend (1970a), p. 197.

[173] Feyerabend (1970a), p. 197.

[174] The reference to Kuhn's ideology is also in the letters Feyerabend wrote to Kuhn in 1960–1962, commenting on an early draft of *The Structure of Scientific Revolutions*, and published as Feyerabend (1995a): see especially pp. 355, 360 and 367–368; see also Feyerabend (2006), pp. 614–618 and 619–620 (it is, he says "one of the most important topics, not only for philosophy, but quite in general": *ibidem*, p. 613).

[175] Feyerabend (1970a), pp. 197–198.

[176] Feyerabend (1970a), p. 198.

[177] The ambiguity of presentation is the second chief objection Feyerabend raises already in his above mentioned 1960–1962 letters: see Feyerabend (1995a), pp. 355–356, and (2006), pp. 614–616, where Kuhn is accused of willingly confusing the descriptive level with the prescriptive (or normative) one, dragging the reader into his own ideology, without respecting his possibility to critically dissociate himself from Kuhn's views.

[178] Feyerabend (1970a), p. 199.

with 'organized crime'; and every statement he has written about the 'individual scientist' applies with equal force to, say, the individual safebreaker".[179]

Just as in the case of Popper, the source of major disagreement is normal science.[180] Already referring to "The Function of Dogma in Scientific Research",[181] Feyerabend lamented the fact that Kuhn seems to be coming to the conclusion that the characteristic features of science as we pursue it today are subordinate to certain unity of doctrine and to a drastic narrowing of critical debate. If Kuhn acknowledges to his interlocutor that science is not a wholly monolithic enterprise, he nonetheless stresses what he calls the "nearly independent" character of the different branches of science, each of which is guided by its own paradigm and inquires into its own specific problems. But Feyerabend retorts that those crises which put an end to the kingdom of a paradigm often depend on the interaction among these allegedly divided parts of mature science. Furthermore, if the scientific enterprise were as monolithic as Kuhn portrays it, there would be no room for the emergence of competing theories.[182]

More important, however, is the second part of his paper, where Feyerabend discusses Kuhn's *functional* argument for normal science:[183] it would constitute a

---

[179] Feyerabend (1970a), p. 200. Kuhn denies the scientific character to the pre-paradigmatic phase since it does not have the peculiarities of normal science: although the people involved in this kind of research are scientists, their activity is "something less than science" (Kuhn (1963a), p. 355). In the review of the volume of proceedings of the 1961 Oxford Symposium on the History of Science Feyerabend regards this as a "purely semantic argument that *condemns* an activity because it is not *customary* to apply a certain word to it" (Feyerabend (1964), p. 251), thus deeming it valueless. Therefore, Kuhn would not be able to provide a new criterion of demarcation because he does not take properly into account the *aim* of science and the question whether normal science is able to attain such an aim or not (there is a Popperian flavour in this remarks by Feyerabend, to which Kuhn replies in his (1983c)).

[180] See also Feyerabend (1978b), p. 204.

[181] Kuhn (1963a). Just like Toulmin (see his (1970), p. 39), Feyerabend very closely associates this paper with *The Structure of Scientific Revolutions*, and both these with Kuhn (1970a).

[182] The problems leading to Einstein's special theory of relativity, Feyerabend notices, could not have arisen without the tension that existed between Maxwell's theory and Newtonian mechanics. Nor was it possible to use Brownian motion for a direct refutation of the phenomenological second law of thermodynamics: the kinetic theory had to be introduced from the start. See the discussion in Feyerabend (1965a), section VI, pp. 175–176: from the microscopic point of view, a Brownian particle is a perpetual motion machine of the second kind and its existence refutes the phenomenological second law of thermodynamics. It therefore belongs to the domain of relevant facts for this law. Could this relation between the law and the Brownian particle have been discovered in a direct manner, i.e. by an investigation of the observation consequences of the phenomenological theory, without borrowing from an alternative account of heat? The answer is negative: a "direct" refutation of the second law that considers only the phenomenological theory and the "fact" of Brownian motion is impossible. As is well known, the actual refutation was brought about via the kinetic theory and Einstein's utilization of it in the calculation of the statistical properties of the Brownian motion.

[183] Kuhn often tries something like functional explanations of scientific practice: see, for example, his (1961a), (1963a), (1964) and (U-1990b). Nancy Cartwright notices that "all

necessary presupposition of revolutions. The force of persuasion of this argument, that aims at describing normal science as a positive phase, lies on two assumptions: first, that revolutions are desirable; and second, that the peculiar way in which normal science leads to revolutions is desirable as well.

According to Feyerabend, Kuhn cannot regard the changes brought about by a revolution as improvements – and the reason is the very incommensurability thesis Kuhn recognizes: if a revolution brings about some changes, but not some actual improvements, then revolutions are not desirable. Furthermore, according to Kuhn (as portrayed by Feyerabend, of course) scientists would commit themselves to the "pedestrian activity"[184] of normal science *ad nauseam*, and would give it up only when problems become too big. But Feyerabend argues that it is possible to bring about revolutions also along another and better path: through the *proliferation* of different theories, that is, through the creation of competing theories, rivalling the dominant one.[185] He advances the *principle of tenacity*, that is, "the advice to select from a number of theories the one that promises to lead to the most fruitful results, and to stick to this one theory even if the actual difficulties it encounters are considerable".[186] Having adopted such principle, "we can no longer use recalcitrant

---

functional explanations have a dubious logic, but they do often bring out instructive aspects of the custom in question": (1983), p. 143.

[184] Feyerabend (1970a), p. 201.

[185] Kuhn forcefully stresses the necessity of mental rigidity since it alone can allow the individuation of an important anomaly and properly highlight it (indeed, his "The Essential Tension: Tradition and Innovation in Scientific Research" (Kuhn (1959a) was a paper delivered at a conference whose participants were supposed to reflect on the ways to foster scientific research by promoting divergent thinking: Kuhn thought the opposite was the case). Therefore, in the final analysis, its adoption alone "will in the end lead to the overthrow of the very same paradigm to which the scientists had restricted themselves in the first place" (Feyerabend (1964), p. 252; see also Feyerabend (1970a), pp. 201–202). Furthermore, for Feyerabend "This is the main reason why the rejection, by mature science, of pre-paradigmatic battle of ideas is defended by Kuhn not only as a historical fact, but also as a reasonable move" (Feyerabend (1964), p. 252). Feyerabend agrees that this is a good argument in favour of paradigms: we need a certain amount of dogmatism since it provides scientists with a guide in the exploration of a nature too complex to be inquired into at random. He writes: "The massive dogmatism I have described is not just a *fact*, it also has a most important *function. Science would be impossible without it*" (Feyerabend (1975), p. 298). We need a guide to discriminate what is relevant from what is not and to individuate the most fruitful areas of research. However, what Feyerabend questions is that Kuhn's arguments are strongly in favour of theoretical monism. The psychological function of a single paradigm, on which Kuhn so much insists, its function as background against which the *Gestalt* of an anomaly stands out so vividly, could very well be played by a *plurality* of theories as well. Ultimately, Kuhn's theoretical monism is not justified by the commitment to one single paradigm. Lakatos thinks along the same lines: "The history of science has been and should be a history of competing research programmes (or, if you wish, 'paradigms'), but it has not been and must not become a succession of periods of normal science: the sooner competition starts, the better for progress. 'Theoretical pluralism' is better than 'theoretical monism': on this point Popper and Feyerabend are right and Kuhn is wrong" (Lakatos (1970), p. 155, emphasis suppressed).

[186] Feyerabend (1970a), p. 203.

facts for removing a theory, *T*, even if the facts should happen to be as plain and straight-forward as daylight itself. But we can use *other theories*, *T'*, *T"*, *T'"*, etc. which *accentuate* the difficulties of *T* while at the same time promising means for their solution. In this case the elimnation of *T* is urged by the principle of tenacity itself. [...] Proceeding in accordance with such principle is *one* method of precipitating revolutions. It is a *rational* method".[187] Kuhn speaks of dogmatic and authoritarian features of normal science:[188] but if "normal" activity is so monolithic, where do alternatives – that is, theories competing with the stablished paradigm – come from? Kuhn, however, acknowledges the multiplicity of theories and indeed attributes a function to it: they make refutations possible and, most importantly, they bring about revolutions.[189] All this leads Feyerabend to suspect "that normal or 'mature' science, as described by Kuhn, *is not even a historical fact*".[190] Kuhn's theoretical monism, Feyerabend concludes, is not only false – both fom the descriptive and the historical point of view – but also methodologically undesirable.

The interplay between tenacity and proliferation is "an essential feature of the actual development of science. [...] it is not the puzzle-solving activity that is responsible for the growth of our knowledge but the active interplay of various tenaciously held views".[191] If science wishes to develop ideas and use rational means for the elimination of even the most fundamental conjectures, it must use a principle of tenacity together with a principle of proliferation. "It must be allowed to *retain* ideas in the face of difficulties; and it must be allowed to introduce *new ideas* even if the popular views should appear to be fully justified and without blemish".[192]

Most importantly, tenacity and proliferation are the only features that may bring about what Feyerabend regards as the highest value:

> This value does not exclude the institutionalized forms of life (truth; valour; self-negation; etc.). It rather encourages them *but only* to the extent to which they contribute to the advance of some individual. What is excluded is the attempt to 'educate' children in a manner that makes them lose their manifold talents so that they become restricted to a narrow domain of thoughts, action, emotion. Adopting this basic value we want a methodology and a set of institutions which enable us to lose as little as possible of what we are capable of doing and which force us as little as possible to deviate from our natural inclinations.[193]

---

[187] Feyerabend (1970a), p. 205. The criticism of Kuhn's monism is one of the key issues also in Feyerabend (1995a): see especially p. 367.

[188] See especially Kuhn (1963a).

[189] This is one of the main point of agreement between Feyerabend and Kuhn. Feyerabend himself recalls how this very point was first highlighted during the lectures on scientific method Popper gave at the London School of Economics between 1948 and 1952. Drafts of these lectures, together with Popper's notes and other course material are kept in the Popper Archives at the Hoover Institution on War, Revolution and Peace, Stanford University.

[190] Feyerabend (1970a), p. 207.

[191] Feyerabend (1970a), p. 209.

[192] Feyerabend (1970a), p. 210.

[193] Feyerabend (1970a), p. 210.

What Feyerabend values most are "the happiness and full development of an individual human being".[194] On the one hand, proliferation means that "there is no need to suppress even the most outlandish product of the human brain. *Everyone may follow his inclinations* and science, conceived as a critical enterprise, will profit from such activity".[195] On the other, tenacity means that "one is encouraged not just to follow one's inclinations, but to develop them further, to raise them, with the help of criticism (which involves a comparison with the existing alternatives) to a higher level of articulation *and thereby to raise their defence to a higher level of consciousness*".[196] The scientific enterprise described by Kuhn, concludes Feyerabend, is not only ill-conceived and non-existent – but its defence is also incompatible with the humanitarian outlook Feyerabend puts at the very centre of his approach to philosophy and life.[197]

The theory of science that should replace Kuhn's is Lakatos'. The latter's picture is a synthesis of two discoveries: "First, it contains Popper's discovery that science is advanced by a critical discussion of alternative views. Secondly, it contains Kuhn's discovery of the function of tenacity which he has expressed, mistakenly I think, by postulating tenacious *periods*. The synthesis consists in Lakatos' assertion [...] that proliferation and tenacity do not belong to *successive* periods of the history of science, but are always *copresent*".[198] Therefore Feyerabend, in agreement with Lakatos, proposes a relation of *simultaneity* and the interaction of different factors. He speaks of the normal *component* and the philosophical *component* of science, not of the normal period and the period of revolution.[199]

Feyerabend does not stop here, however. After criticizing Kuhn following Lakatos' point of view, he now wishes to defend him against Lakatos who, with his insistence on standards, concedes too much to Popper's orthodoxy, thus considerably reducing the revolutionary import of his own model. And he does so by showing that science is an enterprise more *irrational* than Lakatos and Kuhn himself (at least

---

194 Feyerabend (1970a), p. 210.

195 Feyerabend (1970a), p. 210.

196 Feyerabend (1970a), p. 210. "The interplay between proliferation and tenacity also amounts to the continuation, on a new level, of the biological development of the species and it may even increase the tendency for useful *biological* mutations. It may be the only possible means of preventing our species from stagnation" (*ibidem*).

197 Not differently from his teacher, Popper, Feyerabend's theory of rationality – that is, his personal solution to the problem of rationality: how do we wish to live our lives? – has a profoundly ethical nature. See Gattei (2002a) and (2002b).

198 Feyerabend (1970a), p. 211. "Proliferation sets in already *before* a revolution and is instrumental in bringing it about. [...] Proliferation does not *start* with a revolution; it *precedes* it. A little imagination and a little more historical research then shows that proliferation not only *immediately precedes* revolutions, but that it is there *all the time*. Science as we know it is not a temporal succession of normal periods and of periods of proliferation; it is their juxtaposition" (*ibidem*, p. 212).

199 "It seems to me that such an account overcomes many difficulties, both logical and factual, which make Kuhn's point of view so fascinating but at the same time so unsatisfactory" (Feyerabend (1970a), p. 212).

as Lakatos sees him) may think.[200] From his confrontation with Kuhn and above all with Lakatos, Feyerabend's conclusion is that the only permissible principle is an instruction *against method*: "anything goes".[201]

The last issue Feyerabend explores in his paper concerns incommensurability, "a point of Kuhn's philosophy", he declares at the very beginning, "which I wholeheartedly accept".[202] In particular, Kuhn and Feyerabend agree that new theories, however better and more detailed than the previous ones, are not always able to cope with *all* the problems the previous theories had been able to account for. It is the so-called "Kuhn-loss", according to which "new theories, while often better and more detailed than their predecessors were not always rich enough to deal with *all* the problems to which the predecessor had given a definite and precise

---

[200] The perusal of the various criticisms Feyerabend raises against Lakatos goes beyond the scope of the present work, which focuses on Kuhn. I limit myself to highlighting only a fundamental objection Feyerabend raises, that of the *time limit*. The evaluation standards of Lakatosian methodology are effective only when they are combined with a time limit (indeed, what may at first look like a degenerating problem-shift may with time turn into a progressing one): without this limit it is no longer possible to regard the decision to work on a research programme as rational – beyond that limit it would be illegitimate (and therefore irrational) to go on working on a programme in its degenerating phase. One need not make dogma a virtue – as Kuhn said – to recognize that the programmes that marked modern science had *for a certain period of time* showed a decrease of empirical content as regards the rival programmes they eventually ended up replacing (as an example, Kuhn himself cites the takeover of old Aristotelian physics by the new Galilean physics). However, the introduction of a time limit bears devastating consequences for the standards Lakatos wants to defend: either they are vacuous (one does not know when to apply them), or they can be criticized on grounds very similar to those which led to Lakatos' criticism of Popper's naïve falsificationism, and which led to the introduction of these very standards in the first place (see Feyerabend (1970a), p. 215). See also the lively exchange in Lakatos, Feyerabend (1995, 1999).

[201] See Feyerabend (1975), p. 28. Apparently (given Feyerabend's passion for singing, opera and acting, this is indeed most plausible) the source of this slogan is Cole Porter. The first refrain of the song "Anything Goes", from his homonymous musical, goes like this: "In olden days a glimpse of stockings / Was looked on as something shocking. / Now, heaven knows… / Anything goes!". *Anything Goes* opened at the Alvin Theatre in New York, on 21 November 1934 and turned out to be the fourth longest running musical of the 1930s. In 1987 it was revived at the Vivian Beaumont Theatre with Patti LuPone in the leading role and in a revised book by Timothy Crouse and John Weidman; the 1936 screen version starred Ethel Merman and Bing Crosby.

[202] Feyerabend (1970a), p. 219. Indeed, Feyerabend recalls how they both introduced the term, quite independently, in 1962 (in Kuhn (1962a) and in Feyerabend (1962a), respectively): "I still remember marvelling at the pre-established harmony that made us not only defend similar ideas but use exactly the same words for expressing them" (Feyerabend (1970a), p. 219). Feyerabend will refer to such pre-established harmony again, several years later, in a paper whose ideas – he realizes after reading the epilogue of Hoyningen-Huene (1989a) – are "very similar to, and almost identical with, Kuhn's as yet unpublished, later philosophy" (Feyerabend (1989a), p. 405, n. 26).

answer. The growth of knowledge or, more specifically, the replacement of one comprehensive theory by another involves losses as well as gains".[203]

[203] Feyerabend (1970a), p. 219. Revolutions, according to Kuhn, bring about both an increased ability to solve problems and some "losses": loss in the ability to explain some phenomena, in the first place; but also loss of some (authentic) scientific problems, due to the narrowing of a certain scientific discipline; and communication difficulties among professional of different scientific disciplines. Revolutions in science (unlike those in mathematics: see Giorello (1992); but see also the papers in Gillies (ed.) (1992) and Corry (1993)) entail progress at the price of some regress: "Copernicus destroyed a time-honoured explanation of terrestrial motion without replacing it; Newton did the same for an older explanation of gravity, Lavoisier for the common properties of metals, and so on" (Kuhn (1962a), p. 157; see also pp. 66, 103–109, 148–149, 167, 169 and 170). "In the transition from an earlier to a later theory, there is often a loss as well as a gain of explanatory power" (Kuhn (1961a), p. 211, but see also pp. 211–213): this point is central for Kuhn's entire model for the growth of science: "In fact, it is largely the necessity of balancing gains and losses and the controversies that so often result from disagreement about an appropriate balance that make it appropriate to describe changes of theory as 'revolutions'" (*ibidem*, n. 48; see also p. 208, and Kuhn (1970a), p. 20, (1976b), p. 192, and (1992), p. 120). And again: "Entry into another culture does not simply expand one's previous form of life, open new possibilities within it. Rather, it opens new possibilities at the expense of old ones, exposing the foundations of a previous life form as contingent and threatening the integrity of the life one had lived before. Ultimately the experience can be liberating, but it is always threatening" (Kuhn (1984), p. 368). One of Kuhn's favourite example of loss – recurrent both in *The Structure of Scientific Revolutions* and in other works – refers to the chemical revolution: in the context of phlogiston theory, a metal was regarded as a compound of a specific component (the "calx") and phlogiston; since phlogiston was assumed to be present in all metals, the theory could explain why they resembled one another to a much greater extent than the corresponding calces (what we would now call "oxides"). The oxygen theory, by contrast, considers metals to be elementary, and thus lacks any resources to account for their similarity (it cannot appeal to the shared possession of phlogiston, that is). Thus, according to Kuhn, the adoption of phlogiston theory reopened an empirical problem that had been considered settled before (see Kuhn (1977c), p. 323; see also Kuhn (1962a), pp. 132 and 157; a similar discussion is to be found also in Toulmin (1961), pp. 89–94). However, as Andrew Pyle made me observe, the oxygen theory explains the shared properties of calces in terms of their being metal oxides, and these similarities remain unexplained in terms of the phlogiston theory. Likewise for non-metals: phlogistonists can explain the similarities of phosphorus, sulphur and carbon in terms of a shared principle (phlogiston), while they are unable to explain the similarities among their corresponding acids since they regard them as simples. With Lavoisier phosphorus, sulphur and carbon begin to be seen as simples, and so their similarities can no longer be accounted for; by contrast, scientists were now able to explain the qualitative similarity of acids in terms of their composition. The situation, in other words, is perfectly symmetrical: both theories have their (different) lists of simples, and therefore each one cannot provide answers to some questions the other can provide, and vice versa. At a closer look, the great majority of Kuhn's scattered remarks on alleged "losses" through a scientific revolution are simply general claims, with no supporting evidence or worked-out examples, as the situation would require. Among the passages from Kuhn's works I have been able to dig up, the most interesting one is that of Newton's theory, that took to the relinquishment of any ability to explain gravitation, which Descartes and Cartesians had retained. Yet this could be better described as a shift in what counts as an explanation in physics – that is, only a part, however important, of Kuhn's

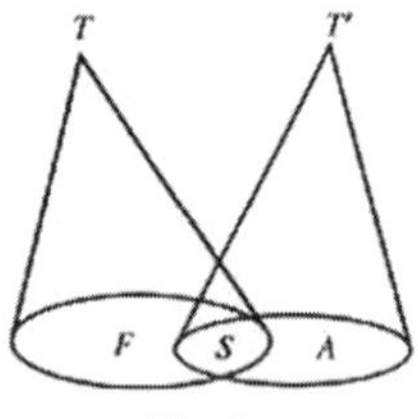

Fig. 1

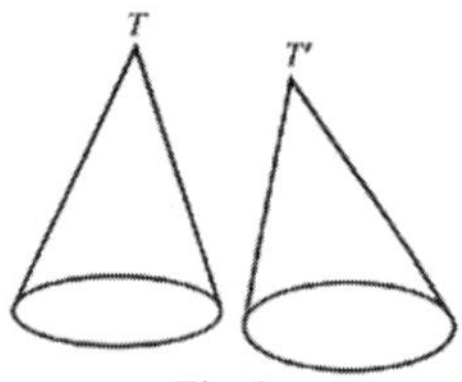

Fig. 2

Feyerabend explains:

> *T* is superseded by *T'*. *T'* explains only why *T* fails where it does (in *F*); it also explains why *T* has been at least partly successful (in *S*); and makes additional predictions, (*A*). Now if this scheme is to work then there must be statements which follow (with, or without the help of definitions and/or correlation hypotheses) both from *T* and from *T'*. But there are cases which invite a comparative judgement without satisfying the conditions just stated. The relation between such theories is as shown in *Fig. 2*. A judgement involving a comparison of content classes is now clearly impossible.[204]

Even more important, seen against the background of Popper's thought, "*T'* cannot be said to be either closer to, or farther from, the truth, than *T*".[205]

It is interesting to see, on this matter, also Lakatos' point of view. In a loose slip of paper inserted in his own copy of *Criticism and the Growth of Knowledge*, he schematizes his own position, as regards those of Kuhn and Feyerabend:

And he adds in a note: "competing paradigms: there is always a neglected merit in the defeated theory".[206]

After refuting some of the criticism of the incommensurability thesis, and particularly of the desirability, or simply possibility, of incommensurable theories,[207] Feyerabend turns to Popper, who referred to this thesis in his own paper.[208]

KHUN FEYERABEND LAKATOS

Incommensurable theories, he says, though not comparable as to their contents, can nevertheless be refuted "by reference to their own respective kinds of experience

wide characterization of "regress through revolutions" in terms also of loss of authentic problems and communication breakdown. On this issue, see also Laudan (1977) and (1990a), Hoyningen-Huene (1989a/1993), pp. 260–261, Preston (1997a), pp. 87–98, Gattei (2000b), pp. 328–329, and Carrier (2002b), particularly p. 56.

[204] Feyerabend (1970a), p. 220, and (1978b), p. 182.

[205] Feyerabend (1970a), p. 220, and (1978b), p. 182.

[206] The drawing, together with the handwritten note, are in Lakatos Archive (10.4). See also Pera (1982a), ch. 4.

[207] See Feyerabend (1970a), pp. 222–229, and (1978b), pp. 185–192.

[208] See Popper (1970), pp. 56–58.

(in the absence of commensurable alternatives these refutations are quite weak, however)".[209] Nor, *pace* Popper, "is it possible to make a judgement of *verisimilitude* except within the confines of a particular theory.[210] None of the methods which Popper wants to use for rationalizing science can be applied and the one that can be applied, refutation, is greatly reduced in strength. What remains are aesthetic judgements, judgements of taste, and our own subjective wishes".[211] Does this amount to a slip in that very subjectivism Popper wishes to attack? According to Feyerabend, "an enterprise whose human character can be seen by all is preferable to one that looks 'objective', and impervious to human actions and wishes. [...] Secondly, matters of taste are not completely beyond the reach of argument".[212]

These remarks trouble Popper's (objective) world 3[213] and take Feyerabend to the conclusion that

---

209 Feyerabend (1970a), p. 227, and (1978b), pp. 199.

210 "These results, that shed new light on the role of *methodology*, bear consequences also in the field of *cosmology*. They show that a certain form of realism, further than being too limited, is also in contrast with the actual practice of science. [...] Now realism can be regarded as a *special theory* of the relationship between man and the world, in its turn subject to developments and improvements, or as a *presupposition of scientific knowledge* (and of knowledge in general). It seems that the majority of contemporary professional realists, and among them also the stern pope of critical rationalists, Sir Karl Popper, understand realism in this sense. They are dogmatists" (Feyerabend (1978b), pp. 201–202).

211 Feyerabend (1970a), pp. 227–228; see also Feyerabend (1978b), pp. 199–200. These remarks closely resemble some of Kuhn's central points: his rejection of verisimilitude and his discourse about "values", the third component of the disciplinary matrix. On truth, see Kuhn (1962a), pp. 170–171, (1970c), pp. 205–207, (1991a), p. 8, and (1993a), p. 330, as well as Gattei (2002a), (2002b) and (2003); as to values, see Kuhn (1962a), pp. 153–159, (1970a), pp. 20–22, (1970b), pp. 241 and 261–262, (1970c), pp. 184–186, 199 and 205–206, (1971a), pp. 145–146, (1977c), pp. 321–325, (1980a), pp. 189–190, and (1983c), p. 209.

212 Feyerabend (1970a), p. 228, and (1978b), pp. 200–201. Right against the very heart of methodological anarchism – the "all-pervasive character of theoretical assumptions", as Feyerabend describes it in 1962, according to which "scientific theories are ways of looking at the world and their adoption affects our general beliefs and expectations, and thereby also our experiences and our conceptions of reality" (Feyerabend (1962a), p. 45) – Popper fought his first battles, when he contrasted the criterion of falsifiability to Marxism, psychoanalysis and analytic psychology that pervaded the eyes of their enthusiastic advocates. To the inductivist myth of a science that rests "upon solid bedrock" he opposed the image of science "like a building erected on piles" (Popper (1935, 1959), p. 111), to the irrationalism myth of paradigm (or framework) Popper opposed critical discussion. An enemy of both logical positivists and irrationalists, he upheld *doxa* when the former exalted *episteme*, and when the former started crying "down with the method!", he kept on supporting methodological rules.

213 "Popper and Lakatos think that the solution of these problems is relatively easy. They refuse to deal with 'mob psychology' within the theory of science and stubbornly affirm that *all* science bears an essentially rational character. [...] Unfortunately, the scientist has to deal also with the world of matter and that of thought. Or, better, he does not have to deal 'also', but *exclusively* with these worlds. And the rules that allow to move easily and without problems in the 'third world' cannot at all be used to solve experimental, theoretical, social and psychological problems that appear in the first and in the second" (Feyerabend (1978b), p. 177).

the attempt to judge cosmologies by their content may have to be given up. Such a development, far from being undesirable, changes science from a stern and demanding mistress into an attractive and yielding courtesan who tries to anticipate every wish of her lover. Of course, it is up to us to choose either a dragon or a pussy cat for our company. I do not think I need to explain my own preferences.[214]

---

[214] Feyerabend (1970a), p. 229.

Chapter 3

Incommensurability

> Phil[osophy] lets everything *loose* – it relativizes the universe – like the Copernican system, it cuts fixed points – and what was quietly resting it turns into something floating. It teaches the relativity of all foundations.
>
> *Novalis (Friedrich von Hardenbergh)*

The controversy over incommensurability dates back to 1962, when the thesis was first stated in print by its two major advocates, Thomas Kuhn and Paul Feyerabend. In "Explanation, Reduction, and Empiricism", while criticizing the model of reduction between scientific theories advanced by logical positivists, Feyerabend claims that successive theories can be reciprocally incommensurable, that is, not comparable with respect to their content and claims about the world. In *The Structure of Scientific Revolutions* Kuhn ascribes to incommensurability a major role in his theory of the development of science as a sequence of revolutionary transitions from paradigm to paradigm.[1]

However, as we have seen, if 1962 was the year in which the incommensurability thesis came to the fore of epistemological debate, it had been in the air for quite some time. In fact, in advancing the thesis both Feyerabend and Kuhn rely on previous developments in the philosophy and history of science, and also in philosophy in general. From several points of view, then, the incommensurability thesis is the upshot of the philosophical climate produced between the late 1950s and the early 1960s. These years saw the establishment of the history of science as a professional discipline, the influence of *Gestalt* psychology on the philosophy of perception, the rapid decline of Logical Positivism, the wide influence of Wittgenstein's later writings and Quine's attack on the distinction between analytic and synthetic propositions.[2]

Nevertheless, although child of its times, incommensurability remains a characteristic feature of the new movement in the philosophy of science emerging in the late 1950s. Together with the thesis of the theory-ladenness of observations, the rejection of the conceivability of a single scientific method (fixed and established once and forever) and the insistence on the relevance of the history of science for the philosophy of science, incommensurability is one of the key theses of what has become known as the "new philosophy of science", that is, "post-positivistic (or historical) philosophy of science". As we saw in the previous chapter, besides Kuhn

---

1 The secondary literature on incommensurability is enormous (to have a rough picture, see Hoyningen-Huene (1989a/1993), p. 207, n. 58), but Kuhn himself remarks that "virtually no one has fully faced the difficulties that led Feyerabend and me to speak of incommensurability" (Kuhn (1983a), p. 34).

2 For an overview, see Brown (1977) and Kordig (1971).

and Feyerabend, the other protagonists of this movement are Norwood Russell Hanson, Michael Polanyi and Stephen Toulmin.[3]

Different Ways of Understanding Incommensurability

The term "incommensurability" derives from the standard employment of this concept in geometry and mathematics: two quantities are said to be incommensurable if there is no common measure whole units of which divide both of them. The application of this mathematical concept to competing scientific theories involves a stretching of the concept that leaves considerable room for alternative interpretations.

The discussion of the incommensurability of scientific theories rarely proceeds in accordance with the mathematical concept of incommensurability. On the contrary, the trend is to frame the comparison in terms of concepts and considerations of semantic or broadly epistemological nature. In fact, discussions of incommensurability are nearly always phrased in terms of incomparability of the contents of alternative scientific theories, or meaning variance of scientific terms, reciprocal translation difficulties of the vocabularies employed by different theories, or else the absence of shared standards of theory appraisal. All this gives rise to the problem of the relationship between the concept of incommensurability *stricto sensu*, that is, the absence of a common unity of measure, and the very many variations on this theme, understood as *lato sensu* as possible. The point is whether incommensurability is a single kind of relationship among scientific theories, the various aspects of which are nothing but components, or different points of view; or, rather, whether incommensurability consists of several different elements together, such as the content incomparability among alternative scientific theories or the absence of shared criteria and evaluation standards each of which, individually taken, gives rise to incommensurability.[4]

---

[3] See Kuhn (1991a), pp. 90–91. Once it was customary to highlight the sharp contrast between post-positivistic (or historical) philosophy of science and Logical Positivism's. However, an increasing number of studies of the history of the philosophy of science in the twentieth century invite a reassessment of the relationship between the two approaches. Indeed, some studies suggest that Logical Empiricism has more in common with Kant and English empiricism than is usually thought (see Coffa (1991), Friedman (1993) and (1999), Parrini (1980), (1995, 1998) and (2002)). Other studies show that the logical positivist "double-language" model contains, *in nuce*, the thesis of meaning variance (see English (1978)). Still some others highlight the very warm welcome given to *The Structure of Scientific Revolutions* by Rudolf Carnap, in his capacity as editor of the *International Encyclopedia of Unified Science* (see Reisch (1991) and Friedman (2001), pp. 18–19 and 41–43): in order to explain it, striking parallels have been drawn between the views of Carnap and Kuhn (see Earman (1993), Irzik, Grünberg (1995), Irzik (2002) and (2003)). At the very least, these studies suggest that the incommensurability thesis can no longer be regarded as a thesis which epitomizes the differences between diametrically opposed positivist and post-positivist positions.

[4] Paul Hoyningen-Huene and Howard Sankey, who more than anybody else worked on these issues in the past years (first separately, and then jointly organizing an international conference on incommensurability – *Incommensurability (and related matters)*, Hannover, 13–16 June 1999: see Hoyningen-Huene, Sankey (eds) (2001) – hold themselves different positions. Hoyningen-Huene regards incommensurability as a compound relationship made

Therefore, the literature on incommensurability concerns very different issues:[5] some authors speak of conceptual change and of intelligibility of alternative conceptual frameworks; others deal with scientific realism and the continuity of reference of theoretical concepts; still others focus on the rationality of theory choice in science and reflect on the actual availability of objective criteria for theory appraisal.

The necessity of referring to a plurality of issues is largely due to Kuhn's and Feyerabend's own original discussions. On the one hand, in "Explanation, Reduction, and Empiricism" Feyerabend understands incommensurability as the absence of logical relationships between theories due to the semantic variance of the terms employed, which involves the scientist's inability to objectively assess and compare the contents of rival, competing theories.[6] On the other hand, in *The Structure of Scientific Revolutions* Kuhn interprets incommensurability as a complex relation among paradigms that involves methodological, semantic as well as perceptual components.[7] According to Kuhn, different paradigms make use of different evaluation criteria and refer to different sets of scientific problems; scientists' vocabulary varies during the revolutionary transition from one paradigm to another; different scientific communities, supporting different paradigms, see the world in different ways and, perhaps, even inhabit different worlds.[8]

Some Precedents

Both in its general form and in its specific applications (such as to the couple classical mechanics-special theory of relativity) the incommensurability thesis certainly does not come out of the blue.[9] We can trace its historical precedents in the epistemological debate at the turn of the twentieth century, in the contraposition of the radical "fractural conventionalism" of Edouard Le Roy (1870–1954), on the one hand, and of the moderate and continuist conventionalism of Jules-Henri Poincaré

---

of several components, such as meaning variance and the lack of shared standards for theory appraisal, that express different facets of the same thing (see, for example, Hoyningen-Huene (1990), p. 488). On the contrary, Sankey denies that such a "unified" approach is possible, at least in Kuhn's case (see Sankey (1993a), pp. 760–765 and (1994a)), and regards semantic incommensurability as a form of incomparability of content (see Sankey (1997c), p. 428). See also Sankey (1999) and Sankey, Hoyningen-Huene (2001).

[5] See, for instance, Hoyningen-Huene, Sankey (eds) (2001), pp. 303–316, or Kuhn (2000a), pp. 367–400.

[6] See Feyerabend (1962a), pp. 62–69 and 92–93.

[7] See Kuhn (1962a), pp. 148–150.

[8] See Kuhn (1962a), ch. X. In some of his later works Kuhn narrows incommensurability to the semantic relationships among terms of different theories, highlighting a special similarity between incommensurability and the indeterminacy of translation remarked by Quine. However, he then distinguishes his own position from Quine's, speaking of the impossibility to achieve a perfect translation of groups of terms belonging to the particular vocabulary of one theory (see particularly Kuhn (1983a) (1983b), (1989a), (1989b), (1990) and (1991a)).

[9] See Giedymin (1968), (1970) and (1971).

(1854–1912) and Pierre Duhem (1861–1916), on the other.[10] In the 1930s Kazimierz Ajdukiewicz (1890–1963) resumed Le Roy's theses within the context of a pragmatic conception of language and meaning, also mentioning classical mechanics and the special theory of relativity as instances of non-intertranslatable languages.[11] Another striking anticipation of the incommensurability thesis is that of Carnap: his meaning holism, deriving from his theory of linguistic frameworks, explicitly involves the idea that in the translation from one language to another the content of an empirical statement is not always preserved.[12] Here I will consider, albeit very briefly, two important – but perhaps less well-known, at least in this respect – precursors.[13]

Frank P. Ramsey

In the summer of 1929 Frank Plumpton Ramsey (1903–1930) wrote a brief paper, "Theories", posthumously published by his friend Richard B. Braithwaite in 1931, and subsequently republished a few times. Not satisfied with the traditional explanation of the status of ever more abstract concepts (such as field of force, electric current and, later, electrons and photons) that physicists had been using during the nineteenth century, Ramsey advances a new explanation. If the philosophical tradition within which Ramsey had grown (and Bertrand Russell in particular) thought that the abstract concepts of physics should be defined in terms of the naturally visible phenomena, Ramsey realizes that that would not have worked: indeed, if things were so, theoretical concepts would mean something only when

[10] See Le Roy (1899–1900), (1900a), (1900b), (1901a) and (1901b); Poincaré (1902), (1905), (1908) and (1913); Duhem (1906, 1914) and (1996). See also Giedymin (1974), (1982), (1991) and (1992) and Zahar (2001).

[11] See Ajdukiewicz (1949) and (1978). On the relationship between Ajdukiewicz's conventionalism, the Polish school of philosophy, the Vienna Circle and some key features of the "new philosophy of science", see Giedymin (1971), (1973), (1974), (1975), (1977), (1978), (1982), (1991), (1992), Szaniawski (ed.) (1989), Woleński (1989) and (1999), Coniglione (1990) and (1996), Coniglione, Poli, Woleński (eds) (1993), Sinisi, Woleński (eds) (1995), Ginzburg (1998), pp. 136–170, and Kijana-Placek, Woleński (eds) (1998).

[12] The remarkable similarities between Carnap's and Kuhn's views will be examined in detail below, in ch. 5: see especially pp. 208–210, where I deal with incommensurability.

[13] Ramsey 1929 paper, "Theories", though sketchy, is widely believed to contain important insights into the structure and functioning of scientific theories. However, little attention has been paid to the way in which it anticipates some key issues of the debates that would take place some three decades later, and especially the incommensurability thesis. As Hugh Mellor remarks, "Nowhere, perhaps, does Ramsey more tantalizingly anticipate later literature than in his brief discussion of theories […]. His treatment of theoretical terms as existentially bound variables has indeed been noted by philosophers of science; but many of the consequences he drew from this treatment have had to be laboriously rediscovered": Mellor (1978), p. 4. See also *ibidem*, pp. 4–5, and Mellor (1990), pp. xx–xxi. On the other hand, the case of Frola is particularly interesting because his reflections do not represent an isolated case, but appear within a wider context – that of the Centro di Studi Metodologici (see below, n. 33) – that was strongly influenced by the ideas of Logical Positivism. Both Ramsey's and Frola's views are therefore indicative of the philosophical climate that dominated the first half of the twentieth century and in whose fertile soil Kuhn's ideas plunged their roots.

they are used in explanations, and there would be no possibility to develop a science by employing new and different uses of those very theoretical concepts – as in fact had happened in the course of the entire development of science. According to Ramsey's idea, sentences referring to theoretical concepts should not be directly translated into sentences referring to observables. Theoretical concepts play a role within very complex sentences – the so-called Ramsey-sentences[14] – that contain both theoretical terms and observables. In other words, Ramsey proposes to cast the whole content of a theory in the form of a single sentence in second order language which, in a sense, eliminates all the theory's theoretical (non-observational) terms. A treatise of physics would consist in one, single, very long sentence that would have the form of a story. A treatise on electrons, for instance, would start with something like: "there are things we call electrons that ..." and would go on describing the properties of these objects.[15]

In "Theories" Ramsey sets out "to describe a theory simply as a language for discussing the facts a theory is said to explain".[16] In order to do so, he constructs a simple example of a theory with two universes of discourse, "the primary system" and "the secondary system". The propositions belonging to the former system represent "the facts to be explained"[17] and are truth-functions of quantifier-free expressions formed from predicate or function symbols A, B, C, ..., and a suitable number of individual constants as names of the individuals belonging to the primary system. The secondary system is an expansion of the primary system: new predicate or function symbols (α, β, γ...) are introduced to form truth-functional propositions

---

14 The term was introduced by Carl G. Hempel in his (1958), p. 216. The idea of treating a theory's theoretical terms as existentially bound variables appears for the first time in Ramsey (1925), pp. 8–11 and 27.

15 Ramsey's method "amounts to treating all theoretical terms as existentially quantified variables, so that all the extralogical constants that occur in Ramsey's manner of formulating a theory belong to the observational vocabulary" (Hempel (1958), pp. 215–216). For example: $(\exists\varphi)\,(\exists\psi)\,[(\varphi x \to (Ax\ Bx))\cdot(\psi x \to Cx)\cdot(\varphi x \to \psi x)]$. In plain words: there are two properties, φ and ψ, otherwise unspecified, such that any object with the property φ also has the observable properties A and B, that any object with the propertyψ also has the observable property C, and that any object with the property φ also has the property ψ. But, Hempel notices, the Ramsey sentence associated with an interpreted theory "avoids reference to hypothetical entities only in letter [...] rather than in spirit" and hence provides "no satisfactory way of avoiding theoretical concepts" (*ibidem*). Indeed, Ramsey himself made no such claim. Rather, as Hempel remarks, "his construal of theoretical terms as existentially quantified variables appears to have been motivated by considerations of the following kind: If theoretical terms are treated as constants which are not fully defined in terms of antecedently understood observational terms, then the sentences that can be formally constructed out of them do not have the character of assertions with fully specified meanings, which can be significantly held to be either true or false; rather, their status is comparable to that of sentential functions, with the theoretical terms plying the role of variables. But of a theory we want to be able to predicate truth or falsity, and the construal of theoretical terms as existentially quantified variables yields a formulation which meets this requirement and at the same time retains all the intended empirical implications of the theory" (*ibidem*, pp. 216–217). See also the discussion in Lewis (1970).

16 Ramsey (1929a), p. 112.

17 Ramsey (1929a), p. 112.

with the individual constants of the primary system. Such predicates or functions can be thought of as the theoretical (or abstract) terms of the theory. They are constrained by a finite number of "axioms", that is, formulae whose predicate or function terms belong exclusively to the vocabulary of the secondary system and whose quantifiers range over the universe of discourse of the primary system.[18] Moreover, "whatever propositions of the secondary system can be deduced from the axioms we shall call theorems".[19] The two systems, or universes of discourse, are linked by a "dictionary", i.e. "a series of definitions of the functions of the primary system *A*, *B*, *C*, … in terms of those of the secondary system α, β, γ […]".[20] General propositions deducible from the conjunction of the axioms and the dictionary are called "laws" and represent empirical generalizations, while singular propositions deducible from that conjunction are called "consequences".[21] The totality of general and singular propositions (that is, the sum of all laws and consequences) in which no extra-logical symbols of the secondary system occur and which are deducible from the conjunction of the axioms and the dictionary Ramsey calls the "eliminant": "it is this totality of laws and consequences which our theory asserts to be true".[22]

Having made these distinctions, Ramsey considers various questions about theories. During the discussion of his answers to these questions in the light of the previous theoretical construction,[23] he raises the problem of the relations among different theories – the problem, that is, of the meaning of phrases like "two contradictory theories", "two equivalent theories" or "one theory contained in (or reducible to) another".[24] He sketches the meaning of these phrases in terms of the content of a theory which, in the light of his previous discussion and definitions, he identifies with what the theory asserts, that is, the totality of laws and consequences of a theory.[25]

Following Jerzy Giedymin,[26] Ramsey's characterization may be summarized and characterized along these lines. Firstly, assuming a theory to be expressed in two languages (primary and secondary system, or observational and theoretical language) and axiomatized by a finite set of axioms on the basis of first order logic, Ramsey claims that the properties of a theory may be best seen if it is reformulated as one single sentence (what Hempel would later label "the Ramsey sentence") obtained from the original theory by a second order existential generalization on all terms of the secondary system (that comprising the theoretical terms). Secondly, such a reformulation replaces theoretical terms by existentially bound variables: in this sense, the Ramsey sentence of a theory may be seen as eliminating theoretical terms (belonging to the secondary system) from the theory in question. Thirdly, the

[18] See Ramsey (1929a), p. 114.

[19] Ramsey (1929a), pp. 114–115, emphasis suppressed.

[20] Ramsey (1929a), p. 115.

[21] See Ramsey (1929a), p. 115.

[22] Ramsey (1929a), p. 115.

[23] See Ramsey (1929a), pp. 119–136.

[24] See Ramsey (1929a), pp. 132–133.

[25] For example: two theories are equivalent if and only if their content – that is, their respective sets of laws and consequences – are equivalent.

[26] See Giedymin (1980), pp. 233–234.

Ramsey sentence expresses the content of the theory it translates: the reformulation of a theory in terms of its Ramsey sentence leaves its empirical content unchanged. Indeed, since all laws and consequences derivable from the original theory are also derivable from the corresponding Ramsey sentence, the empirical content of a theory – what a theory asserts as true, or the totality of laws and consequences deducible from it (the "eliminant", in Ramsey's terminology) – is identical with the content of the corresponding Ramsey sentence. In the fourth place, the relations of equivalence, contradiction, reduction, etc., between two or more theories are defined in terms of the corresponding relations between their empirical contents, or eliminants. Finally, Ramsey's proposal "to express the content of a theory in one single sentence (however complex) and the insistence that questions of the meaning, truth and testing can only be answered within the context of the whole theory, are features of what nowadays is called the *holistic* view of theories".[27]

An immediate consequence that Ramsey draws from this account is that no part, however small, of a theory can be understood without reference to the whole theory. It is therefore impossible to set aside or overlook parts of rival theories only because they do not appear in the theory we support. Writes Ramsey: "If a man says 'Zeus hurls thunderbolts', that is not nonsense because Zeus does not appear in my theory, and is not definable in terms of my theory. I have to consider it as a part of a theory and attend to its consequences, e.g. that sacrifices will bring the thunderbolts to an end".[28] In the same way, if we want to discover whether a part of a theory – such as "Zeus hurls thunderbolts" or "electrons have a certain mass" – is true or not, we cannot simply assess it by considering it individually. As Ramsey writes, "we have to think what else we might be going to add to our stock, or hoping to add, and consider whether [they] would be certain to suit any further additions better than [their negations]".[29]

Another consequence of Ramsey's conception of theories is that rival theories can attribute rather different meanings to shared terms (as in the case of the different meaning attributed to "mass" or "space" by Newton and Einstein), so that there may be no way to directly compare such theories, nor to say whether they are reciprocally incompatible. According to Ramsey, the parts of a theory that contain theoretical variables within the scope of the theory's quantifiers are not "strictly propositions by themselves":[30] their meaning "can only be given when we know to what stock of 'propositions' [they] are to be added".[31] This holistic feature of the meaning of

[27] Giedymin (1980), p. 234. Furthermore, "In so far as in the Ramsey view the sentences of the secondary system of a theory, its theoretical sentences, are not asserted as true, do not enter into its content (although they may be used as a secondary language in which to clothe the theory's content and also as a formal inferential devices) and may be regarded as empirically uninterpreted, in contradistinction to the extra-logical expressions of the primary system, the Ramsey view appears to be a variety of the instrumentalist (or formalist) view of scientific theories" (*ibidem*, pp. 233–234).

[28] Ramsey (1929b), pp. 137–138.

[29] Ramsey (1929a), p. 132. See what Kuhn says of his own experience with Aristotle, for example in his (1977a), pp. xi–xiii.

[30] Ramsey (1929a), p. 131.

[31] Ramsey (1929a), p. 131.

theoretical terms implies, among other things, that rival theories in the same empirical range of phenomena can be incommensurable: "Two theories may be compatible without being equivalent, i.e. a set of facts might be found which agreed with both, and another set too which agreed with one but not with the other. The adherents of two such theories could quite well dispute, although neither affirmed anything the other denied".[32]

Eugenio Frola

Eugenio Frola (1906–1962) was an engineer and mathematician from Turin, who took part in the early meetings of what would have become, in 1946, the Centro di Studi Metodologici ("Centre for Methodological Studies"), an institute that played a major role in the birth of logic and philosophy of science in Italy after the Second World War.[33]

According to Frola, the essential determination of a geometrical object is wholly described by the function that it plays within the theoretical system it belongs to: just as in the game of chess we do not want to know or determine the essence of the various pieces, but rather the rules it must abide by, so in geometry we do not aim at grasping the essence of words like "point", "line", "plane", etc. – rather, we want to rigorously define the rules according to which these words can combine. But if the meaning of "point" is exhausted in implicit definitions, that is, in the rules of its employment, then "a modification of even one single rule totally shifts the meaning

[32] Ramsey (1929a), p. 133. See also Giedymin (1980), Majer (1989), Mellor (1995), Sahlin (1990) and (1997). Ramsey's theory of incommensurability was later resumed by logical positivists in order to highlight the absolute novelty and the substantial incompatibility of the theory of relativity and of quantum mechanics with classical physics. Finally, in the 1960s it was employed in conjunction with the thesis of theory-ladenness and of meaning-variance just to attack Logical Positivism.

[33] The Centro di Studi Metodologici originated from the debates sparked off by Geymonat (1945). Philosopher and mathematician (he was assistant to Giuseppe Peano in the early 1930s) Ludovico Geymonat (1908–1991) studied in Vienna with Moritz Schlick in 1934 and contributed to the diffusion of Logical Positivists' ideas in Italy with his (1934), (1935) and (1936) (in a letter of 26 May 1935, Schlick himself described Geymonat (1935) as "excellent throughout. Undoubtedly, it is the best exposition of our views that has been published so far by a neutral observer": Schlick (1985), p. 298). He was appointed the first chair in the philosophy of science in Italy (in Milan, in 1956) and founded, with other intellectuals, the Centro di Studi Metodologici in Turin in 1947. Apart from Frola, other members include Nicola Abbagnano, Norberto Bobbio, Piero Buzano, Bruno de Finetti, Bruno Leoni, Prospero Nuvoli, Enrico Persico and Francesco Severi. However belonging to different intellectual schools and philosophical orientations, all these scholars shared the ideal of an open, critical, anti-metaphysical rationality, taking issue with the then dominant Crocean idealistic culture (for this reason the movement was also labelled as "neo-rationalistic" or "neo-Enlightenment"), and thus playing a significant role in post-war Italian culture and philosophy (see Abbagnano, Buzano, Buzzati-Traverso, Frola, Geymonat, Persico (1947)). See also Pera (1985) and (1986), together with Pasini, Rolando (eds) (1991) and Minazzi, Petitot (1993).

of the originals[34] – indeed, it creates new originals that are not comparable with the previous ones".[35]

Frola's discussion of implicit definitions is rooted in the works of David Hilbert and Moritz Schlick. The issue is particularly relevant here, since it constitutes a seed of the fact that a term takes its meaning from the theoretical context in which it is employed. In Hilbert's view, terms acquire meaning only by virtue of the axiom system in which they occur and possess only the content that it bestows upon them: they stand for entities whose whole being is to be bearers of the relations laid down by the system. Thus, for example, *The Foundations of Geometry* opens with a system of propositions in which a number of terms (such as "point", "straight line", "plane", "between", "outside of", etc.) have no meaning or content: they acquire one only as a consequence of the system of axioms that defines their mutual relationships.[36] This means that Hilbert's axioms do not presuppose any previously fixed meanings of the primitive terms: indeed, it is the axioms that together provide the required definitions. It follows that a different system of axioms defines a different system of concepts. Therefore, for instance, Euclidean and non-Euclidean geometries cannot be regarded as competing theories or descriptions (of space, of our spatial intuitions, or of anything else): they constitute *different conceptual schemes*.

In his *General Theory of Knowledge* Schlick takes issue with Kant's account of geometry as synthetic truths known by *a priori* intuition. Schlick's (and, more generally, the Vienna Circle's) anti-metaphysical and empiricist bent reacted against both the appeal to *a priori* intuition and to the claim that synthetic truths might be known without recourse to experience. Schlick's approach was to deny that any problem in the form of explaining synthetic *a priori* truths arises: geometry, while known *a priori*, does not comprise a body of synthetic truths – rather, they are analytic (since, in Hilbert's view, the axioms of a geometry are definitions of their primitive, non-logical, terms). In order to apply geometry empirically, we will need to give each primitive term an empirical interpretation that transforms the axioms into propositions about empirical items.[37] Thus interpreted, axioms are no longer analytic: they will express empirical, synthetic propositions about their empirical

[34] "Originals" is Frola's word for "primitive terms".

[35] Frola (1947), p. 100. Frola taught mathematics at the University of Turin and some of his philosophical remarks are scattered throughout his textbooks. This renders a reconstruction of his thought particularly difficult. However, for what follows I will rely on some explicitly philosophical works, such as Frola (1947) and the articles collected in his (1964). I will also use the guide provided by Geymonat in his (1963) and (1964), as well as in Sassoli (1994).

[36] In his (1899) Hilbert undertook to construct geometry on a foundation whose absolute certainty would not be placed in jeopardy at any point by an appeal to intuition. The task was to introduce the basic concepts – that are in the usual sense indefinable – in such a fashion that the validity of the axioms that treat these concepts is strictly guaranteed. Hilbert's solution was to stipulate that the basic (or primitive) concepts are to be defined just by the fact that they satisfy the axioms.

[37] We could define a "straight line", for instance, as the "path of a ray of light"; or "point" by indicating a "grain of sand".

subject matter. But then they are not necessary either: their truth or falsity is the task of empirical investigation to assess.[38]

However, neither Hilbert nor Schlick were anticipating the theses put forward by Kuhn and Feyerabend several decades later. Both were admitting that Euclidean and non-Euclidean geometries were incommensurable, in a sense – but they thought this view was restricted to mathematics, or to *a priori* formal sciences (like geometry). Frola's contribution lies in the fact that he takes the Hilbert–Schlick idea of implicit definition and says that it does have implications for empirical science: the implicit definitions of closed languages (such as mathematics) have consequences for the open language of physics as well.

Mathematics, according to Frola, is a "language […] closed in an ivory tower" that "does not relate to anything external, to any exterior truth, either absolute or relative".[39] Knowledge exists, in Frola's view, only when it is possible to express it in one language: "to conceive languages as separated and free from any impending external necessity is to put ourselves in a position of independency, of sovereignty; it means breaking a heavy bond, regarding knowledge as liberty, as a human creation".[40] But if every mathematical theory is a particular closed language, Frola's argument seems to entail that there is no relationship between mathematical and ordinary language, just as there is no connection between a formal theory and another one. Nor need we conjecture "the existence of an absolute meta-rationality, *a priori* expressing the norms by which the individual rationalities, deriving from the arbitrariness of precedents,[41] elaborate themselves".[42]

---

[38] The connection between concepts and reality is set up through concrete definitions, for these are the only ones that point to something real, that has an individual existence: they exhibit in intuitive or experienced reality what is to be henceforth designated by a concept. On the other hand, implicit definitions – those by which terms are explained by the system of axioms that employ them – remain in the domain of concepts and have no association or connection with reality: "A system of truths created with the aid of implicit definitions does not at any point rest on the ground of reality. On the contrary, it floats freely, so to speak, and like the solar system bears within itself the guarantee of its own stability" (Schlick (1918/1974), p. 37). Accordingly, the construction of a strict deductive science, such as number theory, has only the significance of a game with symbols (of course, not every set of arbitrary postulates may be conceived as the implicit definition of a group of concepts: the defining axioms must fulfil certain conditions, such as that they are not inconsistent). In geometry, however, and even more in the empirical sciences (such as physics), the reason behind our constructing an edifice of concepts is above all our interest in the relationships among real objects: here our interest "attaches not so much to the abstract interconnections as to the examples that run parallel to the conceptual relations. […] But […] the moment we carry over a conceptual relation to intuitive examples, we are no longer assured of complete rigor" (*ibidem*, pp. 37–38). Implicit definitions do enable us to determine concepts completely and thus to attain strict precision in thinking – but the price for that is the radical separation between thought and reality. For a discussion, see Schlick (1918/1974), pp. 27–39, and Bird (1991), pp. 148–155.

[39] Frola (1947), p. 101.

[40] Frola (1947), p. 102.

[41] Frola is here referring to the arbitrariness of postulates.

[42] Frola (1947), p. 107.

Therefore, in Frola we find the seeds of a (radical) incommensurability thesis between scientific theories in terms of non-comparability, that has consequences also for a realistic conception of the world. The cause of such incomparability is to be traced – and here lies Frola's originality – in the specific aspects of mathematics as such. Indeed, when referring to empirical theories, Frola acknowledges that they are open languages, and this holds particularly for physics.[43] Physics, however, "even if substantially different from mathematics", owes to the latter its very existence, since physics is "applied mathematics", for mathematical terms (numbers and measures, that is), together with "essentially physical terms",[44] are fundamental for the construction of the protocol statements that characterize the empirical basis. And "physical phenomenon, and not only physics, is very closely bound to the language in which it is expressed".[45] This has a deep influence on a possible realistic conception of the world: "[…] in order to construct physics it is not necessary to imagine the actual existence of a nature that physics discovers step by step".[46]

Furthermore, since physical theories are developed within mathematics' closed languages, Frola ends up by eliminating not only realism as a metaphysical pseudo-problem, but also the problem of comparability:

> if we wished to compare Aristotelian physics and contemporary physics and say that the former is false while the latter is true, we would make a very grave mistake; they are absolutely incomparable, nor is there a touchstone upon which we could compare them. Protocols and the laws of our physics have no meaning at all within the language of the other physics, and vice versa. They are the description of two different worlds, not two descriptions of the same one.[47]

Some aspects of Frola's theory are connected to conventionalism, but the upshot is a radical fracturing in which we get rid at the same time both of the metaphysical subjection to external reality and of historical constraints.[48]

Frola is only one voice of the rich epistemological debate that took place in Italy in the late 1950s and early 1960s. In this context, it is worth mentioning also Ludovico Geymonat's philosophical stance, as opposed to Frola's. In *Saggi di filosofia neorazionalistica* (1953) Geymonat reacts against the anti-metaphysical and anti-historical upshot of his friend's view: "As opposed to games, capriciously arbitrary and therefore lacking mutual relationships, scientific theories are connected the ones to the others by *historically determined relationships*".[49] In Geymonat's eyes, Frola's extreme formalism is unable to account for these relationships because it lacks the

---

43 See Frola (1947), p. 107.

44 Frola (1947), p. 107.

45 Frola (1947), pp. 107–108.

46 Frola (1947), p. 108.

47 Frola (1947), pp. 108–109.

48 See also Abbagnano (1947), Giorello (1977), Pera (1984), pp. xi–xii, and Sassoli (1994), pp. 180–186. On this issue and, more generally, on the role of Eugenio Frola and contribution in Italian philosophy of science in the twentieth century, see Geymonat (1963) and (1964), Giorello (ed.) (1977), pp. 161–162, Giorello (1986) and Pera (1986).

49 Geymonat (1953), p. 61.

dynamic perspective that alone can describe rationality in its full articulation. Frola's holistic view ("given the change in the concepts and the deductive rules from one theory to another, it seems impossible that any proposition remains unchanged in the transition, and it seems therefore unconceivable that what was a problem for one theory may turn up to be an identical problem for the other, and be resolved by it"[50]) betrays a real difficulty, that of "making sense of the relationship between different languages".[51] "I am firmly convinced", writes Geymonat, "that such relationships do exist, and indeed represent something fundamental for scientific research. Denying them amounts to denying the reality of science. Providing a rational explanation of them is, however, far from easy".[52] Geymonat's point is that history of science shows that between advocates of rival theories there is never the total impossibility of communication Frola depicts:[53]

> The theories in which science is articulated are not closed at all; on the contrary, they richly communicate with one another and with common language, so that the same proposition – transferred from a more restricted theory to a more general one – gets enriched with new meanings, becomes the source of new developments and reveals with more clarity the profound reason of its validity.[54]

It is from his confrontation with Frola that Geymonat becomes convinced that a purely syntactical analysis does not exhaust the methodological examination of theories: historical and pragmatic considerations have to be taken into consideration as well. "It is undeniable", writes Geymonat, "that in the actual process of formation of a theory, the theory that is elaborated does not always possess the logical closure that characterizes perfect theories".[55] The conflicts that mark the history of science must not be regarded, he says, as the overcoming of the false theory by the true one but, rather, as "route adjustments":

> The new technique[56] does not assimilate the good qualities of the old one, nor turns out to be able to solve all the problems it solved, but, following a different path, it confronts the problems that had not previously been tried, or at least that remain unsolved. It may

---

[50] Geymonat (1953), p. 61.

[51] Geymonat (1960), p. 61.

[52] Geymonat (1953), p. 62.

[53] Indeed, in a later book Geymonat argues how Galileo managed to use fragments of the Aristotelian tradition against the most dogmatic followers of Aristotle's doctrines: see Geymonat (1957). Each time a significant scientific revolution occurs, Geymonat argues, when one theory (or paradigm, or research programme) replaces another, the transition takes place thanks to the work of pioneers, such as Galileo, who are able to move from one language to another. The lesson we draw from history, Geymonat continues, has a theoretical import: the question of the relationships between one theory and another is to be treated with reference to the individual researcher, who can use both languages and, indeed, "in both poses problems and tries to solve them" (Geymonat (1953), p. 50).

[54] Geymonat (1960), p. 104.

[55] Geymonat (1953), p. 50.

[56] Here Geymonat confronts the analogy between theory and technique, by appealing to the notion of "techniques of the reason".

even happen that this change involves a regress, rather than a progress; it may happen, that is, that a new technique results to be rougher than the one it replaced, and despite this it reveals more fruitful in that moment of application, since it better fits the cultural situation of the men that have to employ it.[57]

Here Geymonat proves to be very far from the logical empiricist orthodoxy he had defended in his early works.[58] He seems to be actually anticipating several of the issues that would become central in the debates between Popper, Kuhn, Hanson, Toulmin, Lakatos and Feyerabend. Indeed, in one of his major works, *Filosofia e filosofia della scienza* (1960), he not only frees the philosophy of science from any identification with the Standard View, but actually sets himself the task of widening its scope and object from the statics to the dynamics of theories.[59]

Neurath and Popper

Hanson, Toulmin, Feyerabend, Kuhn and the other critics of the "cumulative" model make their own many of the arguments of "radical conventionalism",[60] and even their use of incommensurability in an anti-Popperian function is not new: already Otto Neurath, criticizing the notion of "empirical basis" in Popper's *Logik der Forschung*, referred to Ajdukiewicz, calling attention on the difficulties of translation: "When a primitive man says: 'The river runs through the valley', he certainly defines the terms in a way that is different from that of the European who goes on using the statement".[61] In so doing, Neurath legitimized the replacement of Popper's *Basissätze* with his own *Protokollsätze*, in which the author of the protocol sentence is recorded; and he secured the stability of observation reports and thus made possible "a connection from people to people, from age to age, from scientist to scientist".[62]

Feyerabend himself, for example, highlighting that learning a theory does not begin with observation culminating later in the theory itself, but always involves

---

[57] Geymonat (1953), p. 70.

[58] I am referring, in particular, to Geymonat (1934), (1935) and (1936).

[59] See Geymonat (1960). Pera highlights the curious destiny of Geymonat: while his (1960) was anticipating and autonomously developing issues that abroad were only ripening, when (in the 1970s) also the Italian philosophical environment would look at these very issues with interest, he had already chosen to move towards other approaches (dialectical materialism) that would prove unfortunate (see Pera (1985), p. 153). On the contrary, other authors read Geymonat's move towards dialectical materialism as a renewed interest in history: see Giorello, Mondadori (1978) and (1992).

[60] The term is Giedymin's: see his (1970) and (1983).

[61] Neurath (1935a), p. 129. According to Dirk Koppelberg the doctrine of incommensurability is already *in nuce* in Neurath (1932a), as a coherent development of his holistic view of the relationship between theory and experience, in which neither is privileged: see Neurath (1932a), pp. 58–59. Such a view is influenced also by the historical reflection on the way in which theories dynamically evolve superseding one another (see Koppelberg (1987), p. 27).

[62] Neurath (1935a), p. 129. It is to be noted, however, that since 1936 Ajdukiewicz gave up his radical conventionalist position, deeming it no longer plausible.

both elements, explicitly draws on Neurath's conception that there is no primary phenomenological language that expresses experience without theoretical residues. In an article that intervenes in the protocol sentences debate, Neurath denied that experiences constitute basic elements that do not require verification and therefore cannot be questioned.[63] Popper recognized that Neurath's point of view was a "notable advance" (as opposed to Carnap's early, later revised views).[64] However, he urges a further step, since we need to find rules that limit the possibility of deleting a protocol statement that contradicts a system arbitrarily, otherwise every system would be equally defensible.[65] This inconvenience may be avoided if we admit some basic (or test) statements that are conventionally accepted as such. Therefore, Popper objected to the naïve empiricist, who thinks he can gather experimental data first, and then move on to the theoretical generalization, that science requires points of view and theoretical problems in the first place, in the light of which experience is oriented and guided: "observations, and even more so observation statements and statements of experimental results, are always *interpretations* of the facts observed; [...] they are *interpretations in the light of theories*".[66] "Science", in other words, "does not rest upon solid bedrock. The bold structure of its theories rises, as it were, above a swamp. It is like a building erected on piles".[67] As can be seen, Popper's teachings,

---

[63] See Neurath (1932a).

[64] See Popper (1935, 1959), pp. 96–97.

[65] See Popper (1935, 1959), p. 97: "Neurath's view that protocol sentences are not inviolable represents, in my opinion, a notable advance. But apart from the replacement of perceptions by perception-statements – merely a translation into the formal mode of speech – the doctrine that protocol sentences may be revised is his only advance upon the theory (due to Fries) of the immediacy of perceptual knowledge. It is a step in the right direction; but it leads nowhere if it is not followed up by another step: we need a set of rules to limit the arbitrariness of 'deleting' (or else 'accepting') a protocol sentence. Neurath fails to give any such rules and thus unwittingly throws empiricism overboard. For without such rules, empirical statements are no longer distinguished from any other sort of statements. Every system becomes defensible if one is allowed (as everybody is, in Neurath's view) simply to 'delete' a protocol sentence if it is inconvenient. In this way one could not only rescue any system, in the manner of conventionalism; but given a good supply of protocol statements, one could even confirm it, by the testimony of witnesses who have testified, or protocolled, what they have seen and heard. Neurath avoids one form of dogmatism, yet he paves the way for any arbitrary system to set itself up as 'empirical science'".

[66] Popper (1935, 1959), p. 107, n*3; see also ch. 1, n. 61. However, while Popper thinks that there is a common empirical basis shared by different theories, Feyerabend and Kuhn deny that, arguing that each theory creates the presuppositions for its own existence and therefore different and competing theories cannot be related to any natural empirical data, legitimized to impartially judge among them.

[67] Popper (1935, 1959), p. 111: "Science does not rest upon solid bedrock. The bold structure of its theories rises, as it were, above a swamp. It is like a building erected on piles. The piles are driven down from above into the swamp, but not down to any natural or 'given' base; and if we stop driving the piles deeper, it is not because we have reached firm ground. We simply stop when we are satisfied that the piles are firm enough to carry the structure, at least for the time being"; see also Popper (1979, 1994), p. 136.

however Feyerabend tends to play the fact down, are actually a major influence on his characterization of the relationship between theory and observation.

Paul K. Feyerabend and Thomas S. Kuhn

In introducing the notion of incommensurability Kuhn and Feyerabend share a purpose and (partially) the methods, but substantial differences divide them nonetheless. For Kuhn, incommensurability has a wider scope, while Feyerabend thinks that only theoretical systems can be incommensurable, and only if interpreted in a certain way. Therefore, Kuhn's and Feyerabend's ideas of incommensurability are different and not interchangeable.

Kuhn, says Feyerabend,

> has observed that different paradigms (*A*) use *concepts* that cannot be brought into the usual logical relations of inclusion, exclusion, overlap; and (*B*) make us see things differently (research workers in different paradigms have not only different concepts, but also different *perceptions*); and, (*C*) contain different *methods* for setting up research and evaluating its results. According to Kuhn it is the *collaboration* of all these elements that makes a paradigm immune to difficulties and incomparable with other paradigms. Incommensurability in the sense of Kuhn [...] is the incomparability of paradigms that results from the collaboration of (*A*), (*B*) and (*C*).[68]

The meaning (*A*) of term is due to the meaning-variance.[69] In its (*B*) meaning, that Feyerabend regards as superseded by empirical considerations,[70] incommensurability bears consequences also for scientists, who end up having different experiences:[71] "scientific theories are ways of looking at the world and their adoption affects our general beliefs and expectations, and thereby also our experiences and our conceptions of reality. We may even say that what is regarded as 'nature' at a particular time is our own product in the sense that all the features ascribed to it have first been invented

---

[68] Feyerabend (1977), pp. 363–364; see also his (1978b), pp. 178–179, (1978c), pp. 65–70, and (1958a). For (*B*) see the discussion in Hanson (1958), ch. 1, where this part "is argued with vigour and many examples" (Feyerabend (1977), p. 363, n. 1).

[69] Toulmin writes: "Men who accept different ideals and paradigms have really no common theoretical terms in which to discuss their problems fruitfully" (Toulmin (1961), p. 57); "the interpretation of an observation language comes from the theory that explains what we observe, and changes as soon as this theory changes" (Feyerabend (1977), p. 364). Feyerabend's favourite example of meaning-variance is that of the concept of "mass", whose meaning varies if employed within the context of Newtonian mechnics or Einsteinian relativity theory. It must be noted that Feyerabend remarks that his own version of the incommensurability thesis follows only from (*A*): "As opposed to Kuhn my own research started from certain problems in area (*A*) and my discussion of these problems was restricted to a fairly narrow domain. [...] When using the term 'incommensurable I always meant deductive disjointedness, *and nothing else*" (Feyerabend (1977), pp. 364–365); see also Feyerabend (1965b) and (1978b), pp. 178–181.

[70] See, for example, Feyerabend (1965c) and (1978b), p. 224.

[71] This is the case, according to Kuhn, of Lavoisier and Priestley, or, according to Hanson, of Kepler and Tycho.

by us and then used for bringing order in our surroundings".[72] In the third meaning, two theories are incommensurable because they adopt specific methods of research and evaluation of solutions:[73] the theories, methods and standards learned by the scientist during his training within a paradigm are inextricably interwoven.[74] Such an entangled weaving of theories, methods and standards makes the incompatibility of two incommensurable theories untranslatable in terms of a logical contradiction.[75]

Therefore, in advancing the incommensurability thesis, both Feyerabend and Kuhn are reacting to the then dominant philosophy of science of the logical empiricist tradition. Although their attack on the "Standard View"'s orthodoxy[76] has several aspects, those I am here most interested in are the rejection of the idea of an observation language neutral with respect to theories and meaningful independently of it (that is, point (*B*) described above). Indeed, if the logical empiricists were ready to accept meaning variance at the level of theoretical terms,[77] they thought that the observation vocabulary was independent of theory. The existence of a neutral observation language would effectively have secured exactly what

---

72 Feyerabend (1962a), p. 29; see also Hanson (1958), ch. 1, and Polanyi: "the two sides do not accept the same 'facts' as facts, and still less the same 'evidence' as evidence. [...] For within two different conceptual frameworks the same range of experiences takes the shape of different facts and different evidence" ((1958), p. 167); and again: "we proceed according to what we expect to be the case" (*ibidem*, p. 161).

73 "Beliefs and valuations have accordingly functioned as joint premisses in the pursuit of scientific enquiries. [...] the general views and purposes implicit in the achievement and establishment of a scientific discovery are its premisses [...]" (Polanyi (1958), p. 161); "[...] the premisses of science determine the methods of its pursuit and vice versa" (*ibidem*, p. 166).

74 Each of these three meanings allegedly bears devastating consequences for Popper's critical rationalism: the first, in particular, precludes the possibility of a comparison based on the empirical contents of two theories, making any criterion of choice and the notion of verisimilitude collapse; the second makes the employment of observations for theory tests highly problematic (to say the least); finally, the third excludes the possibility of an invariant methodology: with the failure of any uniformity in the evaluation standards the notion of scientific progress no longer makes sense. According to Marcello Pera (see his (1981), pp. 203–205), the logical consistency and theoretical survival of critical rationalism seem to be indissolubly linked to the rejection of the thesis of the theory-ladenness of observation from which incommensurability follows. In Pera's view, Popper went so far in his own idea of theory-ladenness of observations to include in his own conceptions – despite his intentions and negations – the very thesis of incommensurability, thus endorsing the irrational premises that were later to be stated as inevitable consequences of the thesis. See the references given above, in ch. 1, n32.

75 This is revolutionary both for the logical positivists' programme and Popper: incommensurability cannot be accounted for in logical terms. That is why Kuhn and Feyerabend promote, besides logic, but with a decidedly greater weight, also hermeneutical, sociological and anthropological methods.

76 See Suppe (1974), pp. 3–118, and (1977), pp. 617–632. See also Feigl (1970) and Hempel (1965a), (1970) and, for a self-critique, his (1989).

77 See the discussion of Carnap's views in English (1978); see also Newton-Smith (1981), p. 152.

incommensurability questioned, that is, a semantic common ground that allowed for the confrontation of theories. Against this principle of Logical Positivism, Kuhn and Feyerabend hold that observation is not, in itself, an independent source of meanings for terms: in fact, the meanings of observational terms depend on theory. The incommensurability thesis, then, is rooted in the rejection of the existence of a shared observation language that may leave room for a decision among competing theories. Incommensurability can therefore be characterized as the negation of the existence of a neutral language, independent of theory, by which the contents of rival, competing theories can be expressed and therefore a comparison, and a choice, among them, can be decided.

Crossing paths

Paul K. Feyerabend is two years younger than Kuhn: he was born in Vienna in 1924.[78] When he was a youth he studied physics, philosophy, literature and theatre, besides singing, opera and acting (to which he added also Italian). After taking part in the Second World War on Germany's side, when he was seriously wounded (he would be partially paralysed and an invalid for the rest of his life), Feyerabend attended physics lectures at the University of Vienna, where his teachers were Hans Thirring, Karl Pribram and Felix Ehrenhaft. In 1948, on the occasion of the first international summer school of the Österreichisches Kollegium in Alpbach, in Tyrol, he met Karl Popper "and other dignitaries".[79] On the model of the Wiener Kreis, in 1949 he set up the Kraft Kreis, a university circle centred on the figure of Viktor

---

[78] More information on Feyerabend's life can be found in his autobiography, Feyerabend (1995b), but see also Hoyningen-Huene (1997b). For a comprehensive critical introduction to his philosophy see Corvi (1992) and Preston (1997a). For detailed critical examinations of his philosophy see Duerr (ed.) (1980), Munévar (ed.) (1991), Oberheim (2006), Pera (1982a, 1996) and (1984), Preston, Munévar, Lamb (eds) (2000), Farrell (2003), Stadler, Fischer (eds) (2007), and Tambolo (2007). See also Giorello (1976a), (1976b) and (1979), together with Gillies (1993, 1995).

[79] Feyerabend (1995b), p. 72. No doubt, Karl Popper is the philosopher who influenced Feyerabend most profoundly, always remaining a critical reference for him. Several years after their first meeting in Alpach, Feyerabend would write: "I admired his freedom of manners, his cheek, his disrespectful attitude towards the German philosophers [...], his sense of humour (yes, the relatively unknown Popper of 1948 was very different from the established Sir Karl of later years) and I also admired his ability to restate ponderous problems in simple and journalistic language. Here was a free mind, joyfully putting forth his ideas, unconcerned about the reaction of the 'professionals'" (Feyerabend (1978c), p. 115; see also his (1979, 1980), pp. 202–203, and (1958a), pp. 25–26). Feyerabend is concerned not only with the philosophy of science, but also with every form of human knowledge. Notwithstanding what he would say later, ever more frequently and decidedly as years went by, his approach to philosophy and knowledge as a whole is deeply influenced by Popper. Indeed, Feyerabend's output can be read as an attempt to answer the very questions Popper tried to answer, that is, the possibility and modes of learning and knowledge acquisition. They are questions that capture what Popper singled out as the core problem of epistemology: the growth of knowledge (see Popper (1935, 1959), pp. 15–19).

Kraft, the last member of the Vienna Circle to remain in the Austrian capital.[80] In that same year Ludwig Wittgenstein gave a lecture at the Circle; two years later, in 1951, Feyerabend planned to study with Wittgenstein in Cambridge and applied for a bursary. However, Wittgenstein died before Feyerabend got to England and so he decided to go to London and study with Popper.[81] After spending two years (1952–1953) at the LSE Department of Philosophy, Logic and Scientific Method, he declined Popper's invitation to become his research assistant (Joseph Agassi took the post in his stead) and went back to Vienna, where he was assistant to Arthur Pap, who was then trying to give new life to the Vienna Circle's key doctrines. In 1955, with the help of Popper and Erwin Schrödinger, he was appointed to a lectureship at the University of Bristol. There he taught until 1958, when he joined the Department of Philosophy of the University of California at Berkeley, where he was appointed guest professor. He was offered a permanent post the following year and in 1962 he became full professor.

Thomas S. Kuhn was born in Cincinnati, Ohio, in 1922.[82] In 1940 he entered Harvard University, where he attended the lectures of Philipp Frank and Percy W. Bridgman. He graduated in physics and started working at the Harvard Radio Research Laboratory. His researches on radar countermeasures for war airplanes took him to Europe for a certain period. Back to Harvard, in the fall of 1944, he continued his work at the Radio Research Laboratory, where he met with John H. Van Vleck (a Nobel prize winner for physics in 1977, for his fundamental theoretical investigations of the electronic structure of magnetic and disordered systems), his subsequent Ph.D. supervisor.[83] After receiving his doctorate in 1949 he started teaching courses of General Education under the supervision of James Bryant Conant, Harvard's president: in preparing his lectures on the history of science he had the fundamental insights that would later develop in his philosophy of science. At the meetings of the Harvard Society of Fellows he got to know Willard Van Orman Quine, Stanley Cavell, George Sarton and I. Bernard Cohen. In 1950 he attended Popper's William James Lectures and for the first time he became acquainted with

---

[80] Kraft was one of Feyerabend's examiners for his doctoral dissertation. Analogously to the Vienna Circle, the Kraft Circle "set itself the task of considering philosophical problems in a nonmetaphysical manner and with special reference to the findings of the sciences" (Feyerabend (1966), pp. 3–4).

[81] See Feyerabend (1995b), p. 86, and Agassi (1993a), ch. 4.

[82] More details on Kuhn's life and intellectual development are available in a long interview he gave a few months before his death, Kuhn (I-1997a). See also Hoyningen-Huene (1997a) and (1997c), Buchwald, Smith (1997), Heilbron (1998), Andresen (1999), Gattei (2000c) and Andersen (2001), pp. 1–7; an intellectual biography of Kuhn is currently being written by Keay Davidson. Critical expositions of his philosophy are Buzzoni (1986), Hoyningen-Huene (1989a/1993), Giordano (1997), Bird (2000), Gattei (2000b), Andersen (2001), Sharrock, Read (2002), Marcum (2005) and Preston (2008). See also Giorello (1976a) and (1976b), and Gillies (1993, 1995). Valuable material is also to be found in Fuller (2000) and Gattei (ed.) (2003).

[83] Kuhn's Ph.D. dissertation is his (1949), out of which came also his (1950), (1951a) and Kuhn, Van Vleck (1950).

critical rationalism.[84] It is at this point that Charles Morris invited him to prepare a monograph on the history of science for the *International Encyclopedia of Unified Science*, that he was editing with Carnap, Frank, Jørgensen and Rougier: thus, Kuhn started thinking of *The Structure of Scientific Revolutions*, that would eventually be published in 1962. In 1957 he published *The Copernican Revolution* and accepted the post he was offered from the University of California at Berkeley. There he taught until 1964, working both in the Department of History of Science and in the Department of Philosophy. In Berkeley he renewed his friendship with Stanley Cavell and met Paul Feyerabend, with whom he started a most fruitful confrontation on philosophical and historical issues. During one of their long conversations, they discovered they were employing the very same term – incommensurability – to refer to issues related to scientific progress.[85]

Feyerabend read Kuhn's works for the first time in 1959 and the two of them started discussions the following year. There began two years of intense confrontation and intellectual exchange.[86] In 1964, also following Carl Gustav ("Peter") Hempel's advice, Kuhn moved to Princeton, and since then he met with Feyerabend only a few times. The last one in June 1985, when Kuhn, who was in Paris for some lectures, was invited by Feyerabend to spend a few days in Zurich (Feyerabend was then teaching at the Eidgenössische Technische Hochschule). For three days, the old friends revived intense discussions both on philosophical and personal matters. Despite the promise to meet again, things were to take a different turn, and they were never able to see each other again.

While Feyerabend was already well-known as a philosopher of physics, at the very beginning of the 1960s Kuhn's name was known only within the restricted

---

[84] Popper's ten William James Lectures were delivered from February 16 to April 27, 1950, and ranged from the philosophy of science to the philosophy of the social sciences, politics and ethics. They are still unpublished but can be found in Popper Archive (39.4)–(39.14).

[85] See Kuhn (I-1997a), pp. 297–298. Both introduce the term in 1962, respectively in Kuhn (1962a), p. 103, and Feyerabend (1962a), p. 47. See also Kuhn (1983a), p. 669 and 684, n. 2, and Feyerabend (1958a), pp. 31–36 and (1978b), pp. 178–182. The first occurrence of the idea of incommensurability (but not the term) in Feyerabend's published writings dates back to his works on complementarity: see Feyerabend (1958a), p. 31, (1958b), p. 83, and (1961b), p. 388; see also the different reconstructions offered by Feyerabend in the second and third edition of *Against Method*: (1975, 1988), pp. 228–230, and (1975, 1993), pp. 211–213. Interestingly, as Paul Hoyningen-Huene notices (see his (1989a/1993), p. 207, n. 57) Wolfgang Wieland also employs the term "incommensurability" in 1962, and with very similar meaning: "Indeed, it may happen that reciprocally contradictory statements are actually totally incommensurable; this happens when they try to answer incommensurable questions" (Wieland (1962), p. 30 and, analogously, pp. 37 and 45)).

[86] "Some of which were carried out in the now defunct *Café Old Europe* on Telegraph Avenue and greatly amused the other customers by their friendly vehemence" (Feyerabend (1970a), p. 198, n. 2). A good example of such "friendly vehemence" is documented in the letters published as Feyerabend (1995a) and (2006), which date back to those very years, 1960–1962.

community of historians of science.[87] Feyerabend helped him to be known also among philosophers, where he had already acquired a "controversial reputation". Between 1960 and 1961 Feyerabend often referred to a "forthcoming" book by Kuhn containing several examples drawn from the history of science that would support his own theses.[88]

The two thinkers shared several things[89] and Feyerabend spoke of a "pre-established harmony"[90] between them. Both shared the rejection of the dominant tradition in the English-speaking world, the Logical Empiricism that later contributed to the birth of analytic philosophy.[91] Feyerabend nourished a critical attitude towards this tradition from his discussion of the empirical basis of science (the so-called "protocol sentences debate") and from his apprenticeship under Karl Popper. Whereas Kuhn's scepticism towards the logical empiricist tradition is rooted in his studies of the history of science, begun in 1947: according to Kuhn, the actual history of science did not fit the normative picture philosophers had developed. His later meeting with Popper, in 1950, proved crucial: Popper's critical rationalism always remained a constant critical reference for him. Finally, both Kuhn and Feyerabend had a solid scientific training: Feyerabend held a Master's in astronomy, Kuhn a Ph.D. in theoretical physics.

However, one thing put them on a par in the eyes of the international philosophical community: their simultaneous (and, as we have seen, not completely independent) introduction of an important but highly controversial notion, that would become the centre of lively philosophical discussions following the publication of their works.[92] The confrontation – or the clash – on this issue is far from resolution (in his later

---

[87] History of science was still an underestimated discipline: suffice it to say that in the nearly 4200 pages of the *Encyclopedia of Philosophy* (published in eight volumes in 1967) there is no trace of it, nor of Kuhn's own work, that with *The Structure of Scientific Revolutions* took it to the centre of the epistemological debate. On the contrary, Feyerabend's views are discussed in an article on the philosophical consequences of quantum mechanics (Hanson (1967), that refers to Feyerabend (1957), (1962a) and (1962b)); also, he is author of four articles, respectively dedicated to Boltzmann, Heisenberg, Planck and Schrödinger: Feyerabend (1967a), (1967b), (1967c) and (1967d). He should have written the article dedicated to the philosophical implications of quantum mechanics; however, the task was eventually undertaken by Hanson. Feyerabend's original and most interesting article is still unpublished: see "Philosophical Problems of Quantum Theory", in Feyerabend Archive (11-12-3). Feyerabend's contributions to the philosophy of physics are being collected in Feyerabend (forthcoming).

[88] See, for example, Feyerabend (1961a), p. 61.

[89] For a comparison, see Preston (1997a), especially pp. 87–98, and Hoyningen-Huene (2000).

[90] See Feyerabend (1970a), p. 219, and (1989a), p. 405, n. 26.

[91] See also Hacker (1996), especially chs. 1 and 3, and Baker, Hacker (1985).

[92] Such discussions reached their peak in the 1960s and 1970s, but cooled down considerably in the following two decades. In 1999 an international conference was organized in Hannover in an attempt to recompose a debate and outline possible lines of development: the most significant contributions were published in Hoyningen-Huene, Sankey (eds) (2001).

years, Kuhn himself undertook the task of writing a whole book on it[93]) and has never lost vigour, indeed fuelling debates in many areas, for example the controversies over rationality, progress and realism. It is the notion of incommensurability: as Feyerabend once put it, "apparently everyone who enters the morass of this problem comes up with mud on his head".[94]

The heated discussions on incommensurability partially derive from the fact that Kuhn and Feyerabend themselves have often been interpreted as advocates of a radical thesis of incommensurability (or incommunicability). If things were so, however, the development of science would turn out to be totally arbitrary, since any rational decision procedure would be lacking. But this is not their intention (even if Feyerabend's rhetoric often gave the impression that it was exactly so). They aimed, first and foremost, to provide strong arguments against what they took to be a widely held, but nonetheless simplistic, view our understanding of scientific change.

Paul K. Feyerabend

Just as the contextual theory of meaning[95] is rooted in Feyerabend's interpretation of Wittgenstein's philosophy so, as John Preston notices, his incommensurability thesis is inspired by the Cambridge philosopher Elizabeth Anscombe, who had been very

---

[93] The announced goal of Kuhn's unfinished book was to clarify and defend an idea first advanced in *The Structure of Scientific Revolutions*, namely, the idea that "the normal-scientific tradition that emerges from a scientific revolution is not only incompatible but often actually incommensurable with that which has gone before" (Kuhn (1962a), p. 103). In fact, Kuhn (U-1982-) is a project dominated by one topic: "incommensurability and the nature of the conceptual divide between the developmental stages separated by what I once called 'scientific revolutions'" (Kuhn (1993a), p. 228). Kuhn began the book – "the grandchild of *Structure*, since the child was still-born" (quoted in Buchwald, Smith (2001), p. 464) – in the early 1980s, but never managed to complete it. Initially titled *Scientific Development and Lexical Change*, the book later became *Words and Worlds. An Evolutionary View of Scientific Development*, and received his final title (*The Plurality of Worlds. An Evolutionary Theory of Scientific Development*) shortly before Kuhn's death in Cambridge, Massachusetts, on 17 June 1996. The manuscript, of which only five chapters were completed (the rest is fragmentary), circulated among a restricted group of Kuhn's friends and colleagues. James Conant (grandson of James Bryant Conant, who initiated Kuhn to the history of science and to whom *The Structure of Scientific Revolutions* is dedicated) and John Haugeland, both from the University of Chicago, are now editing it. The various phases of the manuscript are documented in Kuhn (U-1980), (U-1984), (U-1987), (U-1990a) and (U-1990b); (U-1987), together with Kuhn (1989a), contains the most significant points of Kuhn's projected book.

[94] Reported in Hoyningen-Huene (2000), p. 104. "Dealing with incommensurability we enter in territory studded with traps, pitfalls, false alarms, in which rhetoric plays a far more important role than anywhere else" (Feyerabend (1978b), pp. 178–179).

[95] Feyerabend describes the contextual theory of meaning, that applies to any kind of language (and not only to observation language), in the following terms: "a statement will be regarded as observational because of the *causal context* in which it is being uttered, and *not* because of what it means. [...] All we need in order to provide a theory with an observational basis are statements satisfying this pragmatic property. [...] Their meaning they obtain from the theory to which they belong" (Feyerabend (1965a), pp. 198–199; see also pp. 179–181.

close to Wittgenstein.[96] Feyerabend met Anscombe during a lecture on Descartes he gave at the Österreichisches Kollegium and discussions with her extended over months:

> On one occasion which I remember vividly Anscombe, by a series of skilful questions, made me see how our conception (and even our perceptions) of well-defined and apparently self-contained facts may depend on circumstances not apparent in them. [...][97] I conjectured that such principles would play an important role in science, that they might change during revolutions and that deductive relations between pre-revolutionary and post-revolutionary theories might be broken off as a result. I explained this early version of incommensurability in Popper's seminar (1952) and to a small group of people in Anscombe's flat in Oxford (also in 1952 with Geach, von Wright and L.L. Hart present) but I was not able to arouse their enthusiasm on either occasion.[98]

The fifties

The first occurrence of the idea of incommensurability (though not of the term itself) is in Feyerabend's early works on the philosophy of physics, in his discussion of complementarity:

> a theory may be found whose conceptual apparatus, when applied to the domain of validity of classical physics, would be just as comprehensive and useful as the classical apparatus, without coinciding with it. Such a situation is by no means uncommon. [...] the concepts of relativity are sufficiently rich to allow us to state all the facts which were stated before with the help of Newtonian physics. Yet these two sets of concepts are completely different and bear no logical relation to each other.[99]

Feyerabend is here attacking the idea that the meaning of observation language is determined by pure observation. In a body of knowledge no part can be appraised individually, since each one is connected to others;[100] therefore, there is no theory

---

And also: "sense-data cannot be separated from the process of their description" (Feyerabend (1960), p. 37, emphasis suppressed). See also Preston (1997a), p. 102.

[96] Feyerabend recalls her as "a powerful and, to some people, forbidding British philosopher who had come to Vienna to learn German for her translation of Wittgenstein works" (Feyerabend (1978c), p. 114; see also (1979, 1980), p. 201). Anscombe "had a profound influence" upon Feyerabend, "though it is not at all easy to specify particulars" (*ibidem*). She gave him manuscripts of Wittgenstein's later writings and discussed them with him. The passage that follows refers to one of these occasions. See also Feyerabend (1978c), pp. 67, n. 114, and (1979, 1980), pp. 65–66.

[97] Feyerabend is here referring to Whorf's "covert classifications": see also his (1975), pp. 223–230.

[98] Feyerabend (1978c), pp. 114–115; see also (1979, 1980), pp. 201–202, and (1995b), pp. 92–93. For what follows, see the detailed analysis in Preston (1997a), ch. 6.

[99] Feyerabend (1958b), p. 83; see also his (1961c), pp. 387–388, where he argues that within the Aristotelian conceptual scheme, Galileo's or Descartes' law of inertia "does not make sense, nor can it be formulated" (p. 387). For the outline of Feyerabend's position in the next sections, I will substantially follow Corvi (1992) and Preston (1997a).

[100] This is the core of the holistic theory of knowledge, according to which the identity of a single piece of knowledge is not given once and forever, nor is it unalterable, but it is

without observation, nor observation without theory.[101] The meaning of an observation statement is determined neither by the pragmatic conditions in which a language is used, nor by the phenomenon that makes us assert it is true. On the one hand, against what he labels the "principle of pragmatic meaning" Feyerabend remarks that the regularity of linguistic rendering in observation contexts does not determine meaning: "however well behaved and useful a human observer may be, the fact that in certain situations he (consistently) produces a certain noise, does not allow us to infer what this noise means".[102] Against the "principle of phenomenological meaning", on the other hand, Feyerabend holds that immediate experience, associated with the use of observation statements, does not determine the meaning but, at most, constitutes the cause of the statement.

Ever since 1958, however, Feyerabend states what would remain a firm point of his thought, something which, as years would go by, he would draw consequences from, both on the epistemological and the practical level. He holds, that is, that the so-called observation statements are the outcome of an interpretation, one out of many possible ones. He writes: "the interpretation of an observation language is determined by the theories which we use to explain what we observe, and it changes as soon as those theories change".[103] And as there is a link between the stability thesis and the meaning-invariance thesis, so there is one between the just mentioned thesis and incommensurability: "I interpreted observation languages by the theories that explain what we observe. Such interpretations change as soon as theories change. I realized that interpretations of this kind might make it possible to establish deductive relations between rival theories and I tried to find means of comparison that were independent of such relations".[104]

The sixties

In the early 1960s Feyerabend widened the scope of his attack on empiricism by turning it into a thorough attack of the reductionist account of the relationship between rival theories.[105] According to reductionism, the relationship between a superseded theory and the one that supersedes it can be twofold: either the superseded theory reduces to the other one by a process of logical derivation, or it is explained by it.

---

determined by its position within the whole. Together with Wittgenstein's, Feyerabend's own holistic views may have played a role in shaping Kuhn's philosophical stance in this respect.

[101] The second assumption highlights that theoretical presuppositions are not always manifest in the control procedures. Lakatos, in his (1961) and (1963–64), speaks of "hidden lemmas".

[102] Feyerabend (1958a), p. 22.

[103] Feyerabend (1958a), p. 31, emphasis suppressed. Feyerabend ascribes such idea, which forms the core of his scientific realism, to Wittgenstein, but also to Galileo and other scientists. This thesis is logically entailed from the more general contextual theory of meaning, that applies to any languages, not only to observation language.

[104] Feyerabend (1978c), p. 67.

[105] See particularly Feyerabend (1962a) and (1965a).

Following this model, successive theories in the same field are more general theories that comprise the previous theories in the same domain.[106]

In Feyerabend's eyes the reductionist account involves two key assumptions. Since the reduced theory must be deducible from the reducing one, such an account assumes a "consistency condition": "Only such theories are then admissible in a given domain which either *contain* the theories already used in this domain, or which are at least *consistent* with them inside the domain".[107] And given the fact that, in order that a theory is coherent, a univocal vocabulary is required, it is also assumed a "condition of meaning invariance": "meanings will have to be invariant with respect to scientific progress; that is, all future theories will have to be framed in such a manner that their use in explanations does not affect what is said by theories, or factual reports to be explained".[108] The latter condition, independently of the former, is further supported by the empiricist assumption of the existence of a theory-neutral observation language.

Feyerabend's main objection to these two reductionist approaches is that the condition of meaning invariance is violated during some important changes of theory. As he writes at the beginning of "Explanation, Reduction, and Empiricism",

> What happens [...] when a transition is made from a theory *T′* to a wider theory *T* (which, we shall assume, is capable of covering all the phenomena that have been covered by *T′*) is something much more radical than incorporation of the *unchanged* theory *T′* (unchanged, that is, with respect to the meanings of its main descriptive terms as well as to the meanings of the terms of its observation language) into the context of *T*. What happens is, rather, a *replacement* of the ontology (and perhaps even of the formalism) of *T′* by the ontology (and the formalism) of *T*, and a corresponding change of the meanings of the descriptive elements of the formalism of *T′* (provided these elements and this formalism are still used). This replacement affects not only the theoretical terms of *T′* but also at least some of the observational terms which occurred in its test statements. That is, not only will description of things and processes in the domain in which *T′* had been applied be infiltrated, either with the formalism and the terms of *T*, or if the terms of *T′* are still in use, with the meanings of the terms of *T*, but the sentences expressing what is accessible to direct observation inside the domain will now mean something different. In short, introducing a new theory involves changes of outlook both with respect to the observable and with respect to the unobservable features of the world, and corresponding changes in the meanings of even the most 'fundamental' terms of the language employed.[109]

---

[106] The target of Feyerabend's attack is a Nagel's classic (1961), but see also Hempel, Oppenheim (1948) and Nagel (1949). For a brief description of this view, see Suppe (1974), pp. 53–56, and (1977), pp. 619–624.

[107] Feyerabend (1965a), p. 164. In his (1962a) Feyerabend speaks of a "principle of deducibility", according to which "explanation is achieved by deduction in the strict logical sense", (p. 46).

[108] Feyerabend (1965a), p. 164. "According to the principle of meaning invariance, an explanation must not change the meanings of the main descriptive terms of the explanandum" (Feyerabend (1962a), p. 46.

[109] Feyerabend (1962a), pp. 44–45.

The claim that in some cases there is an actual change of meaning during the transition from one theory to another implies that in these cases older theories cannot be logically derived from those that supersede them. Feyerabend holds the meaning variance of theoretical terms by considering the crucial differences in the way in which basic concepts are defined within a certain number of competing theories, and employs his criticism of the idea of a neutral observation language to support the claim that such meaning variance extends over observation language as well. Feyerabend's idea that meaning changes with the change of theories suggests a view according to which the meaning of the terms employed by a theory is determined by the context within which they appear and varies with the varying of such a context:

> the meaning of every term we use depends upon the theoretical context in which it occurs. Words do not 'mean' something in isolation; they obtain their meanings by being part of a theoretical system. Hence if we consider two contexts with basic principles that either contradict each other or lead to inconsistent consequences in certain domains, it is to be expected that some terms of the first context will not occur in the second with exactly the same meaning.[110]

But Feyerabend's position as to the theoretical dependence of observation terms is not limited to the remark that their meaning is determined by the context in which they are employed. Indeed, Feyerabend develops such considerations while advancing, at the same time, his view of realism. By saying that meaning of observation terms depends on theory in which they are employed Feyerabend defends "a realistic interpretation of scientific theories" according to which theories provide their observation terms with meaning:[111] "A realist [...] wants to give a unified account, both of observable and of unobservable matters, and he will use the most abstract terms of whatever theory he is contemplating for that purpose. He will use such terms in order either to *give* meaning to observation sentences or else to *replace* their customary interpretation".[112] Therefore, according to Feyerabend, the meaning of observation terms does not depend upon the theory simply in virtue of the context, but rather because realistically interpreted theories provide the observation terms they employ with their meaning.

Feyerabend's idea seems to be that the meaning of observation terms is determined by the theory exactly because theory aims at describing reality, and because the ontology of a theory that undertakes such a task bears implications for the nature of observed entities. In other words: given the fact that meaning does not derive either from experience or from the conditions of application, the meaning of an observation term, as it is employed within a theory, depends on the way in which the theory describes the entities to which the term refers. Therefore, according to Feyerabend, the meaning of observation terms depends on the theoretical context in

---

110 Feyerabend (1965a), p. 180.

111 See Feyerabend (1962a), pp. 51–53.

112 Feyerabend (1975), p. 279.

the sense that it depends on the picture of observable entities a theory depicts of its own field of application.[113]

The seventies

Feyerabend introduces the concept of incommensurability while arguing that the impetus theory is not reducible to Newtonian mechanics. He considers a version of the law of inertia formulated in terms of impetus and shows how it cannot be reduced to Newtonian mechanics, since the concept of impetus cannot be properly correlated to Newtonian concepts. Indeed, according to Feyerabend, the notion of impetus depends on the Aristotelian principle according to which all movement is the product of the continuous action of some kind of force.[114] According to his reconstruction, impetus was thought of as "a kind of inner principle of motion":[115] it is "the force responsible for the movements of the object that has ceased to be in direct contact, by push, or by pull, with the material mover".[116]

Then, Feyerabend argues that "the 'inertial law' […] of the impetus theory is incommensurable with Newtonian Physics in the sense that the main concept of the former, the concept of impetus, can neither be defined on the basis of the primitive descriptive terms of the latter, nor related to them via a correct empirical statement".[117] The conclusion is that the reason why the term "impetus" cannot be defined in Newtonian terms is that the concept of impetus presupposes the fact that a continuous motion requires a cause. Since within Newtonian mechanics, inertial movement is not regarded as subject to any force, the concept of impetus depends on a principle which is incompatible with Newton's basic assumptions.[118]

Taking his cue from the impetus theory,[119] Feyerabend constructs his first general characterization of incommensurability, according to which the conceptual apparatus of a new theory T is incommensurable with that of a (previous, or competing) theory T' if and only if three conditions hold true: the primitive descriptive terms of T cannot be defined by means of T'; there are no "bridge-laws" linking two sets of

113 See Feyerabend (1965a), p. 170: "For example, we may change our ideas about the nature, or the ontological status (property, relation, object, process, etc.) of the color of a self-luminescent object without changing the methods used for ascertaining that color (looking, for example). Clearly, such a change is bound profoundly to influence the meanings of our observational terms". What matters, to Feyerabend, is highlighting a point: what influences the meanings of the terms is not the whole theory, but only a part of it, that is, the basic principles. In other words, the meaning of theoretical terms depends on their connection to certain fundamental theoretical laws or postulates. See also Feyerabend (1958a), section VI.

114 See Feyerabend (1962a), pp. 62–69.

115 Feyerabend (1962a), p. 65.

116 Feyerabend (1962a), p. 65.

117 Feyerabend (1962a), p. 76.

118 This shows that it is not the whole theoretical context that influences the meaning of theoretical terms, rather, only specific parts of theories.

119 It is to be noted, however, that this is not Feyerabend's only example: he applies the same argument to other case-studies from the history of science, such as the concept of mass in Newton's dynamics and Einstein's theory of relativity (see Feyerabend (1965a), pp. 168–172).

primitive descriptive terms, which result to be correct and explicable consistently with T; the principles of T are incompatible with those of T'. These conditions apply to several pairs of theories that have been employed as instances of explanation and reduction: "Many (if not all) such pairs on closer inspection turn out to consist of elements which are incommensurable and therefore incapable of mutual reduction and explanation".[120]

The immediate upshot of Feyerabend's argument for incommensurability is that a theory cannot be reduced, by deductive assimilation, to another one. Since the terms employed to express them have different meanings, no statement of one theory can be derived from the other one. Since the classes of consequences of those theories do not share common members, Feyerabend concludes that they are "deductively disjoint"[121] and "a judgement involving a comparison of content classes is now clearly impossible".[122]

The first problem raised by the incommensurability thesis is that allegedly incommensurable theories offer alternative descriptions of the phenomena in the same domain. It is not clear how such theories can actually contradict each other in the same domain of phenomena while being, at the same time, logically unrelated: in what sense can incommensurable theories offer alternative descriptions of the very same things if nothing of what one of them affirms is negated by the other? Feyerabend seems to hold that this problem can be avoided, or at least minimized, by limiting incommensurability to general theories. Indeed, he highlights that incommensurability is restricted to "general theories, or non-instantial theories",[123] in cases in which, some structural similarities notwithstanding, some universal principles of one system do not appear in the other: "the problem of incommensurability arises only when we analyse the change of comprehensive cosmological points of view – restricted theories rarely lead to the needed conceptual revisions".[124]

The reason for such a restriction seems to be that general theories do not share a common observation language, while theories of a lower level of generality can be compared by referring to an observation language, given that there is "a background theory of greater generality that provides a stable meaning for observation

120 Feyerabend (1962a), p. 77.

121 Feyerabend (1978c), p. 67.

122 Feyerabend (1970a), p. 220.

123 Feyerabend (1962a), p. 44. "To circumvent the difficulty that arises when we want to say that incommensurable theories 'speak about the same thing' I restricted the discussion to non-instantial theories [...] and emphasized that mere *difference* of concepts does not suffice to make theories incommensurable in my sense" (Feyerabend (1978c), p. 68, n. 118).

124 Feyerabend (1975), p. 284. The question can be tackled also from another point of view: we could actually see incommensurability as an excellent stimulus to look for possible translations and therefore *use Feyerabend against Feyerabend*, arguing that we attempt ever new translations exactly because there are incommensurable schemes. Analogous examples can be drawn from literature: see, for examples, the problem of translating texts like Raymond Queneau's *Les fleurs bleuers* (1965), *Petite cosmogonie portative* (1954) or *Exercices de style* (1947). In presenting *an* Italian translation of these texts, authors like Italo Calvino or Umberto Eco explicitly declare that rather than attempting an actual translation from the originals, they prefer to take up Queneau's challenge and play his own game in another language.

sentences".[125] However, by restricting incommensurability to general theories, the problem cannot be avoided: in fact, if they refer to whatever exists within a certain domain, then they will have to refer to (at least) some of the same things.

Feyerabend attempts to answer this objection by speaking of the comparison between incommensurable theories on the basis of crucial experiments. The fact that theories are subjected to crucial tests is problematic for his account: since incommensurable theories do not share common statements, there can be no prediction asserted by one and denied by the other. Feyerabend's position is based on a "pragmatic theory of observation"[126] according to which "observational sentences are distinguished from other statements not by their meaning, but by the circumstances of their production".[127] Feyerabend distinguishes between an uninterpreted statement and a statement expressed by the basic statement on the basis of a certain interpretation, so that the same statement can express different assertions. In so doing, observation statements can continue to be applied with the same pragmatic conditions even if their meanings vary with the varying of the theoretical context.[128] Thus, while incommensurable theories do not share any observation statements,

> *there is still human experience as an actually existing process*, and it still causes the observer to carry out certain actions, for example, to utter sentences of a certain kind. [...] This is the *only* way in which experience judges a general cosmological point of view. Such a point of view is not removed because its observation *statements* say that there must be certain experiences that then do not occur. [...] It *is* removed if it produces observation sentences when observers produce the *negation* of these sentences. It is therefore still judged by the predictions it makes.[129]

Therefore, an observation statement to which two incommensurable theories attach different meanings can still constitute the report of a crucial test that may confirm one theory and refute the other.[130]

The pragmatic account of an observation explains how incommensurable theories can be applied to the same empirical domain, and be subjected to crucial tests on the basis of the same experimental procedure. This means that incommensurable theories *can* actually compete in providing the description of a shared group of

---

[125] Feyerabend (1965a), p. 214. Observation statements cannot judge theories, with the exception of theories with a low level of generality, that share the same principles upon which the chosen observation language is based. In order to criticize a theory it will be necessary to appeal to alternative theories, not to observation, since every experience belongs to a determined theory, being framed within certain theoretical presuppositions (as opposed to others).

[126] Feyerabend (1965a), p. 212.

[127] Feyerabend (1965a), p. 212.

[128] See Feyerabend (1965a), pp. 197–198.

[129] Feyerabend (1965a), pp. 214–215.

[130] It is not even necessary that there is a single shared observation statement as the result of a test. The same experimental result may be described by theories employing a thoroughly different terminology and nonetheless be in support of one of them and contradict the other: see Feyerabend (1975), pp. 281–283.

phenomena, if there are such experimental procedures that the statements produced by their application confirm one theory and bring the other into discredit.

Thomas S. Kuhn

Kuhn's reflection on the nature of the scientific enterprise may be divided into three phases. The first extends from the mid-1950s to the early 1960s and culminates with the publication of *The Structure of Scientific Revolutions* in 1962.[131] As a consequence of the huge debate following the publication of this book and the confrontation with Popper and his pupils during the 1965 Bedford Colloquium, Kuhn developed and refined his own ideas, reasserting some of them and softening others. Kuhn's major works in this second period are "Logic of Discovery or Psychology of Research?" (1965), "Reflections on My Critics", "Postscript – 1969" and "Second Thoughts on Paradigms" (all written in 1969), "Notes on Lakatos" (1970), "Objectivity, Value Judgement, and Theory Choice" (1973);[132] the collection *The Essential Tension* (1977) virtually closes this second phase. In the following years, Kuhn published several works and contributed to many conferences and symposia, both of history and philosophy of science. His later philosophical works (1980s–1990s) mark the various steps along which evolved the third phase of his thought, for which, regrettably, Kuhn never provided a comprehensive exposition.[133] The underlying theme of Kuhn's philosophical reflection is the notion of incommensurability, which I am here going to persue in the various phases of its development.[134]

The Structure of Scientific Revolutions, 1962

In his *magnum opus* Kuhn employs "incommensurability" in order to characterize the kind of relationship that links two different traditions of normal science,

---

[131] Kuhn's major philosophical works in this period are his (1959a), (1961a), (1962a), (1962b), (1963a) and (1963b), both written in 1961, and (1964), written before *The Structure of Scientific Revolutions*. Kuhn's early philosophical outlook is particularly relevant also in his (1957).

[132] Respectively, Kuhn (1970a), (1970b), (1970c), (1971a), (1974c) and (1977c). Other important works of this second period are (1968a), (1969) (written in 1966), (1971b), (1971c), (1975a) and (1976a), (1976b) and (1977b) (written in 1968). To these we might add Kuhn (1978), the most coherent application of Kuhn's idea of how history of science should be done and of what elements it should highlight. In the fourth section of his (1984), later reprinted as a postscript to the second edition of the book, Kuhn explicitly highlights the links between his philosophical views and his historical work, sketching a brief overview of the close parallelism between the history and the philosophy of science. In his (1980b), (1980c) and (1984) Kuhn also replies to the various criticism raised against his book, albeit completely ignoring Agassi (1983).

[133] The most significant points of Kuhn's later philosophical parabola are his (1979a), (1979b), (1980a), (1981), (1983a) and (1983b), (1983c), (1986), (1989a) and (1989b), (1990), (1991a), (1991b), (1992) and (1993a). Particularly important are also some unpublished papers and series of lectures, such as (U-1980), (U-1984), (U-1987), (U-1990a) and (U-1990b), and his last, unfinished book, (U-1982-).

[134] In particular, I shall analyse the first two phases in the present chapter, while I will devote the next one to focusing on the third phase (Kuhn's "linguistic turn").

before and after a scientific revolution.[135] He considers three aspects. The first is *methodological* and follows from the fact that different paradigms refer to different sets of problems and employ different methodological standards to assess solutions: "the proponents of competing paradigms will often disagree about the list of problems that any candidate for paradigm must resolve. Their standards or their definitions of science are not the same".[136] The second aspect is *semantic* in nature and is due to the variation of the conceptual apparatus employed by different paradigms: "Since new paradigms are born from old ones, they ordinarily incorporate much of the vocabulary and apparatus, both conceptual and manipulative, that the traditional paradigm had previously employed. But they seldom employ these borrowed elements in quite the traditional way. Within the new paradigm, old terms, concepts, and experiments fall into new relationships one with the other".[137] Finally, the third involves *ontological* elements: "the proponents of competing paradigms practice their trades in different worlds. One contains constrained bodies that fall slowly, the other pendulums that repeat their motions again and again. In one, solutions are compounds, in the other mixtures. One is embedded in a flat, the other in a curved, matrix of space. Practicing in different worlds, the two groups of scientists see different things when they look from the same point in the same direction".[138]

a) Methodological incommensurability During a scientific revolution, then, there is a change in the set of problems that can and have to be confronted by researchers. The problems whose solutions played a fundamental role for old research traditions can disappear, or be abandoned as obsolete or even not (or no longer) scientific, while new questions become relevant, questions that had never been seen as such before the revolution, or whose solutions were obvious in the previous research traditions. Together with the problems, criteria and standards that solutions must satisfy in order to be regarded as scientifically acceptable often change as well.[139] As

---

[135] See Kuhn (1962a), pp. 3–4, 101–103, 147–150 and 164–165. On pp. 149–150 and 156–158 Kuhn speaks of incommensurability between paradigms, arguing that "the world of [a scientist's] research will seem, here and there, incommensurable with the one he had inhabited before" (p. 112; see also pp. 3–4). The notion of incommensurability described on pp. 147–150 is richer than the one introduced on pp. 101–103: here, in fact, it refers only to pairs of problems and standards before and after a revolution, while later in the book Kuhn, besides linking concepts and procedures, introduces the description of the relationship between different worlds as the most fundamental feature of incommensurability. The transition to a richer notion of incommensurability apparently takes place on p. 112, where Kuhn applies it to the relationship between different worlds.

[136] Kuhn (1962a), p. 148.

[137] Kuhn (1962a), p. 149.

[138] Kuhn (1962a), p. 150. To the "world change" thesis is devoted the whole of ch. ten of Kuhn's book, titled "Revolutions as Changes of World View": see, in particular, pp. 111, 117, 118, 121, 135 and, later on, p. 150. See also Kuhn (1962b), p. 175.

[139] See Kuhn (1959a), p. 234, (1961a), pp. 211–212, (1962a), pp. 5–6, 52, 84–85, 102–110, 140–141 and 147–150. Clearly, the methodological component of the incommensurability thesis derives from paradigms understood as exemplars, or exemplary problem solutions.

can be seen from the examples chosen by Kuhn,[140] these kinds of change are mostly due to the change of the phenomenal world and for this reason cannot be regarded as successive phases of a better and better approximation to the true description of reality, during which a progressive elimination of errors is brought about.

Kuhn criticizes previous accounts of science as a cumulative enterprise – understood not as a mere accumulation of new problems, solutions and standards, rather, as the idea of those who interpret the change and replacement of old problems and standards with the change and replacement of something that had never been science proper. Historiographical inquiry shows, according to Kuhn, that in the transition from paradigm to paradigm there is no simple accumulation, either of scientific problems or of the specific techniques employed to solve them, or in the solutions that get accepted from time to time (indeed, Kuhn says, accumulation is supplemented by an actual increase of explanatory power). In fact, there is no external unit of measurement that allows us to establish any progress or regress in standards: they simply change, and change as the paradigm changes: "The attempt to explain gravity, though fruitfully abandoned by most eighteenth-century scientists, was not directed to an intrinsically illegitimate problem; the objections to innate forces were neither inherently unscientific nor metaphysical in some pejorative sense. There are no external standards to permit a judgment of that sort".[141]

The lack of "neutral" or super-paradigmatic standards, the fact that every standard for the assessment of problems, methods and instruments (either conceptual or technical) is valid only from the point of view of the particular paradigm that frames it, immediately leads to the incommensurability thesis: every appraisal can be made only by sharing the fundamental paradigmatic presuppositions that decide what is a problem, a solution and an acceptable solution, and what is not.

As to this point, Kuhn insists on the analogy between political and scientific revolutions. Political revolutions are inaugurated by a growing sense, often restricted to a fraction of the society, that existing institutions have ceased adequately to meet the problems posed by the very social reality they have in part created. In much the same way, scientific revolutions are inaugurated by a growing sense, again often restricted to a narrow subdivision of the scientific community, that the existing paradigm has ceased to provide the adequate tools in the exploration of an aspect of nature to which that very paradigm had led the way.[142] In both cases, the sense of malfunction that can lead to crisis is prerequisite to revolutions. Most importantly, since the competing parties in a political revolution "differ about the institutional matrix within which political change is to be achieved and evaluated, [and] acknowledge no supra-institutional framework for the adjudication of revolutionary difference, the parties to a revolutionary conflict must finally resort to the techniques of mass persuasion, often including force".[143] Much in the same way, the choice

[140] The most important ones are the development of dynamics from Aristotle, through Descartes and Newton, to the eighteenth century; Lavoisier's revolutionary transformation of chemistry; the development of Maxwell's electromagnetic theory.

[141] Kuhn (1962a), p. 108.

[142] See Kuhn (1962a), pp. 92–94.

[143] Kuhn (1962a), p. 93.

between competing paradigms "proves to be a choice between incompatible modes of community life. Because it has that character, the choice is not and cannot be determined merely by the evaluative procedures characteristic of normal science, for these depend in part upon a particular paradigm, and that paradigm is at issue".[144]

b) Semantic incommensurability New paradigms are born from old ones and ordinarily incorporate much of the vocabulary and apparatus, both conceptual and manipulative, the old research traditions had previously employed. However, the inherited elements are seldom employed in the same way after a revolution: "Within the new paradigm, old terms, concepts, and experiments fall into new relationships one with the other. The inevitable result is [...] a misunderstanding between the two competing schools".[145] It is conceptual change, or meaning-variance, both from an extensional point of view (some objects leave a concept's extensional domain to enter a new one),[146] and from an intensional point of view (what changes, that is, is the concept employed when varying the characteristics of the objects that are its referents).[147]

According to Kuhn, in the transition from one paradigm to another some fundamental terms radically change their meaning because they "fall into new relationships one with the other",[148] change their "conditions of applicability"[149] and, more generally, are differently employed. Copernicus' adversaries, who thought it foolish that the earth moved, were not, strictly speaking, wrong, since the meaning

[144] Kuhn (1962a), p. 94.

[145] Kuhn (1962a), p. 149. The classical example is the passage from Newton's conception of the world to Einstein's: "To make the transition to Einstein's universe, the whole conceptual web whose strands are space, time, matter, force, and so on, had to be shifted and laid down again on nature whole. [...] Communication across the revolutionary divide is inevitably partial" (*ibidem*; see also pp. 64–65, 102–103, 114–115, 128–129, 130–133, 142–143 and 149–150. Just as in the case of Feyerabend, who originally developed the idea of incommensurability against the reductionists' view (according to which previous theories get deductively absorbed by the ones that supersede them: see Feyerabend (1962a), pp. 62–69), so goes also Kuhn's argument: for both of them, incommensurability is not related only to the difference of the concepts underlying the two theories, but it also involves the dependence of the meaning of observation terms on the theory in which they are employed. Later on (for example in his (1970b)), however, Kuhn would claim he had always understood the meaning-variance as only partial, and this attenuates the parallel between their semantic interpretations of incommensurability. See Preston (1997a), ch. 6, and Sankey (1993a).

[146] See Kuhn (1962a), pp. 114–115, 128–129 and 130–134. See also Kuhn (1970b), pp. 273–274, (1970c), pp. 200–201, (1979b), pp. 203–204, (1981), p. 8, (1989a), pp. 85–86, and (1990), p. 313.

[147] According to Kuhn this conceptual change prevents the possibility of deriving the laws of nature effective within one paradigm in another one. Indeed, the laws that obtain in Einsteinian physics are not at all similar to Newton's ones, despite their similarities. This is because, Kuhn argues, the formulation of the Einsteinian version of these laws employ relativistic concepts, representing space, time and mass as Einstein sees them.

[148] Kuhn (1962a), p. 149.

[149] Kuhn (1970b), p. 266.

they ascribed to the "earth" involved, among other things, that it conserved its stationary position at the centre of the Universe.[150]

The lack of appreciation of the meaning change of terms in the passage from one theory to another implies a "logical lacuna" in the neopositivistic thesis according to which it is possible, given some appropriate restrictions, to derive the preceding theory from the following one: in order to derive the statements of Newtonian mechanics from those of relativistic mechanics we must assume statements like "$(v/c)^2 << 1$",and in so doing we get a set of propositions formally identical to Newton's laws of motion, of gravitation, and so on. However, Kuhn remarks, the laws thus obtained are not Newton's laws: they become such only if they "are interpreted in a way that would have been impossible until after Einstein's work",[151] since the quantities that are in them are relativistic quantities, not classical ones: "the physical referents of these Einsteinian concepts are by no means identical with those of the Newtonian concepts that bear the same name. (Newtonian mass is conserved; Einsteinian is convertible with energy. Only at low relative velocities may the two be measured in the same way, and even then they must not be conceived to be the same.)".[152]

In other words, if we do not change the (relativistic) definitions of the variables and parameters that appear in the statements derived from Einstein's theory, these statements cannot be regarded as Newtonian, while if we change them – something which is possible only by transforming them in the light of successive knowledge, that is, under the guide of Einstein's theory of relativity – we cannot say we actually and logically derived Newton's laws. This, according to Kuhn, explains the reason why Newton's laws seem still applicable to nature – why, for instance, an automobile driver acting as though he lived in a Newtonian universe is justified – but does not at all show that Newton's laws are a particular (or limiting) case of Einstein's:[153] "it is not only the forms of the laws that have changed. Simultaneously we have had to alter the fundamental structural elements of which the universe to which they apply is composed".[154]

---

150 See Kuhn (1962a), pp. 149–150.

151 Kuhn (1962a), p. 101.

152 Kuhn (1962a), p. 102.

153 It was Feyerabend especially who insisted on meaning-variance, highlighting that, in case of universal theories (with all-inclusive ontologies), it is not even possible to state an empirical hypothesis that links sentences to their respective theories. In the case of classical and relativistic mechanics, for instance, when we employ classical terms "we assume a universal principle that is suspended by relativity, which means it is suspended whenever we write down a sentence with the intention to express a relativistic state of affairs. Using classical terms and relativistic terms in the same statement we both use and suspend certain universal principles which is another way of saying that such statements do not exist: the case of relativity vs. classical mechanics is an example of two incommensurable frameworks" (Feyerabend (1975), p. 276; see also pp. 280–281). Each fact of Newtonian mechanics, for instance, "presumes that shapes, masses, periods are changed only by physical interactions and this presumption is suspended by the theory of relativity" (*ibidem*, p. 271).

154 Kuhn (1962a), p. 102. This passage underscores an important convergence between Kuhn's and Feyerabend's versions of incommensurability. For just as in Feyerabend's case

c) Ontological incommensurability[155] Finally, "the third and most fundamental aspect of incommensurability of competing paradigms"[156] entails that scientists see the "same" objects in different ways. All see the very same world, "in some areas they see different things, and they see them in different relations one to the other".[157] The passage from one world to another cannot be forced by logic or by a neutral experience (a so-called *experimentum crucis*): "Like the gestalt switch, it must occur all at once (though not necessarily in an instant) or not at all".[158] The world changes through a revolution and new knowledge is gained in a non-cumulative manner.[159]

This third feature of incommensurability, as introduced in *The Structure of Scientific Revolutions*, requires a detailed analysis. Normal science develops under the guidance of rules drawn from exemplary problem solutions of concrete problems. Such exemplary problem solutions (or exemplars, in short) play a constructive

---

(see his (1962a), pp. 62–69) Kuhn's argument against the derivation of Newton's laws from Einstein's is directed against the reductionist picture of the dynamics of scientific theories. Since such a derivation turns out to be impossible due to the conceptual disparity between the two theories, Kuhn's notion of incommensurability seems to be coinciding with Feyerabend's own – at least so they appeared to Dudley Shapere (see his (1983), p. 83), for example, and to Israel Scheffler ((1967), pp. 49–50). Such alleged coincidence may be further backed by the fact that Kuhn combines the affirmation of conceptual disparity with the negation of the existence of a neutral observation language (see Kuhn (1962a), pp. 126–129). Indeed, all this suggests that for Kuhn, as well as for Feyerabend, incommensurability does not consist simply in the different key concepts of different theories, but also involves the dependence of the meaning of observation terms on the theory that employs them. However, in his later writings Kuhn claims he simply meant to say that part of the vocabularies of incommensurable theories differ in their meanings: see Kuhn (1970b), pp. 266–268. In his (1974d) he very clearly says he only meant to highlight "'some difference in some meanings of some words they have in common' is the most I have ever intended to claim" (p. 506). This attenuates the parallel between Kuhn's and Feyerabend's views of semantic incommensurability, since it suggests that observation language, even if not theoretically neutral, semantically varies only partially in the passage from one theory to another. Kuhn expresses a further dissociation from Feyerabend in the 1980s, when he introduces the notion of "local incommensurability" (see his (1983a)).

155 I refer to this component under the rubric of "ontological incommensurability" because it concerns the thesis of "world-change" through revolutions. However, this is just a label, and I do not wish to attach to it any actual metaphysical claim in the strong sense of the word.

156 Kuhn (1962a), p. 150. Indeed, Kuhn devotes the whole ch. X of the book to it.

157 Kuhn (1962a), p. 150.

158 Kuhn (1962a), p. 150. "The transfer of allegiance from paradigm to paradigm is a conversion experience that cannot be forced. Lifelong resistance, particularly from those whose productive careers have committed them to an older tradition of normal science, is not a violation of scientific standards, but an index to the nature of scientific research itself" (*ibidem*, p. 151). See also Worrall (1990), (1995), (2000), where the cases of David Brewster and Joseph Priestley are treated, and (2002).

159 See, Kuhn (1961a), p. 208: "innovations in scientific theories are not simply additions to the sum of what is already known. Almost always (always, in the mature sciences) the acceptance of a new theory demands the rejection of an older one. In the realm of theory, innovation is thus necessarily destructive as well as constructive". See also Kuhn (1962a), p. 66.

function: by virtue of the similarity and dissimilarity relations they involve, they build the world the scientists inhabit.[160] Since during the periods of normal science the exemplary problem solutions are not questioned, their contribution to the constitution of the world remains constant: there is no change in the region of the phenomenal world relevant for a given scientific community, and the knowledge of it cannot but grow by accumulation. Not so through revolutions: for the characteristic feature of revolutionary periods is that the world changes throughout a revolution.[161] There is a change in the region of the phenomenal world (that is, the woven web of similarity and dissimilarity relations) with which a given scientific community deals and in which it carries out its research activity. For after a revolution the phenomenal world may comprise phenomena and entities that were absent in the world preceding the revolution. Even during the periods of normal science it was possible to discover new phenomena and new entities, but such discoveries were not surprising and did not involve the revision of the body of (explicit or implicit) knowledge of the researchers. But

> during revolutions scientists see new and different things when looking with familiar instruments in places they have looked before. It is rather as if the professional community had been suddenly transported to another planet where familiar objects are seen in a different light and are joined by unfamiliar ones as well. [...] In so far as their only recourse to that world is through what they see and do, we may want to say that after a revolution *scientists are responding to a different world.*[162]

Practising their trades in different worlds, scientists supporting different paradigms see different things when they look in the same direction with the same instruments. However, this does not mean, Kuhn specifies, "that they can see anything they please. Both are looking at the world, and what they look at has not

---

[160] See Hoyningen-Huene (1989a/1993), pp. 159–160 and ch. 5; see also Gattei (2000b), pp. 299–308 and 311–319. In his (1961) Toulmin refers to a paradigm as a set of "examples of a standard type", examples used "as objects of comparison" (p. 52). Shortly afterwards, he speaks of "relating the unfamiliar to the familiar", "relating the anomalous to the accepted" and "relating the phenomena to our paradigms" (Toulmin (1961), p. 60).

[161] See Kuhn (1962a), pp. 6, 61, 85, 102–103, 106, 111, 117, 118, 120, 121, 134–135, 141, 143 and 150, (1970c), pp. 193 and 201, (1974c), p. 309 n. 18, (1977a), p. xxiii, (1979b), pp. 206–207, (1983a), p. 52, and (1986), p. 33. See also Feyerabend (1978c), p. 70, and (1978b), p. 202: "We need to acknowledge that incommensurable theories", Feyerabend writes, "present different worlds, and that the passage from a theory to another determines the transformation of a world into another". Feyerabend is here referring to a "change of *conditions*", and in this sense "we say that a change of the universal principles determined a change of the world. In this way the idea of an world independent from our knowing activities disappears from general philosophy and is limited to a few theories. Therefore, our knowing activity determines the structure of the world even where it is most steady; it eliminates gods and replaces them with aggregates of atoms in a void space" (*ibidem*).

[162] Kuhn (1962a), p. 111, emphasis added. During a revolution some numerical data may also change, since some old quantitative expectations related to some phenomena may be replaced by new ones, or new quantitative expectations may arise where before there were none.

changed. But in some areas they see different things, and they see them in different relations one to the other".[163]

Kuhn, in other words, ascribes to "world" two different meanings: on the one hand, different paradigms determine different "worlds", that is, the domains subject to scientific inquiry; on the other, "world" is the invariant reality, however unknowable and directly inaccessible (he usually comprises in this latter meaning of the word also the world of everyday experience, the one Kuhn describes as "outside the laboratory"[164]). In Kantian terms, we might say that, on the one hand, there is a sense in which the world changes after a revolution while, on the other, it remains the same. The explanation is that Kuhn uses "world" both in the sense of *noumenon*, and in the sense of *phenomenon*. The noumenal world exists independently from paradigms, languages and minds: this explains Kuhn's statement that after a revolution the world remains the same.[165] However, such a world is unknowable, indescribable and unfathomable, as opposed to the phenomenal world constituted by paradigms. Different paradigms structure the world in different ways, imposing different similarity and dissimilarity relations, and giving rise to what Paul Hoyningen-Huene called "the plurality of phenomenal worlds".[166] This is why Kuhn says that the world changes when paradigms change: while there is only one single noumenal world, there are several phenomenal worlds, different from one another, each constituted by a different paradigm.[167]

The change of the worlds of possible experience (or of what theories refer to) must therefore be understood only as a partial change. Indeed, Kuhn clarifies that "successive theories are incommensurable (which is not the same as incomparable) in the sense that the referents of some of the terms which occur in both are a function of the theory within which those terms appear".[168] And, even more explicitly: "Some sort of communication goes on, and we must learn to understand it by making something out of phrases like 'partial communication', or 'preservation of reference for certain terms, although some of the referential apparatus has itself changed'".[169]

Kuhn wavers at length between the characterization of revolutions as changes of the world or as changes in the way we look at the world. Following Hanson, he speaks of *Gestalt*-switches, a term he draws from psychology to highlight a dramatic

[163] Kuhn (1962a), p. 150.

[164] Kuhn (1962a), p. 111; see also his (1970c), pp. 201–202.

[165] See Kuhn (1962a), pp. 111–112, 118, 120–121, 128–129, 134–135 and 150.

[166] See Hoyningen-Huene (1989a/1993), pp. 36–42.

[167] This explanation was not available to Kuhn until the later phase of his philosophical development. Having never received a proper philosophical training, he lacked the terminology that would have made clearer some of the passages of *The Structure of Scientific Revolutions* (oddly enough, however, those who misunderstood or misrepresented Kuhn's position were all philosophers). But by the time he realized he could easily hold that after a paradigm shift the world was both the same world and another world, he had already stopped talking about paradigms and characterizing revolutions as paradigm shifts. This was in 1969.

[168] Kuhn (1979b), p. 204.

[169] Kuhn (1974b), p. 409.

and irreversible world-change.[170] In his later works, Kuhn reviews this qualification. Individual scientists can experience these switches, but the term becomes illegitimate if it gets applied to the true subject of the scientific revolution, that is, the community of professionals of a given scientific discipline. A switch takes some time to occur, and although some members of the community can experience it, not all of them will do that at the same time. Revolutions, in other words, are a phenomenon much more complex than the metaphor of the *Gestalt*-switch indicates.

The sixties and seventies

The characterization of incommensurability in *The Structure of Scientific Revolutions* is at the same time strong and vague, and this made Kuhn (partly) modify his position as years went by.[171] Ever since 1969 he no longer spoke of incommensurability between paradigms, but among theories, terms, vocabularies and languages: incommensurability therefore began to characterize the differences in the process of acquisition of concepts belonging to different theories. In so doing, Kuhn considerably shrank the scope of the incommensurability thesis: the changes in the problems and standards, together with the idea of world-change, were from then on associated with the changes of meaning.[172]

The upshot is a new version of the concept of incommensurability, substantially reduced with respect to the one offered in *The Structure of Scientific Revolutions*: while in 1962 it presented three aspects, among which Kuhn had not yet developed the mutual relationships, incommensurability is now limited to only one of those aspects, the semantic one. The other two, the change in the problems and standards of their possible solutions, and the change of the phenomenal world, are now closely linked to meaning change. Kuhn would later complete the narrowing process by explicitly saying that he never understood the incommensurability thesis as implying that all the concepts employed in rival theories actually change their meanings in the transition to a new theory.[173] What he had always had in mind, he now says, is a

---

170 However, just as for the parallel between scientific progress and biological evolution (see below, ch. 4, pp. 177–180), there is an important difference: while for *Gestalt* pictures there is perfect symmetry between the two possible interpretations, and it is possible to switch back and forth freely and easily between them, in actual practice the switch is irreversible: "the scientist does not preserve the gestalt subject's freedom to switch back and forth between ways of seeing" (Kuhn (1962a), p. 85). Once a scientist succeeds in conceiving space as curved, instead of flat, or realized the existence of oxygen as the reason for combustion, he is no longer able to see a Newtonian flat space, or imagine the presence of phlogiston as in pre-Lavoisier times.

171 Kuhn would interpret these changes always as improvements on his early positions, not as a withdrawal of his most radical points.

172 See Kuhn (1970b), pp. 266–269, (1970c), p. 198, (1971a), p. 146, (1976b), pp. 188–190 and 189, n. 20, (1979b), pp. 203–204, (1983a), pp. 34–35 and 48–49, (1983b), p. 714.

173 See Kuhn (1983a), p. 36: "the term 'incommensurability' functions metaphorically. The phrase 'no common measure' becomes 'no common language'. The claim that two theories are incommensurable is then the claim that there is no language, neutral or otherwise,

sort of "local incommensurability" in which meaning change refers only to "a small subgroup of (usually interdefined) terms".[174]

In 1969 and the 1970s Kuhn regards two theories as incommensurable only if there is no "language into which at least the empirical consequences of both can be translated without loss or change".[175] Such a language would be a neutral observation language, neutral at least with respect to the two theories in question – but such a language cannot exist, since "In the transition from one theory to the next words change their meanings or conditions of applicability in subtle ways: Though most of the same signs are used before and after a revolution – e.g. force, mass, element, compound, cell – the ways in which some of them attach to nature has somehow changed".[176] Kuhn bases incommensurability upon his scepticism about the possibility of a (relatively) neutral observation language. Such a scepticism was already expressed in *The Structure of Scientific Revolutions*, but there it is not related to incommensurability.[177]

At the beginning of the 1970s Kuhn supported the notion of incommensurability between two theories with the notion of untranslatability: in the process of learning a theory we acquire sets of similarity and dissimilarity relations that constitute the world of the would-be professional in a certain discipline. Such a learning process is not cognitively neutral, but already contains a certain knowledge of the world. Two different theories are therefore phrased within conceptual frameworks with different empirical instances. Accordingly, reciprocal translation becomes impossible without a loss (or modification) of meanings: there is no conceptual system able to allow for all the empirical consequences of different theories. In short, as Kuhn writes in the early 1980s, "if two theories are incommensurable, they must be stated in mutually untranslatable languages".[178]

Although the possibility of accomplishing a translation between successive theories remains at the centre of Kuhn's philosophical concerns throughout the 1970s, he does not provide a detailed analysis of the difficulties that it may present. He simply offers a general indication of the cause and scope of such difficulties. Kuhn explains that the translation, whether between theories or languages, is problematic because "languages cut up the world in different ways".[179] Different theories employ different "ontological categories"[180] to classify objects in their domain of application. During a transition from one theory to another classification schemes change:

> One aspect of every revolution is, then, that some of the similarity relations change. Objects which were grouped in the same set before are grouped in different sets afterwards

---

into which both theories, conceived as a sets of sentences, can be translated without residue or loss".

[174] Kuhn (1983a), p. 36.

[175] Kuhn (1970b), p. 266; analogously, p. 268. See also Kuhn (1970c), p. 201, (1974b), p. 410, (1976b), pp. 190–192, and (1979b), pp. 203–204.

[176] Kuhn (1970b), pp. 266–267; analogously, (1970c), p. 198.

[177] See Kuhn (1962a), pp. 126–129 and 145–148.

[178] Kuhn (1983a), p. 34. See also Sankey (1990).

[179] Kuhn (1970b), p. 268.

[180] Kuhn (1970b), p. 270.

> and *vice versa*. Think of the sun, moon, Mars, and earth before and after Copernicus; of free fall, pendular, and planetary motion before and after Galileo; or of salts, alloys, and a sulphur-iron filling mix before and after Dalton. Since most objects within even the altered sets continue to be grouped together, the names of the sets are generally preserved.[181]

The notion of incommensurability does not ban the possibility of a confrontation (and therefore a rational choice) between theories.[182] If different lexicons give rise to different object domains, such difference of domains – and of problems within them – is not global, since incommensurability of lexicons is only local. Therefore, the empirical potential of incommensurable theories may be compared: two theories may empirically intersect and contradict each other.

The original version of the incommensurability thesis, expressed by Kuhn in *The Structure of Scientific Revolutions*, involved, in the revolutionary transition between two different periods of normal science, a general change in the set of problems that can and must be confronted by researchers, a conceptual change (or meaning variance) and a change in the very world scientists deal with. The picture that emerges from Kuhn's successive reflections, between the late 1960s and the 1970s, combines these elements in a more coherent manner.

Feyerabend's criticism of Kuhn (1960–1962)

If about incommensurability Kuhn's and Feyerabend's views are similar, more significant are their differences as to scientific methodology. Indeed, Feyerabend writes:

> while I thought I recognized Kuhn's *problems*; and while I tried to account for certain *aspects* of science to which he had drawn attention (the omnipresence of anomalies is one example); I was quite unable to agree with the *theory of science* which he himself proposed; and I was even less prepared to accept the general *ideology* which I thought formed the background of his thinking. This ideology, so it seemed to me, could only give comfort to the most narrowminded and the most conceited kind of specialism. It would inhibit the advancement of knowledge.[183] And it is bound to increase the anti-humanitarian tendencies which are such a disquieting feature of much of post-Newtonian science.[184]

---

181 Kuhn (1962a), p. 275.

182 Even in its original mathematical sense (which is very close to its etymological meaning), incommensurability does not imply impossibility of comparison: "lack of common measure does not make comparison impossible. On the contrary, incommensurable magnitudes can be compared to any required degree of approximation" (Kuhn (1983a), p. 35). See also Kuhn (I-1997a), p. 298.

183 See also Feyerabend (2006), p. 614: "I think that these problems are very important because their solution, one way or another, will have a tremendous influence upon the way we think, the way we act, and thereby upon our well-being. [...] I think that some of your ideas, when published, may have a disadvantageous influence".

184 Feyerabend (1970), p. 197.

Contrary to Nagel,[185] Kuhn cannot be charged with presenting a cumulative account of the growth of scientific knowledge: more than anybody else he insisted on the non-cumulative character of scientific revolutions. But Feyerabend contested the fact that although he offered very good arguments for the existence of paradigms, he failed to motivate his specific version of theoretical monism, according to which every scientific discipline, in its mature phase, is dominated by *one single* paradigm.[186] And this, argues Feyerabend, for two reasons: one historical (or descriptive) in character, the other with a functional (or prescriptive, or normative) role.

The historical issue Feyerabend raises refers to Kuhn's picture of science and scientific practice. He crystallizes it by saying that "normal science is *monistic*; crises are *pluralistic*".[187] As he had done a little earlier in some letters to Kuhn, Feyerabend accuses him of not having offered a mere description of actual scientific practice, but of having willingly suggested an evaluation of the phases through which science moves, advancing the idea that normal science periods are desirable, while crises and revolutions are undesirable – or, at least, they are desirable only as long as they lead to an actual improvement of knowledge and a new period of normal science. In Feyerabend's eyes, Kuhn ends up by identifying mature science with normal science: "In short, it is believed that science is essentially normal science. Crises are embarassments, periods of confusion which should be passed through as quickly as possible".[188] Feyerabend is worried that the conservative attitude inplicit in Kuhn's description of normal science has influenced both historiography of science and the methodology of scientists. The case of quantum mechanics provides "a marvellous example of the way in which philosophical speculation, empirical research, and mathematical ingenuity can jointly contribute to the development of physical theory":[189] physicists try to construct a theory that satisfied their positivistc ideal, a theory that described reality but to which it was impossible to give a realistic interpretation. In so doing they acted following the enchanting influence of that empiricism which, when applied to the history of science, has monistic outcomes very similar to the ones described by Kuhn's picture of the scientific enterprise. Indeed, the attempts of those physicists gave way to the "monolithic" quantum theory of the 1930s.

The years 1960–1961 see the period of more intense frequentation and confrontation between the two philosophers, and to these years date two letters containing several critical remarks raised by Feyerabend against a draft of *The Structure of Scientific Revolutions*.[190] They are important both from the historical and the philosophical point of view: historically, they are more direct and lively than the published confrontation between them;[191] philosophically, Feyerabend's letters

---

[185] See particularly Nagel (1961).

[186] I am here following John Preston's detailed discussion of theoretical monism in his (1997a), ch. 5.

[187] Feyerabend (1962b, 1967), p. 136. See also Preston (1997a), pp. 90–92.

[188] Feyerabend (1962b, 1967), p. 137.

[189] Feyerabend (1962b, 1967), p. 138.

[190] Feyerabend (1995a) and (2006).

[191] See, in particular, Feyerabend (1970a) and (1978b), and Kuhn (1970b).

anticipate many of the issues later raised in the public debate that has grown up around Kuhn's works, including some misunderstandings. Furthermore, it is interesting to notice how Feyerabend is still very much on Popper's side, advocating some of the core ideas of his critical rationalism.[192] The progressive rejection of Popper's philosophy will only begin later, with "Explanation, Reduction , and Empiricism" (1962) but will become irreversible only at the beginning of the 1970s.[193]

Besides the very many detailed points Feyerabend raises, there are a few general criticisms that will constitute the central issues for Feyerabend's "official" published reply to Kuhn, in *Criticism and the Growth of Knowledge*. In particular, Feyerabend suspects that underlying Kuhn's arguments is an ideology covered up by history – an ideology that might be of help only for the most narrow-minded and conceited forms of specialism: that is the reason behind the title of Feyerabend's paper, "Consolations for the specialist". In Feyerabend's eyes such ideology would tend to increase the anti-humanitarian tendencies of science. It is the chief target of his criticism: Feyerabend sees himself as a stern opponent of all dogmatic tendencies in science, philosophy and (later on) politics, but also as a champion of the humanitarian values of freedom and everybody's right to develop his own individuality.[194]

In *The Structure of Scientific Revolutions* Kuhn outlines a general model for the development of science: there are some regularities, he argues, in the succession of the different phases of scientific research. In the first place, Feyerabend questions the adequacy of some features of Kuhn's pattern, contradicting Kuhn from a historical, or descriptive, point of view. Secondly, Kuhn discusses the functional roles the different elements of the scientific enterprise play in scientific progress. For example, Kuhn characterizes periods of normal science as those in which a kind of dogmatic attitude is manifested and explains the reasons why such a tendency, paradoxically, does not constitute an obstacle to the growth of science – rather, it not only helps and favours it, but proves to be its necessary prerequisite. Kuhn, in other words, positively values certain aspects of science on the basis of their functional role for scientific progress. Again, Feyerabend questions Kuhn's appraisal of these elements, contradicting Kuhn also from a methodological, or normative point of view. Finally, in *The Structure of Scientific Revolutions* the descriptive and prescriptive elements are not clearly separated, and often alternate with each other. In hindsight,[195] Kuhn explains such ambivalence arguing that in certain conditions descriptive statements about the development of science directly involve normative statements concerning the actual ways of operating in science. In contrast, Feyerabend sees Kuhn's way of presenting his ideas as a tricky tool employed to deceive the reader with ideological stances, without allowing room for critical detachment and evaluation. Therefore,

---

[192] See, for example, Feyerabend (1995a), pp. 358, 359, 360, 361, 363, 366–368, 373 and 376–377; and (2006), pp. 616 ("The question which is still in need of being answered is why you direct your attention to paradigms rather than to method") and 624–627.

[193] See, for example, Feyerabend (1965d), a wholly favourable review of Popper (1963a).

[194] More on this in the next section.

[195] In his replies to criticism in *Criticism and the Growth of Knowledge*, Kuhn (1970b).

Feyerabend's criticism tackles Kuhn from three points of view: historical-descriptive, methodological-prescriptive and concerning the manner of presentation.

At the roots of these three criticisms is the notion of normal science, that is, the phase in the development of science characterized by wide consensus and dogmatic commitment. Such consensus is based on concrete scientific results, that are so convincing and possess such a big heuristic potential that a tradition of research, within a discipline, can be modelled on them, and that in analogy with them it is possible to determine what might constitute a problem and what form a possible solution might have. Exemplary problem solutions constitute the core of the paradigms: during the periods of normal science they cannot be questioned and therefore cannot be subjected to test or verification. This dogmatic attitude notwithstanding, periods of normal science do give rise to scientific revolutions, during which a number of exemplary problem solutions which were deemed valid till then, become questioned and replaced by new exemplars.

First, Feyerabend questions the historical existence of normal science. He thinks that there is no historical differentiation between periods of theoretical monism and those in which new and rival theories proliferate. Feyerabend offers but one historical example in support of his claim,[196] an example that he takes to prove the coexistence of theoretical proliferation and research carried out within a given frame. On the one hand, this can be seen as a weak point in Feyerabend's argumentation but, on the other, it can also be regarded as a valid support to his thesis, since it expresses how difficult it is to justify or refute, by referring to concrete historical examples or case-studies, conclusions concerning regularities in the history of science.

Feyerabend's second criticism concerns the evaluation of normal science. Like other critical rationalists, such as Popper and Watkins,[197] he looks at it with true horror.[198] The reason for this profound dislike is the dogmatic (or quasi-dogmatic) element that characterizes the periods of normal science. For Kuhn, this dogmatic element is functional to the growth of knowledge and the progress of science, while for a critical rationalist (and for Feyerabend in particular) it is absolutely unacceptable, since science is, and must be, an essentially and fundamentally critical enterprise. For Popper and his pupils doing science means devising bold hypotheses, as clear-cut and falsifiable as possible. One of the hinges of critical rationalism is that if somebody decides not to test his hypotheses any more, he ceases to do science.[199] If Kuhn positively values the dogmatic element of normal science he violates this principle. And such a violation, in Feyerabend's eyes, is anti-humanitarian, since "Progress has always been achieved by probing well-entrenched and well-founded forms of life with unpopular and unfounded values. This is how man gradually freed

---

[196] See Feyerabend (1995a), p. 357; see also p. 366.

[197] See particularly Popper (1970), pp. 51–56, and Watkins (1970).

[198] See Feyerabend (1970a), pp. 201–210, (1978b), pp. 158–168, and (1995a), pp. 361–362.

[199] "The game of science is, in principle, without end. He who decides one day that scientific statements do not call for any further test, and that they can be regarded as finally verified, retires from the game" (Popper (1935, 1959), p. 53).

himself from fear and from the tyranny of unexamined systems".[200] The "normal" element of normal science is nothing but a conservative and anti-humanitarian constituent.

Feyerabend rejects Kuhn's theoretical monism, that is, his stressing the claim that "it is only by concentrating *on a single paradigm*, by trying to fit nature into it despite all apparent difficulties, that scientific progress is achieved".[201] Indeed, Feyerabend firmly believes that

> considering a set of mutually inconsistent but factually adequate theories *increases* the empirical content of any element of the set and this for the simple reason that many tests presuppose the existence of an alternative! […] If this is the case then we must make the decision: what do we prefer, increased empirical content of the theories we possess or that unanimity of research and the close fitting produced by it in the periods which you call the normal periods. History cannot help us in this decision. Many scientists seem to prefer the latter alternative […]. But advance of knowledge, so I would have thought, has nothing to do with membership in communities (Wittgenstein notwithstanding).[202]

But what really makes Feyerabend furious is Kuhn's style of presentation. He realizes the necessity of a point of view, or an "ideology", that provides the background of any presentation, the starting point for a reading of the historical facts. Without such an ideology living "would be both impossible and inadvisable",[203] and any historical account offered without a point of view "if it would be possible, would be the most drab and uninteresting affair imaginable".[204] Furthermore, he does not pretend "that in history a nice distinction can be drawn between what is regarded as a factual support, and what is regarded as interpretation according to some point of view".[205] In other words, Feyerabend acknowledges a methodological, evaluative point of view, on the basis of which Kuhn judges some elements of science as rational and other as irrational. But he charges Kuhn with not rendering explicit his point of view, so that readers can be made aware of the existence of other points of view and of other possibilities for an evaluation: "points of view *can* be made explicit, and it *is* possible to write history in such a manner that the reader is always aware of the one's

---

200 Feyerabend (1970a), pp. 209–210.

201 Feyerabend (1995a), pp. 355–356. Feyerabend insists on this issue particularly in his second letter: see *ibidem*, pp. 372–373, 375–376, 377–378 and 385.

202 Feyerabend (1995a), p. 356. And again: "I think that very often anomalies can be discovered […] not by further and further elaborating a given paradigm, but by elaborating an *alternative* paradigm and producing with its help testable predictions which, if they are confirmed, show that the first paradigm is in trouble. Your insistence upon faithfulness *to one and only one paradigm* is bound to result in the elimination of otherwise very important tests and is bound in this way to reduce the empirical content of the paradigm you want to be accepted" (*ibidem*, p. 365). "[…] alternatives should be considered as they are necessary to accelerate whatever paradigm is in the center of attention […]: *alternatives are* both used, and needed; and they are needed as it is only with their help that it is possible to find anomalies in whatever theory is being held at a special moment" (*ibidem*, p. 366).

203 Feyerabend (1995a), p. 355.

204 Feyerabend (1995a), p. 355.

205 Feyerabend (1995a), p. 355.

ideology or point of view *as well as of the possibility of an alternative interpretation of the historical facts*. That is, history can be written in such a manner *that what is factual and what is reasonable appear as two clearly distinct affairs*".[206] On the contrary, in Kuhn's case, it seems that the appraisal of historical facts is immediately drawn from facts themselves: "What you are writing is not just history. It is *ideology covered up as history*":[207]

> What I do object to most emphatically is the way you present this belief of yours; you present it not as a *demand*, but as something that is an obvious consequence of historical facts. Or rather, you do not even talk about this belief, you let it as it were emerge from history as if history could tell you anything about the way you *should* run science (it *does not* imply ought!). It is this bewitching way of representation to which I object most, the fact that you take your readers in rather than trying to persuade them. This manner of presentation you share with Hegel and Wittgenstein; and with all those who say, when engaging in a political enterprise, that 'history will be our judge'.[208]

---

[206] Feyerabend (1995a), p. 355. "Your 'history'", Feyerabend remarks, "is ideologically infected double talk" (*ibidem*, p. 360). Feyerabend rails at Kuhn's ambiguous discourse at various times, especially on pp. 366–368. In particular, he writes: "you do not write history plain and simple, but […] you present an ideology, and a very questionable monolithic ideology […] in the covers of history. In this respect you are really very similar to those who point to history in order to justify their crimes. You are a mystic, an irrationalist. And by this I mean that you not only hold certain beliefs (conservative character of normal science) but that you are not prepared to let these beliefs speak for themselves; you rather present them in a manner which suggests that they are facts and thereby force people to swallow them without criticising them. What are you afraid of? Are you afraid that people will oppose at once when your beliefs are presented to them in their proper form, viz. as demands as to how science ought to be run? […] Again, it is this kind of *double-talk* to which I object most. You really are like a witch doctor. That is, I do not object to your findings, but to the manner in which you represent them – as if they were (α) indisputable […] and (β) unescapable; neither of which is the case" (*ibidem*, pp. 367–368). "In sum: it seems to me that you do not tell history as what it is: a series of accidents combined with struggle for power etc. etc. You perceive some inherent reason in it. But in order to perceive inherent reason in a series of accidents one must distort reason itself" (*ibidem*, p 371: in this passage is eminently clear the influence of Popper's thought: see, in particular, his (1945, 1966), vol. II, ch. 25, and (1957b)). See also Feyerabend (2006), pp. 614–618, (1970a), pp. 198–199, and (1978b), pp. 155–156.

[207] Feyerabend (1995a), p. 355. Again: "[…] doing this you do a disservice both to history and to philosophy. You do a disservice to history because you misuse it to wrap up your own predilections in it without saying so. You *falsify* history just as Hegel falsified it in order to finally arrive at the Prussian state. And you make philosophy irrational by presenting the philosopher not with a doctrine, an ideal concerning knowledge which he would then be able rationally to discuss; you present him rather with a so-called fact (which is not really a fact at all but a misleading report – misleading, I believe, even to you – of your own predilections) which he simply must swallow for facts cannot be different from what they are" (*ibidem*, p. 360; needless say, Popper's criticism of Hegel and historicism is firmly present in Feyerabend's mind; see also p. 356).

[208] Feyerabend (1995a), p. 355.

A few years later, in his "official" reply to Kuhn, Feyerabend would draw a sharp conclusion from Kuhn's repeated double-talk:

> I venture to guess that the ambiguity is *intended* and that Kuhn wants to fully exploit its propagandistic potentialities.[209] He wants on the one side to give solid, objective, historical support to value judgements which he just as many other people seem to regard as arbitrary and subjective. On the other side he wants to leave himself a safe second line of retreat: those who dislike the implied derivation of values from facts can always be told that no such derivation is made and that the presentation is purely descriptive.[210]

Other reasons for confrontation[211]

To the charge of defending a dangerous dogmatism and theoretical monism, and of willingly confusing the descriptive with the prescriptive aspects of the philosophical reflection on the scientific enterprise, Feyerabend added the wider issue of human happiness and the good life – which, if at first it may look like an ethical problem, reveals itself, in Feyerabend's eyes, as an essentially political problem, for it cannot be resolved on the individual level but requires an evaluation of its collective and social dimension. There is no reason why somebody should not live according to his own beliefs and convictions. However, it is important that everybody be aware of the fact that the tradition he belongs to is relative, and therefore he cannot contest the right of other traditions, supporting different values, to assert themselves and pursue their own ways of living.[212] And a good life is definable in terms of the key notion of freedom, which Feyerabend deems the supreme value: spiritual freedom, freedom of thought, freedom from dogma. This vision, to which Feyerabend dedicated the last years of his life, plainly contrasts with Kuhn's picture of "mature" science. "The "closed society" of "normal" scientists is not compatible with the "free society" Feyerabend always hoped for.[213]

The gap between them seems to be unbridgeable. And Feyerabend always kept his original judgement of Kuhn's philosophy unchanged, firmly believing he had understood him. However, in his later years, after reading one of his pupils' dissertation, Feyerabend changed his mind. In the second, revised edition of *Against*

---

[209] Feyerabend also charges Lakatos for this: see his (1981a), p. 47, n. 6.

[210] Feyerabend (1970a), p. 199. See also Popper (1974a, 1976), pp. 193–196.

[211] This section is not strictly related to incommensurability. However, I think it is important to see Kuhn's and Feyerabend's different views of incommensurability within a wider context, about which I give some details here. This section also highlights the clashing metaphysics underlying Kuhn's and Feyerabend's views – the clash between an upholder of dogma, institutionalism and intellectual conformism, and a defender of freedom, intellectual, political, spiritual or otherwise. More than ever, Feyerabend embodies the spirit of Popper's philosophy.

[212] Feyerabend's systematic demolition of the idols of Reason does not result – as somebody has pretended to claim – in irrationalism or nihilism. On the contrary: his radical scepticism (this is how I tend to read his attitude and philosophy) is the design of a free society in which any beliefs, ideas or forms of life have the right to exist and be defended. See, in particular, Feyerabend (1978c), (1979, 1980), (1987a), (1995b) and (1999b).

[213] See Feyerabend (1978c) and (1979, 1980).

Method (1988), in which the dedication "To Imre Lakatos – Friend, and fellow-anarchist" disappears, he wrote:

> I have joined Kuhn in demanding a historical as opposed to an epistemological grounding of science but I still differ from him by being a relativist (Kuhn, if I understand him correctly, rejects apparently relativistic implications of his views) and by opposing the political autonomy of science. But having read Paul Hoyningen-Huene's admirable account of Kuhn's philosophy (*Die Wissenschaftstheorie Thomas S. Kuhns*, Thesis, ETH, Zürich, 1988) and thus having been made aware of the great complexity of Kuhn's thought I am not at all sure that our differences are as great as I often thought they were.[214]

In an article published the following year he took one more step forward, saying that the ideas of this paper "are very similar to, and almost identical with, Kuhn's as yet unpublished, later philosophy".[215] Furthermore, in the third English edition of *Against Method* (1993) he reinforced this opinion.[216] Finally, in a posthumous review published in 1994, Feyerabend describes Paul Hoyningen-Huene's reconstruction of Kuhn's philosophy of science as "a surprisingly coherent and powerful system of thought".[217] A last-minute reconciliation, that would confirm the idea that these two influential philosophers of the twentieth century were not, after all, so distant from each other.[218]

The Critics

The secondary literature on incommensurability is enormous and a synthesis can hardly be attempted.[219] This is comprehensible, given the original divergences

---

214 Feyerabend (1975, 1988), p. 230. See also Hoyningen-Huene (2000).

215 Feyerabend (1989a), p. 405, n. 26. See also the "Postscript on Relativism", in Feyerabend (1975, 1993), pp. 268–272.

216 "[…] our views (i.e. my published views and Kuhn's as yet unpublished recent philosophy) by now seem to be almost identical except that I have little sympathy for Kuhn's attempt to tie up history with philosophical or linguistic, but at any rate with theoretical ropes: a connection with theory just brings us back to what I at least want to escape from – the rigid, though chimerical (deconstruction!) boundaries of a 'conceptual system'" (Feyerabend (1975, 1993), p. 213).

217 Feyerabend (1994), a short review of Hoyningen-Huene (1989a/1993) and Biagioli (1993).

218 To recall the "melancholic dream" with which Jorge Luis Borges concludes one of his most poetic stories, "The end of the story can only be told in metaphors, since it takes place in the kingdom of heaven, where time does not exist. One might say that [Paul Feyerabend] spoke with God and found that God takes so little interest in [epistemological] differences that He took him for [Thomas Kuhn]. That, however, would be to impute confusion to the divine intelligence. It is more correct to say that in paradise, [Tom] discovered that in the eyes of the unfathomable deity, he and [Paul] (the orthodox and the heretic […]) were a single person" ("The theologians", in Borges (1952), p. 207, with the appropriate modifications).

219 Just to mention a few attempts: Buzzoni (1986), ch. 3, Sankey (1994a), (1997a), (1997c) and (1999), Oberheim, Hoyningen-Huene (1997) and Sankey, Hoyningen-Huene (2001).

between Kuhn's and Feyerabend's positions and the very many forms the ensuing debate has taken (a debate that has often ignored the distinction of the different issues involved in the incommensurability thesis). Therefore, I limit myself to a brief overview of those that, after several years, have proved to be the most serious and fruitful attempts to appraise and develop the incommensurability thesis.[220]

"Meaning-variance" and partial communication

Both Kuhn and Feyerabend reject the empiricist idea of the existence of a theory-neutral observation language, whose terms bear a theory-independent meaning.[221] The incommensurability thesis derives from the combination of the theory of meaning variance and that of the theory-ladenness of observation: given the contextual nature of meaning, the vocabularies of different theories cannot comprise meanings shared by both of them. But if it is not possible to express theories by means of a shared vocabulary, their content cannot be directly compared: for in the absence of a shared, semantically neutral vocabulary, it is impossible either that a theory affirms or negates the same thing as another.[222] Theories that, for these reasons, cannot reach an agreement or highlight a disagreement between them on any fundamental issue, are semantically incommensurable – that is, given the semantic variance of the terms in which they are expressed, it is not possible to compare them directly.

What I have just described is the semantic theory of incommensurability in terms of radical meaning-variance, which applies to the whole set of terms employed by a theory. However, it is possible to state a weaker version of the semantic thesis, in terms of partial meaning-variance, when the semantic variance affects only a limited portion (or subset) of the vocabulary employed by theories. The radical version of the meaning-variance theory is usually associated with the *enfant terrible* Feyerabend while its partial version is usually attributed to Kuhn.[223]

The replies to this first version of the incommensurability thesis have followed two main lines. On the one hand, some authors accepted a certain measure of meaning-variance of the terms employed by different conceptual systems, but nonetheless have claimed and advocated the existence of a sort of "overlapping" semantic area

---

[220] See also Sankey (1999) and Sankey, Hoyningen-Huene (2001).

[221] For Kuhn's and Feyerabend's original remarks on meaning-variance and the absence of a neutral observation language see Kuhn (1962a), pp. 101–103, 126–129 and 149–150, (1970b), pp. 266–272, and (1970c), pp. 200–204; Feyerabend (1958a), (1962a), pp. 44–45, 64–68 and 76–77, and (1965a), pp. 168–172 and 179–181.

[222] Indeed, in replying to Dudley Shapere, Feyerabend acknowledges that statements that do not share a common meaning cannot contradict each other (see Feyerabend (1965c), p. 115). As to Kuhn's views on the limits of the possibility of a direct comparison of contents given the lack of a common language, and the impossibility of mutual translation between different theories, see Kuhn (1970b), pp. 266–267, (1976b), 188–190, and (1979b), pp. 203–205. See also Sankey (1992) and (2000b).

[223] See, respectively, Feyerabend (1962a), pp. 68 and 93, (1965b), p. 97, and (1965c), pp. 114–115; and Kuhn (1970b), pp. 266–267, and (1983a), pp. 35–37. For the following survey, see Sankey (1999).

that is sufficient to make comparison possible. On the other hand, other authors have claimed that the idea of untranslatability involved in the semantic version of the incommensurability thesis is incoherent.

First reaction: it is possible to find a common reference

In his *Science and Subjectivity* (1967) Israel Scheffler, by appealing to the Fregean distinction between *Sinn* ("meaning", or "sense") and *Bedeutung* ("denotation", or "reference"),[224] stresses that even if the meaning of a scientific term varies with the theory that employs it, it does not follow that the reference of that term also varies with them. Different terms may differ in their meaning but share their referents, that is, they may co-refer. However, if the terms employed by theories preserve reference (that is, the object they refer to) through variation of meaning, it remains possible to compare the theories with respect to their content. For statements which theories make about the world may agree or disagree with respect to common states of affairs, provided only that their constituent terms refer to the same things, despite variation of meaning.[225]

However, Scheffler's argument, according to which changes of reference do not involve changes of meaning, does not solve the problem. There are several historical cases in which different theories seem to employ the same terms to refer to different objects.[226] Furthermore, examples of radical conceptual changes suggest

---

[224] See Frege (1892).

[225] See Scheffler (1967). The condition by which, in order that statements belonging to semantically variant theories preserve a comparable content, terms refer to the same things, can be understood in a wider sense. For in order that a content comparison is possible, it is sufficient that there is a (partial) overlapping in the extension of terms – it is not necessary that extensions perfectly coincide. As Michael Martin (in his (1971) and (1972)) has shown, the statements expressed by alternative theories may conflict provided that they comprise predicates whose extension are linked by relations of inclusion or intersection. A similar conclusion can be reached also on the basis of Hartry Field's notion of partial denotation (see Field (1973)).

[226] Over the last thirty years of his life Kuhn directed much effort into formulating precise conditions under which two theories would be undeniably incommensurable with one another. His first step, in the late 1960s, was to argue that incommensurability must result when two theories involve incompatible taxonomies. The problem he then struggled with, never obtaining a solution he found entirely satisfactory, was how to extend this initial line of thought to sciences like physics in which taxonomy is not so transparently dominant as it is, for example, in chemistry (see also above, ch. 2, n. 204). In their (2001) historians of science Jed Buchwald and George Smith argue for an interesting case, showing how the evidence for the theory of polarization within ray optics did not carry over to wave optics, so that the case can be regarded as a prototypical case of discontinuity of evidence, and hence of incommensurability in the way Kuhn wanted: "Without any question, the transition from ray to wave optics early in the nineteenth century involved what Kuhn would have originally called a revolutionary paradigm shift. Wave optics requires an entirely different way of thinking about light. Its conceptual framework could never have arisen through a sequence of small changes starting from the conceptual framework of ray optics" (*ibidem*, p. 495); "the transition from ray to wave optics involved a Kuhnian conceptual readjustment if ever there was one" (*ibidem*, p. 477). However, as these authors remark, the evidence for classic geometry optics

the existence, in the history of science, of long periods of referential discontinuity. Therefore, although common referents can very well be more than sufficient to allow for comparison of the contents of different theories, it is still to be shown how it is possible to preserve continuity of reference through the transition from one theory to another, semantically variant with it.[227]

In any case, Scheffler's argument has the merit of calling attention to the issue of reference, highlighting the problem of how terms that belong to different systems can preserve their referents through a process of conceptual content variation that those systems link to them. In particular, it raises the problem of the determination of the referents of terms employed by scientific theories.

Other authors, like Saul Kripke and Hilary Putnam, have given up the Fregean approach for a causal theory of reference: reference, they claim, is determined in a direct manner by the causal relation between speaker and object, and not by the descriptive content that the speaker associates to the terms he employs.[228] From this point of view, the reference is fixed at the initial introduction of the term (by appealing to simplifications, in most cases) and later employments of the same term will have a reference that connects to the original one through a historical concatenation. Since reference is determined independently from descriptive content, the denotation of the terms employed by scientific theories can be preserved notwithstanding the variation of the concepts associated with those terms. But if the reference preserves a stability of its own through the variation of conceptual content, there arises no problem of comparison between different theories, since reference is preserved even if the terms are associated with a different conceptual content, according to the theoretical context they belong to.

New objections have been raised against this proposed solution. It is not plausible to think, in the first place, that the reference of a term is fixed once and for all at the moment of its introduction: this would exclude, in principle, the possibility

---

did carry over to wave optics, notwithstanding the fundamental conceptual readjustment that Fresnel's wave theory required: "The conceptual discontinuity across the transition [...] does not automatically entail an across-the-board discontinuity of evidence" (*ibidem*, p. 495). The question of incommensurability of evidence across revolutionary changes, Buchwald and Smith conclude, is far more complicated than the picture offered in *The Structure of Scientific Revolutions*: "the relationship between the normal-scientific traditions before and after the revolution can be extremely complicated, with the continuity of evidence in some places, discontinuity and hence incommensurability in others, and various gradations between continuity and full discontinuity in still others. Claims about incommensurability will have to be adjudicated only through a detailed historical examination of the science before and after the revolution" (*ibidem*, p. 497). See also Buchwald (1989) and (1992), and Buchwald, Smith (1997).

[227] Some terms whose reference has changed through the history of science are: "atom", "electron", "mass", "planet", "space", "time", "force", "energy", "compound" and "gene" (this is not the proper place to do that, but it would be extremely interesting to analyse also terms and concepts taken from human sciences, such as, for instance, "morality"). For the idea according to which a term changes its reference when the theory that employs it varies, see Kuhn (1962a), pp. 101–102, (1970b), pp. 268–270, and Feyerabend (1965b), p. 98.

[228] See Kripke (1980) and Putnam (1975). See also Putnam (1965), (1972) and (1974).

of a change of references of terms through the history of science.[229] Secondly, if the reference for natural kind terms is fixed in a completely non-descriptive way – that is, by the ostension of samples of a kind – it would be impossible to secure an unambiguous reference to a specific natural kind as opposed to the numerous other kinds instantiated by the sample set. The problem is how to choose an object as a member of a certain kind.[230] Finally, if the reference of theoretical terms is determined by means of a causal relation between observed phenomena and the entities responsible for these phenomena, it would be impossible for theoretical terms not to refer – and yet the lack of reference seems to occur routinely in the history of science.[231]

These difficulties suggest that the causal theory of reference needs be modified to allow the variation of reference after the initial introduction of term, and also to leave room for the role played by descriptive content in the determination of reference. This led to the development of various modified versions of the causal theory, such as the descriptive-causal theory of reference. According to the latter, to the causal relations between speaker and object we must also add at least a minimal descriptive apparatus to determine the referent more precisely.[232] Such a descriptive apparatus, that may include the specification of a natural kind or of the causal role of theoretical entities, is necessary to solve the problem of how to choose an object as a member of a certain kind ("*qua* problem") and to allow that theoretical terms have a reference.

However, these modified versions of the causal theory seem to be unable to achieve such results as to unequivocally and effectively refute the incommensurability thesis, as their proponents aimed to do. For the insertion of a role for description in the process of reference determination, together with the possibility of a reference change after the initial introduction of terms, leave enough room in order for the reference to vary with the varying of the theory that employs it.[233]

---

[229] See Fine (1975). Indeed, there can be changes of reference, sometimes. The question is whether these circumstances coincide with scientific revolutions, for they may refer only to the very early phases of the process of fixing the use of a certain term, or name.

[230] It is the so-called "*qua* problem": see Papineau (1979), Sterelny (1983) and Devitt, Sterelny (1987, 1999).

[231] See Enç (1976), Nola (1980b) and Kroon (1985).

[232] The use of the term "causal descriptivism" varies from author to author. Here I use it in the sense Howard Sankey uses it (see his (1994a), pp. 61–67, and (1997c)). This usage is similar to the sense in which the phrase "causal-descriptive theory" is employed Devitt and Sterelny to refer to theories that are a sort of hybrid between causal theories and descriptive theories of reference (see their (1987, 1999), pp. 96–101). Therefore, the use I adopt here is different from David Lewis' (see his (1984), p. 226), according to whom a causal-descriptive theory of reference provides for the reference to be fixed by a description of its causal relations, and not by the causal relations themselves. See also Sankey, Hoyningen-Huene (2001), pp. xi and xxxi).

[233] For a detailed description of the required modifications of the causal theory of reference and of the consequent implications for what concerns incommensurability, see Sankey (1994a), ch. 2; see also his (1991a), (1991b), (1991c), (1991d).

Second reaction: translation is possible

A second set of objections to the incommensurability thesis targets one of the consequences of the theory of meaning-variance, according to which it is impossible to provide a mutual translation between the vocabularies employed by two incommensurable theories. The radical version of the meaning-variance theory seems to imply that the terms employed to express a theory can be untranslatable into any of the terms employed by a rival one, incommensurable with the former. Taken to its extreme consequences, the meaning-variance theory suggests that there could be even entire non-intertranslatable languages.

The idea of a totally untranslatable language sounded paradoxical to most philosophers: for if it is impossible to translate even a little part of a language, it is not even possible to know whether it is actually a language.[234] If, on the other hand, some sentences are regarded as evidence of the impossibility of a translation, this very fact contradicts the incommensurability thesis, since showing instances within one's own language presupposes a form of translation. Even worse: attempting to understand instances of an untranslatable language seems to imply the actual possibility that some sort of translation is in fact possible, since understanding a foreign language seems to presuppose its trans-lation (that is, its moving from one language to another) into an understandable language.

Ideas of this kind lie at the heart of Donald Davidson's "On the Very idea of a Conceptual Scheme" (1974). But in his article Davidson goes well beyond the semantic version of the incommensurability thesis: to make sense of the idea of a language independent of translation requires a distinction between conceptual schemes and the content organized by such schemes. But, Davidson argues, no coherent sense can be made of the idea of a conceptual scheme, and therefore no sense may be attached to the idea of an untranslatable language.[235]

Some of the issues raised by Davidson can be avoided by considering two points both Kuhn and Feyerabend make about translation. First, since the early 1980s Kuhn developed what he labelled "local incommensurability", that reduces the impossibility of translation to small sets of interdefined terms belonging to rival scientific theories.[236] By restricting the scope of incommensurability to these "local" groups of terms, or even to the specific vocabularies of theories, it is possible to avoid the need to make coherent sense of a totally untranslatable language or of the scheme/content dualism. Secondly, both Kuhn and Feyerabend always attempted to make a clear distinction between translating a language and understanding it:[237] although we might not be able to translate a foreign language into our native language, this does not imply that we are not able to understand it.

The combination of these two remarks yields a refined and improved version of the semantic thesis of incommensurability, in which the ability to translate is

[234] This is certainly Davidson's view. However there may be cases in which such a view seems questionable: even before hieroglyphs were read, for instance, it was quite clear that they represented a language; and the same can be said for Linear B.

[235] See Davidson (1974).

[236] See, for instance, Kuhn (1983a), pp. 35–37.

[237] See Kuhn (1983a) and Feyerabend (1987b).

restricted to the specialized vocabularies within a language, which those sharing different theoretical systems may be able to understand anyway. Davidson's attack is unable to address this refined version.

Further developments

The fact that these two reactions to the incommensurability thesis have been treated separately must not be misunderstood: they do not address completely disjointed aspects of the incommensurability thesis. While the problem of comparison between two theories on the basis of their co-referentiality is different from the problem of the consistent idea of an untranslatable language, the solution (or, better, the attempted solution) of one of these two problems can yield consequences for the solution of the other. For the problem whether the terms employed by one theory refer to the same objects as the terms employed by the other is not distinct from the problem whether the terms of one theory do actually express the same meanings as the other's.

Howard Sankey has developed an approach to semantic incommensurability that deals with both the issue of reference and that of translation. In *The Incommensurability Thesis* (1994) he adopts a modified version of the causal theory of reference that allows for a change of reference after the initial introduction of a term and secures a role for description in the determination of reference. Sankey holds that there may arise problems of translation between two theories due to the different manner and tools by which reference is determined, and defends a partial impossibility of translation in response to the points raised by Davidson. His approach combines acceptance of the first objection to the incommensurability thesis (according to which it is possible to find common references) with rejection of the second one (concerning the impossibility of the mutual translation between rival theories): Sankey claims semantically variant theories can be compared thanks to an overlapping of referents, notwithstanding the (only partially insurmountable) difficulties of mutual translation. Such an approach shows how it is possible, although the two objections address different problems, to grapple with the issues raised by both of them within a unified approach to the semantic theory of incommensurability.[238]

Theory appraisal, rationality of choice and relativism

As to the second version of the incommensurability thesis, I will refer mainly to the developments of Kuhn's thought, for while Feyerabend advocated a variety of well-known issues related to the nature and the limits of scientific method,[239] he did not do so by dealing explicitly with incommensurability. Indeed, unlike Kuhn, who always claimed that incommensurability had one or more methodological aspects, Feyerabend always focused his own treatment of incommensurability on the semantic relations between theories.

---

[238] See Sankey (1994a). For a critical examination of Sankey's work see Hoyningen-Huene, Oberheim, Andersen (1996); Sankey's own rejoinder is in his (1997c), section 8.

[239] To mention but some of his most relevant works, see Feyerabend (1970b), (1975), (1978c), (1979, 1980) and (1987a).

While the term "incommensurable" is usually understood in a semantic way, some authors employed it, and still employ it, in its methodological sense. In this second sense many prefer to deal with the problems raised by the methodological version of incommensurability under the rubric of "rational theory choice" and "relativism" due to the variation of evaluation criteria and methodological standards. However, it needs to be noted that the starting point of the discussion about relativism and rational theory choice is very often Kuhn's statement that criteria and standards for theory appraisal vary with the varying of paradigms.[240]

The methodological theory of incommensurability states that there are no objective and shared criteria or standards for the appraisal and choice of scientific theories: they vary from one theory to another, they change when paradigms change. There are no external or neutral standards that may be employed in the comparative assessment of competing theories. As a consequence, alternative scientific theories may turn out to be incommensurable due to the lack of common methodological standards on the basis of which we can achieve a univocal choice among different theories.

The idea that rival theories may be methodologically incommensurable draws on the rejection of the traditional position according to which there is one uniform and invariant scientific method, employed by the whole scientific community and that constitutes the characteristic feature (the essence, some would tend to say) of science. In *The Structure of Scientific Revolutions* Kuhn claims that the standards for theory appraisal depend on the dominant scientific paradigm of a given period and vary with it. Paradigms "are the source of the methods, problem-field, and standard of solution accepted by any mature scientific community at any given time";[241] "when paradigms change, there are usually significant shifts in the criteria determining the legitimacy both of problems and of proposed solutions".[242] Such criteria, however, do not determine a choice between different paradigms, since such a choice "is not and cannot be determined merely by the evaluative procedures characteristic of normal science".[243] Nor are there independent standards for theory choice since "As in political revolutions, so in paradigm choice – there is no standard higher than the assent of the relevant community".[244] In other words, Kuhn's denial of the availability of standards independent of paradigms gives the impression that the rationality of acceptance of scientific theories is relative to the prior choice of a scientific paradigm – and the latter choice cannot be governed by shared objective standards of theory appraisal.

---

240 See Kuhn (1962a), especially chs. V, IX, X and XIII, together with (1970b), pp. 266–277: such a statement represents one of the constitutive aspects of the incommensurability between paradigms. Another constant point of reference in the discussions of this aspect is Feyerabend's criticism of the existence of a single scientific method, established once and for all: the *locus classicus* is of course Feyerabend (1975). On these issues, see Sankey (1993b), (1994b), (1995a), (1996a) and (1997b).

241 Kuhn (1962a), p. 103.

242 Kuhn (1962a), p. 109.

243 Kuhn (1962a), p. 94.

244 Kuhn (1962a), p. 94.

Kuhn's emphasis on the nature of the fundamental commitment that bonds a scientific community to the dominant paradigm highlighted the conservative component of science. On the contrary, with his principle of proliferation of theories, Feyerabend stressed exactly the opposite elements of change, pluralism and competition.[245] Just like Kuhn, however, Feyerabend claimed that methodological rules and scientific standards are subject to variation and do not remain constant through the history of science. Furthermore, he claimed that all scientific rules advanced from time to time have been justifiably violated in particular periods of the history of science.[246] Accordingly, there is no single, invariant and cogent scientific method that may be applied to all sciences at any time. This does not mean that there is no scientific method, nor that science does not follow any rules – rather, what Feyerabend meant is, quite simply, that "all methodologies, even the most obvious ones, have their limits".[247]

This, I think (despite the floods of ink spilled to argue for the contrary) is the real meaning of the well known slogan "anything goes" – which is not, then, despite Feyerabend's own rhetoric, an instigation to epistemological anarchy, but a polemical way to appeal to history, critical rationality and pluralism. Analogously, *The Structure of Scientific Revolutions* is not meant to be an attack on rationality, but on the current conceptions of realism. Kuhn wishes to question not so much the rationality of theory appraisal and choice, as the epistemic character of the chosen theories, about which we cannot say they are a better approximation to the truth. This reading explains several misunderstandings that interweave in the wide and heated debates on rationality in the 1960s and 1970s.

As it emerges from the works of both Kuhn and Feyerabend, then, there is no fixed set of scientific standards to which we may appeal to univocally determine the choice among conflicting theories. Such a choice cannot be made on the basis of standards and criteria allegedly shared by all parties in the dispute. Kuhn's and Feyerabend's methodological ideas seem therefore to imply an epistemological relativism according to which scientists, in perfectly rational ways, can accept conflicting theories on the basis of conflicting methodological standards. Moreover, in the absence of higher level standards that are able to determine a choice among competing theories, a choice would seem to rest on eminently and inevitably subjective and irrational factors, rather than on objective methodological grounds. These generically relativistic features of Kuhn's and Feyerabend's views of scientific method, theory appraisal and choice have been opposed by many authors, who have subjected their ideas to detailed analyses and sharp criticism.[248]

---

[245] The principle of proliferation recommends that scientists "Invent, and elaborate theories which are inconsistent with the accepted point of view, even if the latter should happen to be highly confirmed and generally accepted" (Feyerabend (1965c), p. 105, emphasis suppressed). On the contraposition between Kuhn's theoretical monism and Feyerabend's pluralism, see Feyerabend (1970a), pp. 205–207, and (1995a), especially pp. 355–356, 372–373, 375–376, 377–378 and 385; see also Preston (1997a), chs. 5 and 7.

[246] See, for example, Feyerabend (1975), pp. 23–25.

[247] Feyerabend (1975), p. 32; see also his (1978c), p. 32.

[248] See, to mention but a few, Lakatos (1970), pp. 177–179, Shapere (1983), pp. 46–47, Siegel (1987), pp. 51–54, Howard Margolis (1993) and Laudan (1996), ch. 5. For a reading

While Feyerabend did little to dispel the charge of relativism, Kuhn tried hard to distance himself from the relativistic implications of his own original position. In his later works he states he never wished to deny that the choice among alternative scientific theories could be a rational process, governed by methodological standards. His aim was simply to highlight that "There is no neutral algorithm for theory-choice, no systematic decision procedure which, properly applied, must lead each individual in the group to the same decision".[249] Lacking such an algorithm, a theory of rational choice inevitably implies a judgement or deliberation. Nor did Kuhn mean to deny the existence of fixed standards of theory appraisal. Indeed, he lists a number of them: accuracy, consistency, simplicity, scope and fruitfulness, to which he adds, from time to time, plausibility, explanatory power, unity of science and above all ability to define and solve as many theoretical and experimental problems as possible.[250] These criteria of choice "function not as rules, which determine choice, but as values, which influence it".[251] Furthermore, Kuhn says, individual scientists "may legitimately differ about their application to concrete cases. […] when deployed together they repeatedly prove to conflict with one another".[252] The outcome is that despite adherence to a shared set of standards, there can arise rational disagreement among scientists who advocate different theories, since each one interprets or gives a different weight to various standards individually taken.

Kuhn's idea of rational disagreement governed by non-algorithmic standards provides a promising solution to the problem of the rational choice among alternative scientific theories, but leaves a question open, which Kuhn was never able to solve satisfactorily. It is the meta-methodological problem of the normative basis of the standards deployable to operate an appraisal and a choice. In *The Structure of Scientific Revolutions* he seems to be founding epistemic normativity on the consensus within the community of experts.[253] Later, however, he tries to naturalize such normative character by founding it on successful scientific practice.[254] Finally, in the 1980s, he proposes a conceptual foundation of scientific norms, whose rationality he thinks analytically secured by the very concept of science.[255]

In the past years the problem of the meta-methodological justification of epistemic norms has been subjected to a detailed analysis by Larry Laudan. He distinguishes between different meta-methodological positions (intuitionist, conventionalist

---

of Kuhn as a "moderate relativist" see Doppelt (1978), while for the view that Kuhn's and Feyerabend's positions on theory choice may contain the elements of a new way of thinking about rationality see Bernstein (1983). See also Andersson (ed.) (1985), Joseph Margolis (1991), Meiland, Krausz (eds) (1982) and Nola (ed.) (1988).

249 Kuhn (1970c), p. 200.

250 Kuhn lists these values, with no claim of completeness, in his (1977c), pp. 321–322. But see also Kuhn (1962a), pp. 153–159, (1970a), pp. 20–22, (1970b), p. 241 and 260–262, (1970c), 184–186, 199 and 205–206, (1971a), pp. 145–146, (1977c), pp. 321–325, (1980a), pp. 189–190, and (1983c), p. 210. See also Feyerabend (1978b), pp. 199–200.

251 Kuhn (1977c), p. 331.

252 Kuhn (1977c), p. 322.

253 See Kuhn (1962a), pp. 94–95.

254 See Kuhn (1970b), pp. 236–237.

255 See Kuhn (1983c).

and naturalist), arguing that in order to answer the relativist challenge we need a naturalistic meta-methodology that founds a normative methodology in the empirical facts related to the means to achieve epistemic ends. In any case, whatever the destiny of Laudan's proposal, the solution to the problems related to the methodological version of the incommensurability thesis will require the development of an adequate meta-methodological theory to secure methodological norms.[256]

Incommensurability and realism

In the light of what has been said so far about the semantic and the methodological versions of the incommensurability thesis, and particularly the difficulties facing the problem posed by the rational choice among competing theories, it could seem that both these versions have an essentially epistemological nature. Indeed, if it is impossible to compare different theories as both to their content and possible common standards then it is unclear how it may be possible to choose one or the other on rational and objective grounds. But the controversy over incommensurability is not limited to the epistemic issues relative to the actual possibility of rational choice. It involves issues of a broader metaphysical scope as well. Semantic incommensurability, in particular, leads to a number of controversial issues on the relationship between theory and reality, which bear special relevance for the wide debate involving realist and antirealist approaches to the philosophy of science.

In the section devoted to the first reaction to the advocates of the semantic version of the incommensurability thesis I mentioned the possible discontinuity of reference in the transition from one theory to another (take, for instance, the case of terms such as "atom", "planet", or "mass").[257] A radical conceptual change may give rise to massive referential discontinuity, such that no term in the new theory refers to an object referred to by the previous theory as well. A referential discontinuity of this kind conflicts with a realist philosophy of science, since realist philosophers hold that successive theories in the same domain typically provide alternative descriptions of the same entities, and that scientific progress consists in the increase of the quantity of known truths with respect to a shared set of entities. But if successive theories no longer refer to the same entities referred to by previous ones in any way, then the realist's idea of scientific progress as the growth of known truths about a shared set of entities becomes untenable.

As we have seen, the referential discontinuity problem results in the problem of determining such reference. The degree to which reference may vary with the varying of theory depends on how much such reference is "sensitive" to the variation of the descriptive content associated to the terms employed. But the problem whether

[256] For Laudan's approach see particularly his (1977), (1984), (1990a) and (1996), ch. 7. For a critical assessment of his theses, see Worrall (1988) (to which Laudan replies in his (1989); Worrall's final rejoinder is his (1989)), Doppelt (1986) and (1990) (Laudan's reply is in his (1990b)) and Siegel (1990). For further comments on normative naturalism in connection to the broader theme of methodological incommensurability see Sankey (1996b) and (2000).

[257] See above, pp. 132–136.

successive theories can preserve their reference to a common domain of entities depends, in a way, on metaphysical issues. For it is not simply a matter of the way in which reference is determined, rather, it is a matter of the ontological status of the entities referred to. The realist holds that the entities to which a theory refers exist independently of the theory, and that the world studied by the natural sciences constitutes an objective reality that exists independently of human thought. But such assumptions could be rejected by those anti-realist philosophers for whom the objects of the referential process and the world studied by science largely depends on human thinking. Some anti-realist philosophers hold that the world and the objects belonging to it are constituted, in the whole or in some parts, by our own theories, concepts and language. They could deny that the terms employed by conceptually varying theories refer to the same objects, since those theories constitute their very domain of reference. In short: the problem whether successive theories actually refer to the same entities referred to by previous ones gives rise to metaphysical questions that create a divide, within the philosophy of science, between advocates of realism and advocates of anti-realism.

The ontological status of the objects referred to by theories is particularly relevant for the incommensurability thesis, since both Feyerabend and Kuhn tended to head in the direction of an anti-realist metaphysics with at least some traces of idealism. While Feyerabend defended scientific realism against the instrumentalist view of scientific theories, some of his occasional remarks do suggest that he might have advocated an actually anti-realist attitude about the relationship between theory and reality:

> [...] a certain form of realism, besides being too narrow, is also in contrast with the actual scientific practice. [...] It seems that the greater part of contemporary professional realists, and among them is, of course, the stern pope of critical rationalists, Karl Popper, intend realism [as a presupposition for scientific knowledge]. They are dogmatists. [...] we have to realize that incommensurable theories present us with different worlds and the transition from one theory to another determines the transformation of one world into another.[258]

Kuhn's case, in this respect, is particularly interesting, since a critical attitude towards realism and a neo-Kantian view of the relation between theory and reality always characterized his works.[259] In *The Structure of Scientific Revolutions* Kuhn appealed to the image of "world change" to describe the transformation involved in

---

[258] Feyerabend (1978b), pp. 201–202. See also Feyerabend (1962a), p. 29, and (1978c), p. 70, where he says that the world changes with theories, and rejects the assumption of "an objective world that remains unaffected by our epistemic activities". Feyerabend favourably welcomed Paul Hoyningen-Huene's neo-Kantian interpretation of Kuhn's philosophy: see Feyerabend (1989a), p. 405, n. 26, and (1994). For further discussions of this aspect, see Devitt (1984, 1991), ch. 9, and Sankey (1994a), ch. 6.

[259] For some of Kuhn's critical remarks on realism see, for instance, Kuhn (1970c), 205–207, (1979b), pp. 206–207, (1991a), pp. 95–96, and (1993a), pp. 243–245. The neo-Kantian tendency of Kuhn's thought is evident, for instance, in his (1979b), pp. 206–207, (1991a), p. 104, and (1993a), p. 245. See also Hoyningen-Huene (1989a/1993), Sankey (1997b), Part II, Bird (2000), ch. 6, Gattei (2000b) and Andersen (2001), pp. 60–64.

paradigm-shifts. On the grounds of past documentation read through the spectacles of most recent historiography one is tempted, Kuhn observed, "to exclaim that when paradigms change, the world itself changes with them", since it is "as if the professional community had been suddenly transported to another planet where familiar objects are seen in a different light and are joined by unfamiliar ones as well".[260] In an analogous way, Kuhn spoke of new entities come into being through paradigm-shifts, as well as of scientists who, supporting different paradigms, inhabit different worlds.[261] For example, wrote Kuhn, "Pendulums were brought into existence by something very like a paradigm-induced gestalt switch",[262] and "Lavoisier [...] saw oxygen where Priestley had seen dephlogisticated air and where others had seen nothing at all".[263] At some point, Kuhn described the idea that "proponents of competing paradigms practice their trades in different worlds"[264] as the "most fundamental aspect of the incommensurability of competing paradigms".[265] Such remarks suggest that underlying Kuhn's reflections on incommensurability is a somewhat idealist conception of the relationship between theory and reality.[266]

However, it is important to notice that the sort of idealism we find in Kuhn is not an extreme idealism, according to which the world is entirely a product of the human mind. In several respects, Kuhn's position allows room for and requires the existence of a reality independent of the mind.[267] In order to understand it, we have to distinguish between the reality from the world of phenomena that we access by means of a theory. Reality is one of these worlds, not the single possible world-in-itself. Thus understood, it preserves some subjective elements, but that does not turn it into an arbitrary construction: the world of phenomena is a particular "reconstruction" of the world-in-itself and retains some of its objective features.[268]

---

[260] Kuhn (1962a), p. 111.

[261] "[...] paradigm changes do cause scientists to see the world of their research-engagement differently. In so far as their only recourse to that world is through what they see and do, we may want to say that after a revolutions scientists are responding to a different world" (Kuhn (1962a), p. 111).

[262] Kuhn (1962a), p. 120.

[263] Kuhn (1962a), p. 118.

[264] Kuhn (1962a), p. 150.

[265] Kuhn (1962a), p. 150.

[266] The first author to notice and deal with Kuhn's idealism is Israel Scheffler: see his (1967), pp. 18–19. See also his (1972) and (1996). See also Hoyningen-Huene (1989a/1993), pp. 267–271, and (1989b), Sankey (1997a), chs. 2–4, and Ghins (1998).

[267] See Brown (1983) and Devitt (1984, 1991), p. 156. The form of idealism we find in Kuhn is a Kantian, or "constructive" idealism: for a discussion of these issues, see Hoyningen-Huene (1989a/1993), pp. 267–271, and Sankey (1997b).

[268] "Underlying all these processes of differentiation and change, there must, of course, be something permanent, fixed, and stable. But, like Kant's *Ding an sich*, it is ineffable, undescribable, undiscussible. Located outside of space and time, this Kantian source of stability is the whole from which have been fabricated both creatures and their niches, both the 'internal' and the 'external' worlds. Experience and description are possible only with the describer and the described separated, and the lexical structure which marks that separation can do so in different ways, each resulting in a different, though never wholly different, form of life. Some ways are better suited to some purposes, some to others. But none is to be

The web of similarity and dissimilarity relations on which the phenomenal world – and, with it, the language by which we refer to it – is founded is a product of history, of the evolution of a certain language and of the particular way to grasp reality inherent in that language. Such a process has no guide, even if it has its point of reference in the community that employs that language. The web of similarity and dissimilarity relations that contributes to the constitution of our image of the world is a property of community, not of individuals: as a consequence, the idealistic element of Kuhn's understanding of reality has a social, not individual character.[269]

While some authors regard Kuhn's image of "world change" as a mere metaphor,[270] others take it more seriously. As I have already mentioned, Paul Hoyningen-Huene interprets such an image in neo-Kantian terms.[271] In his *Reconstructing Scientific Revolutions* (1989) Hoyningen-Huene claims that Kuhn's metaphysical position can actually be seen as a dynamic Kantian position, based on the distinction between the unknowable world-in-itself and the phenomenal world, jointly constituted by the material contribution of the world-in-itself and the conceptual contribution of the knowing human being.[272] Kuhn differs from Kant in allowing that the human

---

accepted as true or rejected as false; none gives privileged access to a real, as against an invented, world. The way of being-in-the-world which a lexicon provides are not candidates for true/false" (Kuhn (1991a), p. 104).

[269] See also above, ch. 2, n. 108. Any network of relations is affected by objective elements: the world-in-itself offers some resistance to our attempts to frame it with our conceptual nets (it makes significant anomalies possible and contributes to the replacement of an old network of relations with a new one). Therefore, it is impossible to separate, within such a network, the objective from the subjective elements, the individual from the reality in-itself. "Can a world that alters with time and from one community to the next correspond to what is generally referred to as 'the real world'? I do not see how its right to that title can be denied. It provides the environment, the stage, for all individual and social life. On such life it places rigid constraints; continued existence depends on adaptation to them; and in the modern world scientific activity has become a primary tool for adaptation. What more can reasonably be asked of a real world?" (Kuhn (1991a), p. 102). See also Hoyningen-Huene (1989a/1993), pp. 262–264 and 267–271, and (1989b), together with Gattei (2000b), pp. 329–332 and 339–344.

[270] Howard Sankey understands it as a metaphor that commits Kuhn to a substantially metaphysical position: see his (1994a), pp. 152–153.

[271] The hypothesis of an interpretation of Kuhn's thought along neo-Kantian lines had already been advanced by Gerd Buchdahl (in his (1969), p. 511, n. 1), when he says that Kant's doctrine that the order of nature is a function of the mind's constructive activity "somewhat cryptically anticipates" Kuhn's views. The same idea is also expressed by Michael Devitt in his (1984, 1991), ch. 9.

[272] In his detailed analysis of Kuhnian metaphysics, Paul Hoyningen-Huene inquires into the relationship between Kant and Kuhn, drawing an interesting parallel between them: see Hoyningen-Huene (1989a/1993), pp. 31–36. Hoyningen-Huene's book (particularly chs. 2–3) advances a new interpretation of Kuhn's philosophy of science, presenting it within a Kantian anti-realistic framework. On the basis of Kuhn's own description of the process of acquisition of empirical concepts provided by Kuhn in his (1974c), Hoyningen-Huene distinguishes the world-in-itself, unknowable and constant through time, and the phenomenal world, a world of appearances constituted by both (external) objective elements and (internal)

conceptual contribution varies with the varying of theories.[273] Due to such conceptual variation, the world that changes in the transition from one theory to another is the scientists' phenomenal world: the world-in-itself is not affected by such change. On the basis of this interpretation, incommensurable theories can be compared in various ways, also appealing to common references.[274] Such references, however, must not be interpreted in a realist sense, as references to an shared domain of objects independent of the mind. For while incommensurable theories can refer to the same objects, these are not mind-independent: such objects belong, so to say, to the intersection of the phenomenal worlds of different alternative theories (whose object domains can partially overlap), that is constituted both by external contributions, from the world-in-itself, and by internal contributions, from individual human subjects.[275]

An alternative interpretation of the "world change" image has been advanced by Ian Hacking, who reads it as an expression of nominalism.[276] He sees Kuhn as a "revolutionary nominalist" for whom the world does not contain underlying natural kinds: the only existing kinds are those resulting from the imposition of classificatory schemes to the world.[277] The individual entities that constitute the world do not change. What changes, in the transition from one theory to another, is rather the system of kinds in which theories classify individual objects.[278] Hacking's nominalistic reading of Kuhn allows for incommensurability in the sense of untranslatability between

---

subjective ones, intrinsic to the knowing subject. When individuals become members of the scientific community they acquire its conceptual network, that is, they enter its particular phenomenal world. During a scientific revolution there is the transition from one paradigm to another, the taxonomy of the community changes and with it changes also the phenomenal world the community refer to. Incommensurability would be a relationship between different theories that describe different phenomenal worlds, not the single, objective real world: thus, incommensurability would be the very basis of Kuhn's idealism.

273 "I am a Kantian with moveable categories": Kuhn (I-1997a), p. 264; see also Kuhn (I-1997b), p. 114.

274 See Hoyningen-Huene (1989a/1993), pp. 218–222.

275 See also Irzik, Grünberg (1998), and Irzik (2000), pp. 637–638. For a critical assessment of Hoyningen-Huene's interpretation, see Sankey (1995b). For a discussion of the possibility that rival theories can share some referents within Hoyningen-Huene's neo-Kantian reading of Kuhn, see Sankey (1997c), pp. 440–441.

276 For a development of Hacking's view, drawing a connection between Kuhn and the empiricist legacy, specifically between his thesis of incommensurability (in particular in its later taxonomic form), and Bastiaan van Fraassen's constructive empiricism, see Bird (2003). Quite interestingly, Bird argues that Kuhn and van Fraassen do not differ as much as might be thought as regards the theory-ladenness of observation. On the possible links between Kuhn's and van Fraassen's philosophies, see also the following remark buy Kuhn: "In philosophy of science, we disagree, I think, about only one thing: the universal translatability of whatever may pass for observation statements. That single disagreement leaves us with almost totally different problematics" (Kuhn (1993b), p. 5).

277 See Hacking (1983), p. 109.

278 See Hacking (1993), p. 306. Kuhn appreciated Hacking's interpretation, but rejected the nominalist stance he is ascribed. Among other things, Kuhn objected to Hacking that what varies with classificatory schemes is not simply the kinds to which objects belong, but the objects themselves: see Kuhn (1993a), pp. 229–230.

kind terms belonging to different theories.[279] It seems to be consistent with the first reaction to the semantic version of the incommensurability thesis, analysed above: indeed, it allows that objects differently classified by theories do not themselves undergo a change in the transition from one theory to another, and this would allow for theories to be confronted thanks to the shared reference to the same entities.

These alternative and, in a way, complementary versions of Kuhn's thought show how the incommensurability thesis is not restricted to epistemic issues, but raises questions whose nature is eminently metaphysical as well. And it is not only the semantic version of incommensurability that paves the way to metaphysical problems related to the realist framework of the nature of scientific progress. For it seems that the major advocates of incommensurability defend metaphysical positions in striking contrast with the basic assumptions of a realist philosophy of science.

Such a metaphysical disagreement involves another important issue that has arisen out of the debate on incommensurability. Given Kuhn's and Feyerabend's anti-realistic tendencies, it is sensible to suppose that they might have based some aspects of incommensurability on anti-realistic assumptions. This indicates that the metaphysical attitude towards reference and the nature of the world studied by science is itself one of the issues in question among the different approaches to the incommensurability thesis. This has two important consequences. First, by assuming that objects referred to by theories are ontologically dependent on or independent of human knowledge, both realists and anti-realists beg the question.[280] Secondly, given the role played by background metaphysical assumptions, the controversy over incommensurability does not constitute simply a dispute between rival approaches within the philosophy of science, but reflects a deeper confrontation between realist and anti-realist metaphysical stances.

Feyerabend and the Return to Ontological Issues

In a 1987 rejoinder to Putnam[281] Feyerabend wrote that incommensurability, as he understood it,

> is a rare event. It occurs only when the conditions for meaningfulness for the descriptive terms of one language (theory, point of view) do not permit the use of the descriptive terms of another language (theory, point of view); mere difference of meanings does not yet lead to incommensurability in my sense. Secondly, incommensurable languages (theories, points of view) are not completely disconnected – there exists a subtle and interesting relation between their conditions of meaningfulness.[282]

---

[279] See Hacking (1993), pp. 294–295.

[280] The possibility of a *petition principii* in discussing incommensurability among realists and anti-realists has led some to advance the hypothesis that such a discussion could be affected by a meta-level of incommensurability: see Hoyningen-Huene, Oberheim, Andersen (1996) and Oberheim, Hoyningen-Huene (1997). For a critical analysis of this position, see Sankey (1997c) and Devitt (2001); see also Devitt (1979).

[281] Putnam (1981).

[282] Feyerabend (1987b), p. 81.

If there were no connections between the concepts of incommensurable theories, then the relationships between concepts could not provide sources for rivalry between them. Rather, Feyerabend thought of an incompatibility relation between the principles that define the concepts of incommensurable theories.

The reason why concepts of one theory cannot be defined within another theory, incommensurable with it, is that the principles required to define them are rejected by the latter. Such basic principles constitute the conditions for the formation of the concepts of a theory, or in order that they have a meaning. And since certain principles are rejected by one theory, it is impossible to form, within it, any concepts based on those very principles.[283]

In order to see how the principles of such theories can be incompatible, we need to highlight the fact that incommensurable theories differ at the level of their ontology.[284] For two incommensurable theories can compete at a deeper level: they have different ontological commitments and actually postulate the existence of diametrically distinct kind of objects in the same domain.[285] Feyerabend repeatedly highlighted that incommensurability is not simply a matter of difference in meaning among theories: "the mere *difference* of concepts does not suffice to make theories incommensurable in my sense. The situation must be rigged in such a way that the conditions of concept formation in one theory forbid the formation of the basic concepts of the other".[286]

One example of the broader (i.e. not only semantic) scope of a theory is the Aristotelian law according to which continuous movements require a continuous causation process.[287] The ontological import of such a principle has two facets: on the one hand, it implies the postulation of the existence of entities of a certain kind; on the other, it describes the behaviour of these entities. Therefore, the law according to which motion requires a cause leads to the postulation of a force, impetus, that sustains the movement of projectiles. Postulating the existence of such an entity means acknowledging the existence of a force that acts on movements that, in the context of Newtonian physics, are not subject to any force. Since inertial movement, according to Newtonian physics, is free from causal influences, the assumption that leads to the postulation of impetus is incompatible with Newton's physics.

The nature of the relation between theoretical principles gets further elaborated by Feyerabend in his discussion of the kind of theoretical change that leads to incommensurability. An objection raised against the idea that any meaning is

---

283 See Feyerabend (1978c), p. 70, and (1978b), p. 202.

284 In arguing that incommensurable theories do not share any statements, Feyerabend underlines that "we very often discover that entities we thought existed did, in fact, not exist. Realizing this, we must *eliminate* and *replace* the terms designating these entities from our factual descriptions" (Feyerabend (1965a), p. 170). See also Feyerabend (1975), pp. 274–276.

285 See Sankey (1994a), pp. 139–152.

286 Feyerabend (1978c), p. 68, n. 18; see also his (1975), p. 269, and (1965b), section 2.

287 Another one is the Newtonian principle according to which "some very fundamental properties of physical objects, such as shapes, masses, volumes, time intervals and so on [...] *inhere* in objects and change only as the result of a direct physical interference" (Feyerabend (1975), p. 275).

dependent on the theoretical context in which a term is employed was the consequence that any alteration of theory would involve a change in meaning. To answer this objection, Feyerabend specified the extension of theoretical change required for incommensurability.

He reserved incommensurability for deep theoretical changes that involve an ontological change and a change of the conceptual framework. In the following passage Feyerabend distinguished the changes that do not affect meaning from those that lead to incommensurability:

> a diagnosis of *stability of meaning* involves two elements. First, reference is made to rules according to which objects or events are collected into classes. We may say that such rules determine concepts of kinds of objects. Secondly, it is found that the changes brought about by a new point of view occur *within* the extension of these classes and, therefore, leave the concepts unchanged. Conversely, we shall diagnose a *change of meaning* either if a new theory entails that all concepts of the preceding theory have zero extension or if it introduces rules which cannot be interpreted as attributing specific properties to objects within already existing classes, but which change the system of classes itself.[288]

In other words, the alterations of a theory within a stable system of concepts does not induce changes of meaning. But a change of theories in which the concepts employed are not able to refer, or that alters the system of classes, involves a change of meaning. More generally, if the kind of entities to which a theory is committed are rejected by successive theories, then we have a change of meaning involving incommensurability.[289]

It is important to notice that the passage I have just quoted implies a more extreme view of the conflict between the basic principles on the basis of which the concepts of theories are defined. For it suggests that in the transition from one

---

[288] Feyerabend (1965b), p. 268. Although this passage does not directly correlate the change of meaning with incommensurability, when the distinction is applied to the concepts of space and time of classical mechanics and general relativity, Feyerabend employs it to diagnose incommensurability (see *ibidem*, pp. 269–271).

[289] It is interesting to notice, also in the case of Feyerabend, an appeal to the Kantian model about the distinction between phenomenon and thing-in-itself. The advocates of theory-ladenness of observation fundamentally repeat that we cannot know reality-in-itself since all of our observations depend on the way in which we look at the world. Yet while for Kant the *a priori* forms of sensibility and intellect were the same for all humans as rational beings, for Feyerabend the conceptual grid is relative to the cultural horizon of the knowing subject and changes depending on whether it is the case of an Achaean warrior, a Hopi indian or a twentieth-century scientist. Although Feyerabend is resistant to any label, since his possible contact points with the great philosophers of the past are set in the context of a very personal interpretation of doctrines that span from Protagoras to Carnap, it is possible to notice how in Feyerabend (and Kuhn) there is the Kantian-flavoured tendency to reduce the object of knowledge to what appears to us, together with the elimination of the thing-in-itself. This combination leads to an idealism opposed to Hegel's, since the reality-in-itself is not replaced by free and autonomous creation of a rational thought (or *logos*), but rather, by the creation of a thought that associates (and sometimes prefixes) rationality with intuition, emotion and instinct.

conceptual system to another there is also a change of references: indeed, if all the concepts employed have an empty extension, and no term of the new theory refers to any member of the old classes, then no terms of any theory refer to the same things. Therefore, besides the mere change of meaning, there seems to be a radical discontinuity between such theories.

In 1989 Feyerabend further specified his point of view, gradating his original uncompromising position and reconstructing the genesis of his own theory. He said he simply wanted to criticize some theses about explanation and reduction, highlighting a feature of scientific change, incommensurability, that those theses were unable to explain and account for. And he added:

> As far as I am concerned *incommensurability is no difficulty for the sciences or, for that matter, for anyone else* – it is a difficulty only for some very naïve philosophical theories and, as these theories were regarded as essential ingredients of a certain type of 'rationality', for this type as well. But it was blown up into a profound feature of all 'creative' thought and it was soon used to provide equally profound reasons for the lack of understanding between cultures and scientific schools. That, to me, is just nonsense. [...] The phenomenon I called incommensurability accounts only for a small part of these misunderstandings and I regard it not only as naïve but as *downright criminal* to blow it up into One Big Monster that is responsible for all the troubles in science and the world at large. Of course, incommensurability is a boon for philosophers and sociologists – and by this I now mean people who call themselves 'philosophers' or 'sociologists' – who like big words, simple concepts and trite explanations and who love giving the impression they understand the deep reasons behind troublesome affairs. *The matter is criminal because it emphasizes difficulties, dwells on them, makes theories about them instead of trying to get out of them.* Different cultures now seem to be doomed to talk past each other just as Einstein seemed doomed forever to misunderstand the wonderful discoveries of the quantum theory. Let us agree that Plato is different from Aristotle but let us not forget that Aristotle spent twenty years at the Academy and certainly learned how to talk the Platonic lingo. Let us also remember that Bohr and Einstein liked each other, often talked to each other and that Einstein *accepted* Bohr's way of defusing his counter-examples. No 'incommensurability' here! Of course, he still had a different metaphysics, but that is not a matter of incommensurability except for the most doctrinaire rationalist.[290]

[290] Feyerabend (1989b), pp. 154–155, emphasis added (apart from the words "downright criminal" and "accepted", which were italicized by Feyerabend).

Chapter 4

Kuhn's "Linguistic Turn"

A man conceived the notion of a vast
Enterprise: to encode the universe
In one book, and with tireless energy
Amassing the dense and lofty pages, he
Emended and recited the last verse.
Grateful to Fortune for all this, he soon
Lifted his eyes and, seeing in the air
A burnished disc, realized with despair
That he had left out just one thing: the moon.
This tale that I've recounted, though a fiction,
May well stand as a figure for the curse
On all of us who suffer the addiction
Of turning our existence into words.
They always miss the essential.

Jorge Luis Borges

As we saw at the end of the previous chapter, in his later years Feyerabend completely abandoned the radical position he had been strenuously advocating in the preceding decades. He clearly highlighted how incommensurability does not represent – for science as well as for philosophy – an insurmountable problem but, at most, a difficulty. For sure, it is a difficulty that can be quite serious at times, but not overwhelming; it demands effort and engagement, but in so doing it welcomes and encourages critical discussion rather than banning it.

In Feyerabend's own words: "[incommensurability] emphasizes difficulties, dwells on them, makes theories about them instead of trying to get out of them". It is remarkable that these words come from Feyerabend, the keenest advocate of the incommensurability thesis, the one who more than any other, in the 1960s and 1970s, had made incommensurability an epistemological and philosophical flag. In so doing, Feyerabend clearly and unmistakeably denounced the very nature of incommensurability, that reveals it to be a *solution*, rather than a *problem* – a too easy way out of a problem, that is: rather than recognizing a difficulty and engaging in the effort to solve it, by appealing to confrontation and the critical discussion of different positions, too often we opt for branding it as an insurmountable problem, and therefore avoid any critical dialogue, moving from the assumption that different positions are simply incompatible and that a rational choice is impossible.

As the years went by, also Kuhn's conception of incommensurability became increasingly narrow, both because of the criticisms he received and the internal modifications of his views. However, the revision of the views he previously held does not lead to the re-evaluation Feyerabend attained. For, contrary to his friend, Kuhn always left incommensurability at the heart of his epistemological reflections and – by trying to specify it – proposed to develop the original thesis along new directions.

On the one hand, such focusing allowed Kuhn to concentrate on what he regards as the key feature of the incommensurability thesis, namely, its semantic-linguistic aspect (that, as we have seen, was under attack by several philosophers of analytic extraction, among which are Israel Scheffler, Willard Quine, Saul Kripke, Hilary Putnam and Donald Davidson). On the other hand, however, such focusing involved a considerable impoverishment of the very notion of incommensurability, which lost two of its most theoretically fruitful and philosophically devastating elements. In fact, from the tripartite incommensurability (in the first phase of Kuhn's philosophical development, in the 1960s), he later came to a more faded notion of it (in the 1970s) and eventually to a decidedly monolithic version of it, to which Kuhn wholly dedicated himself, without being able to achieve a definite position, until the last years of his life.

Beginning in the early 1980s, Kuhn increasingly emphasized the role played by taxonomic lexicons and structured vocabularies in the characterization of scientific revolutions and incommensurability.[1] This later phase of Kuhn's thought – often referred to as "linguistic turn"[2] for the novel attention it pays to the consideration of almost exclusively the linguistic and semantic aspects of scientific theories and conceptual dynamics – can be better understood against the background of some

---

1 The reference texts for this "third phase" of Kuhn's philosophical development are his (1981), (1983a), (1989a), (1990), (1991a), (1991b), (1992), (1993a), (1993b) and (1999), besides (U-1980), (U-1982-), (U-1984), (U-1987), (U-1990a) and (U-1990b). Some traces of the later trend of Kuhn's concerns are to be found already in his (1976b) and (1979b). In the former, in particular, he adopts the conception of scientific theory introduced by Joseph Sneed and Wolfgang Stegmüller (see Sneed (1971, 1979) and (1976), Stegmüller (1973/1976), (1975), (1976) and also (1979a) and (1979b)), and characterizes a theory in terms of a theoretical structure consisting in a set of different applications that share the same group of laws, and of a set of constraints that bond the various applications. This enables Kuhn to link his own notion of exemplar with a non-Carnapian conception of scientific theories: learning a theory implies the ability to employ and handle its various exemplary applications. Moreover, such a knowledge becomes at the same time a tool to control the ways in which the terms of the theory refer to nature. Therefore, the two processes of learning a language of a theory and of learning something about nature go always together. Kuhn's writings in the 1980s and 1990s largely constitute elaborations of this early insight: see also Gattei (2000b), pp. 309–311.

2 The linguistic aspect dominates Kuhn's last, unfinished book. It is to be noted, however, that Kuhn's emphasis on this aspect has nothing to do with Richard Rorty's own "linguistic turn" (see Rorty (ed.) (1967, 1992)): Rorty describes his own position as "linguistic" since the distinctions and other structural relations embedded in a given set of concepts are acquired through learning and the usage of the terms that signify them. Kuhn's position, on the contrary, seems to be closer to the interest for language typical, at the beginning of the twentieth century, of the philosophical tradition that would have later become analytic philosophy.

neo-Kantian themes.[3] Having become convinced that language heavily influences our thinking, our understanding and our very experience of the world, Kuhn thought that structured lexicons are constitutive of the phenomenal worlds and of our possible experiences of them. Kuhn's latest position can therefore be characterized as a sort of post-Darwinian linguistic neo-Kantianism.[4]

From Paradigms to Lexicons

Kuhn's 1969 papers[5] and the following articles he published in the 1970s explained incommensurability in terms of (a sort of) mutual untranslatability of two theories as a consequence of the lack of some (relatively) neutral observation language. Untranslatability is the primary notion required to understand incommensurability: "if two theories are incommensurable, they must be stated in mutually untranslatable languages".[6]

Analysis of the concept of translation

However, such an explication of the incommensurability thesis seems to be contradicting Kuhn's own 1969 papers. Indeed, in these papers he repeatedly speaks of the choice of a theory in terms of a situation involving the translation from one theory's language to another's, and vice versa, despite these translations proving to be problematic.[7] The seeming contradiction disappears by observing that in the 1980s Kuhn employs a different notion of translation. Indeed, in 1969 he writes:

> Though one must know two languages in order to translate at all, and though translation can then always be managed up to a point, it can present grave difficulties to even the most adept bilingual. He must find the best available compromises between incompatible objectives. Nuances must be preserved but not at the price of sentences so long that communication breaks down. Literalness is desirable but not if it demands introducing too many foreign words which must be separately discussed in a glossary or appendix.[8]

In 1982 Kuhn specified how this (usual, everyday) sense of translation, that applies also to literary texts, divides itself in two heterogeneous moments: a moment of translation strictly understood, and an interpretative moment.[9]

In the course of translation in the strict, technical sense, single words or groups of words in the original language are systematically replaced by single words or groups

---

3 See especially Hoyningen-Huene (1989a/1993), Irzik, Grünberg (1998) and Irzik (2003).

4 See Kuhn (1991a), p. 104; see also Kuhn (1993a), pp. 331–332.

5 Kuhn (1970b), (1970c) and (1974c), all written in 1969, a turning-point in Kuhn's intellectual development.

6 Kuhn (1983a), p. 34. See also Kuhn (1981), p. 8, (1983b), pp. 713 and 715, (1984), p. 363, (1989a), pp. 60–61, 74–75 and 76–77, and (1990), pp. 299, 308 and 315.

7 See Kuhn (1970b), pp. 266–270, (1970c), pp. 202–203, and (1977c), pp. 338–339.

8 Kuhn (1970b), p. 266.

9 See Kuhn (1983a), pp. 37–40, (1989a), pp. 59–63, and (1990), pp. 299–301.

of words in the destination language. During this process neither the original language nor the destination one changes, and both texts will bear the same meaning. This kind of translation does not require any "compromises between incompatible objectives": either the translation between two languages is possible, or it is impossible.

The second phase (or aspect, in Kuhn's words) of a translation is the interpretive one. It typically occurs in the case in which historians or ethnographers deal with texts that are partially or wholly unintelligible. In such cases, the process of understanding requires the learning of a more or less substantial part of the language in which the text we want to understand is expressed. In order to achieve the desired stage of understanding, that is, in order to learn how to think in that language, it is not always sufficient to be able to identify the referents of the concepts that were previously unintelligible. It is also necessary to understand how those very referents form the extension of a single concept, and to this end it may turn out to be necessary to know something also of the series of assumptions about the world on which the unintelligible texts or expressions are based.[10]

Once unintelligible words or expressions have become intelligible, it still remains to be seen whether the text just made intelligible is translatable into another language – in the strict, technical sense of translation. Despite the fact that the text may present more or less serious difficulties for such a translation, it is still translatable in the usual, everyday sense. But the latter translation requires compromises akin to those described in the passage quoted above, and also some changes in the destination language, either through the addition of new concepts, or maybe the more or less subtle alteration of the antecedently available concepts.

In this phase of his philosophical development Kuhn regards two theories as incommensurable *only* in the case in which they are formulated in languages that are not translatable in the strict sense. As to the reason why there may exist non-intertranslatable languages in this sense, Kuhn offers a (partial) explanation by introducing the notion of lexicon.

The notion of lexicon

In the 1980s Kuhn employed the concept of lexicon and its structure as a new tool for explicating incommensurability.[11] The position that Kuhn adopted in his later writings is the following: the speaker applies to nature some concepts by determining, on the basis of some criteria, whether a given concept refers to a given object or to a given situation. The criteria employed by every single speaker are not generally universal, that is, are not generally held valid by the entire linguistic community to which the speaker belongs. In principle, two speakers can employ totally disjoint sets of criteria without differing, for this reason, as to the use they both make of concepts. In order to employ a concept unequivocally, the relations between it and the other concepts, which are determined by the speaker's set of criteria, must be identical with the analogous relations holding for the other members of the community (even if they might do so in virtue of other criteria). As a whole, the relations between the

[10] See Hesse (1983), p. 707, and Kuhn (1983a), pp. 40–43, and (1983b), p. 712.

[11] See Kuhn (1983a), pp. 51–53, (1983b), pp. 713–714, (1989a) and (1990).

empirical concepts of a given system, or lexicon, are called the "structure of the lexicon".

The changes in the criteria for the employment of a concept may also propagate within the whole of the linguistic community without changing the structure of the lexicon. According to Kuhn, this is what happens every day during the periods of normal science.[12] By contrast, during revolutions the structure of the lexicon changes and such a change characterizes revolutionary periods.[13] In the latter periods a key role is played by change in the immediate similarity relations that constitute the concepts and the world.[14] Such a change involves a change in the extension of concepts, since the fundamental taxonomies change so as to put the objects previously belonging to the same extension into the complementary extension, and vice versa. It may also lead to a change in the descriptive vocabulary.[15] At the same time, also the knowledge of nature involved in the lexicon changes: we therefore have the "Violation or distortion of a previously unproblematic scientific language", that constitutes "the touchstone for revolutionary change".[16]

Here, then, is the explanation provided by Kuhn of the reason why before and after a revolution languages are not inter-translatable in the strict sense.[17] A necessary condition for translation in the strict sense is a systematic mapping of the concept of the original language into the concepts of the destination languages, so that every concept refers to a concept with the very same extension and meaning. If the lexicons of the original and destination languages have different structures, and if the structural difference is such that the extensions of more or less corresponding concepts are not identical, the conditions required for translation in the strict sense do not hold. Incommensurability, then, holds only when the structure of the world, as it appears through the mediation of the lexical structure, is different from the previous one.[18]

The world-constitutive function of categories

As a consequence of the "linguistic turn", several significant changes affect Kuhn's philosophy, each of which is connected to the notion of taxonomic change. First of all, the terms "paradigm" and "disciplinary matrix" disappear: Kuhn preferred to employ simply the term "theory".[19] It is not simply a terminological change, since

---

12 See Kuhn (1981), p. 19.

13 See Kuhn (1981), pp. 8 and 19–21, (1983a), pp. 51–53, (1983b), pp. 713–714, (1984), p. 364, (1989a), pp. 76–77 and 83–86, and (1990), pp. 313 and 314–315.

14 See Kuhn (1979b), pp. 203–205, (1983a), pp. 48–49 and 51–53, and (1989a), pp. 85–86.

15 See Kuhn (1981), pp. 25–28, (1983a), pp. 48–49, and (1984), p. 363.

16 Kuhn (1981), p. 32; "the central characteristic of scientific revolutions is that they alter the knowledge of nature that is intrinsic to the language itself and that is thus prior to anything quite describable as description or generalization, scientific or everyday" (*ibidem*). See also Kuhn (1983a), pp. 52–53.

17 Kuhn (1983a), pp. 52–53.

18 See Kuhn (1983a), pp. 38–40, 42–43, 48–49 and 51–53.

19 After 1969 Kuhn employs the term "paradigm" only in the restricted sense of "exemplar" or "exemplary problem solution", preferring to speak instead of theories and of

while the notion of paradigm is too wide to allow for talking about the lexicon of a paradigm, it is perfectly meaningful to speak about the lexicon of a theory. Indeed, Kuhn now emphasizes the point that every scientific theory is a taxonomically ordered web of kind terms and kind concepts, some of which are antecedently available (in the sense in which Carl G. Hempel used this expression[20]) with respect to the theory in question.[21]

Secondly, kinds, to which kind terms refer, populate the world and at the same time divide it up into categories that establish mutual relationships that together make up the structure of the lexicon. The no-overlap principle, for example, according to which kind terms that express different species of the same kind cannot share common referents, is an important principle that governs the structure of every single lexicon.

Thirdly, lexicons are the prerequisite for the formulation of scientific problems and for their solutions, for the description of nature and its regularities. Revolutions can therefore be characterized as significant changes in the lexicons of scientific theories. With the varying of lexical categories vary also the criteria employed for the categorization and the way in which objects are distributed in the pre-existent categories. Since different lexicons allow for different descriptions and generalizations, scientific development is necessarily discontinuous.

---

choice among different, competing theories. The label "disciplinary matrix" is abandoned as well, and in his later years he would replace even "scientific revolutions" with "lexical change" or "change of taxonomy".

[20] See, for instance, Hempel (1966), pp. 74–75 and 88, but also the papers collected in his (1965a). Discussing the status of the terms used in a scientific theory, Hempel divides them into two classes, making a distinction between "theoretical terms", which are characteristic of the theory, and "pre-theoretical" or "antecedently available" terms: "while the internal principles of a theory are couched in its characteristic *theoretical terms* ('nucleus', 'orbital electron', 'energy level', 'electron jump') the test implications must be formulated in terms (such as 'hydrogen vapor', 'emission spectrum', 'wavelength associated with a spectral line') which are 'antecedently understood', as we might say, terms that have been introduced prior to the theory and can be used independently of it" (Hempel (1966), p. 75). See also Kuhn (1993a), p. 226.

[21] For Wittgenstein too language does not reflect the given form of the states of things, but is a moment that belongs to an inter-subjective experience and acquires meaning once in it a certain way to organize experience is codified: see Wittgenstein (1953), Part I, §65. According to him concepts have both an operative and a functional nature: in order to understand them we need to grasp them as ways of ordering things, as rules that shape a given linguistic game. To the Platonic image of the concept as a single identical idea in which all things participate, Wittgenstein substitutes the suggestion of an organization of experience following a set of consolidated practices and habits. To the Platonic invitation to abandon individual instances in order to think of the reason why all occurrences of a term designate the very same thing, Wittgenstein contrasts the suggestion to look at the interwoven web of similarities that guide and support our linguistic games (see Kant's image of the "light dove", mentioned above, in ch. 2, n. 8). Concepts are meanings that get constituted in the practice of linguistic games, as a sort of ideal moment that consolidates in the customary usage of words: outside the customary usage shown in the various linguistic games, there is simply no concept.

As a consequence – and this is the fourth significant change in Kuhn's philosophy as a consequence of the "linguistic turn" it underwent – the distinction between normal and revolutionary science now becomes a distinction between activities that require a change of the scientific lexicon and activities that do not require it. Revolutions involve, among other things, new discoveries, that cannot be described within the existing lexical network. Scientists are therefore compelled to adopt a new one. The mentalist description, in terms of *Gestalt* switches, of what happens through a revolution, therefore disappears.

In the fifth place, "incommensurability [...] becomes a sort of untranslatability, localized to one or another area in which two lexical taxonomies differ".[22] Claiming that two theories are incommensurable simply amounts to claiming that "there is no language, neutral or otherwise, into which both theories, conceived as sets of sentences, can be translated without residue or loss".[23] What gives rise to incommensurability and therefore prevents complete communication between theories is the lack of identity among lexical structures.[24] The notion of incommensurability, in the sense of lack of shared standards for assessing and appraising rival scientific theories, is no longer present in Kuhn's philosophy.[25]

Finally, structured lexicons play the function of constituting the world and the experience that was once played by paradigms. They become the relativized version of Kantian categories:

> Though it is a more articulated source of constitutive categories, my structured lexicon resembles Kant's a priori when the latter is taken in its second [that is, 'constitutive of the concept of the object of knowledge'], relativized sense. Both are constitutive of *possible experience* of the world, but neither dictates what that experience must be. Rather they are constitutive of the infinite range of possible experiences that might conceivably occur in the actual world to which they give access.[26]

And, even more explicitly:

> the position I'm developing is a sort of post-Darwinian Kantianism. Like the Kantian categories, the lexicon supplies preconditions of possible experiences. But lexical categories, unlike their Kantian forebears, can and do change, both with time and with the passage from one community to another.[27]

Therefore, the constitutive function of Kant's categories of the intellect is transferred, in Kuhn, to lexical taxonomies. Moreover, while for Kant the categories

---

22 Kuhn (1991a), p. 93.

23 Kuhn (1983a), p. 36.

24 These lexical structures could not be combined into a single taxonomy without violating the no-overlap principle. This is the *function* of the no-overlap principle: the combination of the two lexical structures into a single taxonomy, which would appear an easy way of escaping incommensurability, is ruled out by the no-overlap principle. On the other hand, the *reason* why Kuhn introduced it is the projectibility of kind terms: see below, n. 59.

25 See Hoyningen-Huene (1989a/1993), pp. 206–218, Irzik (2000) and Sankey (1993a).

26 Kuhn (1993a), p. 245.

27 Kuhn (1991a), p. 104.

of the intellect are fixed and universal, for Kuhn lexical categories change through history. This is the reason why, according to him, there is a variety of phenomenal worlds but only one, single noumenal world.[28]

The Linguistic Theory of Scientific Revolutions

In his later years Kuhn wished to clarify his own position from an entirely new point of view with respect to the past, developing an interpretation of the incommensurability thesis that comprises a linguistic theory of scientific revolutions (the theory of kinds), a study of the process of language learning and an analysis of the rationality of scientific development that results in an evolutionary theory of knowledge.[29]

Changes of taxonomy

In the original presentation of his model for the growth of scientific knowledge in *The Structure of Scientific Revolutions* Kuhn greatly emphasized the non-cumulative nature of revolutionary scientific change.[30] An aspect of such a change that assumes particular relevance in the later developments of Kuhn's thought is the change of taxonomy: scientific revolutions are characterized by changes in the taxonomic patterns through which theories classify objects belonging to their domain of application.

Kuhn often employed examples drawn from the history of science, such as astronomical categories like "planet" or "star", or chemical ones, like "compound" or "mixture". These examples indicate that in Kuhn's eyes scientific theories classify objects and phenomena to which they apply in a variety of categories. Such categories contain elements that theories group together on the basis of the proprieties or characteristic behaviours they share. Since theories typically classify their subject domains in a variety of different categories, such theoretical classifications require a taxonomic system comprising multiple categories. Kuhn advanced the hypothesis that scientific revolutions are characterized by the change of such taxonomical systems.[31]

He observed that the criteria defining taxonomical categories change through scientific revolutions. However, since these criteria may vary also during the periods

[28] Notice that for Kant there is a single phenomenal world and a single noumenal world, and that Kuhn does not say anything about the role played by space and time which, according to Kant, are the *a priori* forms of sensorial intuition, mediating between the logical forms of judgement and the chaotic variety of sensations. For a critical discussion of this point, see Irzik, Grünberg (1998).

[29] Howard Sankey labelled it "taxonomic incommensurability": see his (1993a) and (1998). For what follows, see Hacking (1993), Chen (1997), Sankey (1997a) and (1997d), and Irzik, Grünberg (1998).

[30] See Hoyningen-Huene (1989a/1993), Part III, Giordano (1997), pp. 73–90, Bird (2000), pp. 44–63, Gattei (2000b), pp. 316–332, and Andersen (2001), pp. 29–32.

[31] See, in particular, Kuhn (1981). See also Hempel (1965b), which might have influenced Kuhn's views.

of normal sciences, the characteristic features of revolutionary change must lie somewhere else:

> What characterizes revolutions is [...] change in several of the taxonomic categories prerequisite to scientific descriptions and generalizations. That change [...] is an adjustment not only of criteria relevant to categorization, but also of the way in which given objects and situations are distributed among preexisting categories.[32]

Revolutionary scientific change, then, is not restricted to changes in the theories' assertions about the members of shared categories – but it affects the very classification system responsible for the determination of the categories one belongs to, with consequent alterations both of the criteria of classification and of membership to categories.

One of Kuhn's favourite examples is the transition from Ptolemaic to Copernican astronomy: "Before it occurred, the sun and moon were planets, the earth was not. After it, the earth was a planet, like Mars and Jupiter; the sun was a star; and the moon was a new sort of body, a satellite".[33] What was previously classified in a certain way, gets now classified in a different one. But not only objects move from one category to another: sometimes categories themselves change, and a revolution may lead to the introduction of new categories and the deletion of old ones.[34]

The transition to Copernican astronomy very well illustrates what Kuhn was trying to convey. There is, in the first place, a fixed set of entities (in this case, the set of celestial bodies) that constitutes a common domain of objects, shared by different classification systems. Secondly, the change of taxonomy does not turn into an overall change of the taxonomic pattern, since many of the old categories are preserved in the new classification. Thirdly, the change of taxonomy involves

---

32 Kuhn (1981), p. 30. "Normal science, too, alters the way in which terms attach to nature. What characterizes revolutions is not, therefore, simply change in the way referents are determined, but change of a still more restricted sort. [...] roughly speaking, the distinctive character of revolutionary change in language is that it alters not only the criteria by which terms attach to nature but also, massively, the set of objects or situations to which those terms attach." (*ibidem*, pp. 29–30).

33 Kuhn (1981), p. 15. This example, however, may be seen also as not implying a change of taxonomy. For while the astronomers' beliefs in what, for instance, counts as planet do change, their criteria for classifying astronomical objects may well remain the same.

34 A good example, I think, is Kepler's redefinition of the astronomical terminology, with which he tried to make clear the new ways in which traditional astronomical quantities had to be understood: "My dear Fabricius", he wrote on 1 August 1607, "if I were to hand over astronomy *de novo*, in such a way that it was not necessary for me to speak with the ancient words, I would use other ones. I would speak of delay, arc, angle, circle, the designation of the elliptical arc, the measure of delay, the area of the circle" (quoted in Voelkel (2001), p. 205; compare this passage with Kuhn (1962a), p. 149). The necessary change in vocabulary became only one aspect of the difficulty Kepler knew he would face with the reception of his work, particularly the *Astronomia Nova* (1609). On the problem of the reception of revolutionary works in the history of science, with particular reference to Kepler, Galileo and Newton, see the extraordinary works of Guicciardini (1999), Voelkel (2001) and Bucciantini (2003).

the shifting of objects, or groups of objects,[35] from one category to another, as well as the introduction of new categories. Finally, such reclassification involves that entities previously regarded as dissimilar are regarded as belonging to the same category after a revolution.

This taxonomic change bears a series of important consequences on the semantic level. It may happen that important ontological changes, or the introduction of new categories, may involve the introduction of a new vocabulary, semantically different from the one previously in use. However, in several cases the old vocabulary is preserved the taxonomic change notwithstanding, and its terms are therefore subject to meaning change. While the first kind of change refers to the criteria by which a categorical term is applied, the second kind of change may affect the meaning of the term in question. In the case of shift of objects from one category to another, however, the terms that are preserved may undergo a variation in their extension.

Local holism and untranslatability

The semantic variance associated with change of taxonomy is at the roots of Kuhn's new conception of incommensurability. Instead of talking of a total breakdown of communication, Kuhn speaks of difficulties in achieving a proper translation between subsets of (usually inter-defined) terms within theories' particular languages. The difficulties in translation are due to what Kuhn calls a "local holism".[36]

---

[35] The development of Dalton's atomic theory "implied a new view of chemical combination with the result that the line separating the referents of the term 'mixture' and 'compound' shifted; alloys were compound before Dalton, mixtures after" (Kuhn (1970b), p. 269).

[36] A major source for Kuhn's theoretical context view of meaning was undoubtedly Ludwig Wittgenstein, whose views are characterized by a strong attitude towards a holistic view of language. Already in the *Tractatus Logico-Philosophicus* – which is considered as a programme for logical atomism – we find reference to Frege's context principle: names have meaning only in the context of a sentence (see Wittgenstein (1921), 3.3: "Only propositions have sense; only through the nexus of a proposition does a name have meaning"). Wittgenstein's subsequent work seems to be a progressive generalization of the context principle. The Fregean principle is quoted as such at the beginning of *Philosophical Investigation* (Wittgenstein (1953), §§22, 49 and 71) in order to show that words do not have denotation in isolation, but only when used within a language game, a social environment of speech and action: "to imagine a language means to imagine a form of life" (*ibidem*, §19). In a recent rejoinder to Preston (2004) Alexander Bird has argued that despite the fact that Kuhn might have acquired his theoretical context view of meaning from Wittgenstein's local holism about meaning applied to the particular case of theoretical terms, Ernest Nagel seems to have exerted a stronger influence on him: "when we recall Kuhn's relatively thin philosophical training, it is far more plausible to infer that he adopted this view because it was the then predominant view of theoretical meaning in the philosophy of science. For example the view is explicit in Ernest Nagel's hugely influential 1961 book *The Structure of Science*. Kuhn mentions Nagel as one of four friends who were most significant in helping him rewrite his draft of *The Structure of Scientific Revolutions*. (I surmise that the similarity of the titles of Nagel's and Kuhn's books is not mere coincidence)" (Bird (2004), p. 340). Furthermore, as Bird notices, Nagel was also on the advisory committee of the *International Encyclopedia of*

Indeed, unlike the gradual scientific change that characterizes normal science periods, revolutionary change of the taxonomic structure proceeds in a holistic manner. This is due to the shift of objects from one category to another, which is typical of revolutionary taxonomic change:

> Since such redistribution always involves more than one category and since those categories are interdefined, this sort of alteration is necessarily holistic. That holism [...] is rooted in the nature of language, for the criteria relevant to categorization are ipso facto the criteria that attach the names of those categories to the world. Language is a coinage with two faces, one looking outward to the world, the other inward to the world's reflection in the referential structure of language.[37]

Therefore, the holistic nature of taxonomic change is reflected, at the semantic level, through the inter-definition of terms that refer to taxonomic categories. For taxonomic kind terms are defined within an integrated conceptual structure in which a certain number of different concepts are semantically interconnected.

Given the holistic interconnection of terms, it may turn out to be impossible to translate the names of taxonomic categories defined within one theory by recurring to the terms of another. One of the examples Kuhn discusses is the debate between the chemistry of phlogiston and that of oxygen. While the great part of the language employed by the supporters of the phlogiston theory is still in use, "a small group of terms remains for which the modern chemical vocabulary offer no equivalent".[38] Terms like "phlogiston", "dephlogisticated air" and "principle" form a group of conceptually correlated terms that cannot be defined by appealing to the vocabulary of the oxygen theory: "they constitute an interrelated or interdefined set that must be acquired together, as a whole, before any of them can be used, applied to natural phenomena".[39]

Incommensurability is therefore a local phenomenon, the outcome of the difficulties of translation due to the holistic inter-definition of categorical terms, and restricted to small groups of subsets of terms belonging to alternative theories.[40]

Unified Science. At any rate, I think Wittgenstein played a fundamental role in shaping Kuhn's philosophical stance. Of course, he might very well have been influenced from more than one intellectual source, but the influence of Wittgenstein's thought unquestionably underlies the whole evolution of Kuhn's philosophy and becomes most evident and explicit in his later years, after the "linguistic turn" of 1980s–1990s.

37 Kuhn (1981), p. 30.

38 Kuhn (1983a), p. 42.

39 Kuhn (1983a), p. 44. See also James Conant (1957), Musgrave (1976b) and Carrier (2002a).

40 The incommensurability thesis, thus revisited, constitutes an improvement with respect to previous versions. For it makes it possible to reply to Dudley Shapere's objection according to which incommensurable theories cannot rival each other (see Shapere (1983), pp. 45 and 73), since between them there is a common semantic ground sufficient to allow their conflict (see Kuhn (1983a), pp. 35–37). And it also replies to the objections raised by Donald Davidson (1984) and Hilary Putnam (1981), according to which the incommensurability thesis legitimately presupposes the translatability of untranslatable terms: for given the distinction between a semantically stable and shared vocabulary, and a hard core of untranslatable terms,

The theory of kinds

In *The Structure of Scientific Revolutions* Kuhn illustrated the incommensurability thesis by recurring to an analogy drawn from *Gestalt* psychology: during scientific revolutions there takes place a "*Gestalt* switch" as a consequence of which scientists see the world in a different way, or even live in different worlds.[41] The notion of incommensurability implies that scientists face difficulties in appraising competing paradigms since there are no criteria or standards shared by their respective proponents.

Later, Kuhn modified his positions and abandoned the analogy with *Gestalt* psychology, and with it the perceptual interpretation it involved. He developed, in its stead, a metaphor based on language: during scientific revolutions scientists face difficulties of translation when they attempt a discussion of the concepts peculiar of another paradigm, as if they were dealing with another language. Incommensurability thus becomes a sort of untranslatability and is therefore limited to meaning variance.[42]

A further modification aims at retrenching the scope of revolutions. In the early 1980s Kuhn introduced the notion of "local incommensurability",[43] arguing that during a scientific revolution "Most of the terms common to the two theories function the same way in both; their meanings, whatever those may be, are preserved; their translation is simply homophonic. Only for a small group of (usually interdefined) terms and for sentences containing them do problems of translatability arise".[44] Incommensurability thus becomes untranslatability produced by the meaning variance of a small group of terms. More recently, moving forward along this line, Kuhn further limited the scope of incommensurability by introducing the theory of kinds. He explained:

> By now, however, the language metaphor seems to me far too inclusive. To the extent that I'm concerned with language and with meanings at all [...] it is with the meanings of a restricted class of terms. Roughly speaking, they are taxonomic terms or kind terms, a widespread category that include natural kinds, artifactual kinds, social kinds, and probably others.[45]

With this new theory Kuhn redrew the picture of scientific revolutions. Theories classify objects and phenomena to which a series of categories apply, each of which

---

there is no need to translate untranslatable expressions to argue for incommensurability (see Sankey (1990)).

41 See the passages referred to in the previous chapter, n. 161.

42 See Kuhn (1989a), pp. 60–61; see also Partee (1989).

43 In fact, he holds that this is the sense in which he has always meant incommensurability: "Insofar as incommensurability was a claim about language, about meaning change, its local form is my original version" (Kuhn (1983a), p. 36). Kuhn's statement is an indirect reply to the criticism raised in Davidson (1974), Kripke (1980) and Putnam (1981).

44 Kuhn (1983a), p. 36.

45 Kuhn (1991a), p. 92. On this point, see Hacking (1993), p. 290, and Kuhn (1993a), pp. 228–229.

comprises elements that theories group together on the basis of shared characteristic features, or features pointing to a common behaviour. Each theoretical structure is associated with a taxonomic system, and a revolution is the passage from one taxonomic system to another:

> What characterizes revolutions is [...] change in several of the taxonomic categories prerequisite to scientific descriptions and generalizations. That change [...] is an adjustment not only of criteria relevant to categorization, but also of the way in which given objects and situations are distributed among preexisting categories.[46]

Scientific revolutions thus become changes of taxonomy.[47] A revolution produces a new lexical taxonomy in which some kind terms refer to new referents that overlap with those referred to by old kind terms. Prerequisite to complete translation between two different taxonomies are not shared characteristics of individual concepts, but shared lexical structure. By "lexicon" or "lexical structure" (terms that now tend to be preferred to "paradigm") Kuhn meant a structured vocabulary of kind terms that represents a taxonomy of natural kinds: in order for individuals to communicate, it is not necessary that they adopt common criteria of application of terms to the world, but that they operate with homologous lexical structures, that is, with a structured vocabulary that shares the same taxonomic system.[48] Analogously, in order that reciprocal translation from one lexicon to another is possible, theories must share the same lexical taxonomy, otherwise they are "incommensurable". Such a prerequisite aims at securing that terms belonging to intertranslatable lexicons refer to the same natural kind, and not only that they have the same extension: for a set of objects can belong to different natural kinds, and reference to a given natural kind requires that its members are determined as members of a given natural kind, not simply on an extensional basis. This, in Kuhn's later writings, is the characteristic feature of incommensurability. It is not simply the outcome of a failed translation of individual concepts: rather, scientists upholding different paradigms – to use this by now old-fashioned word once more – face incommensurability because they build different lexical structures and, consequently, classify the world differently.

Every scientific theory has its own structured lexicon, that is, a taxonomically ordered network of kind concepts or kind terms (that may be natural, social or

---

[46] Kuhn (1981), p. 30. Such redistribution always involves more than one category, each of which is inter-defined with the others.

[47] In his later years Kuhn repeatedly highlighted this aspect, but already since the early 1970s he claimed that scientific revolutions produce changes in the classificatory systems employed by scientists: "One aspect of every revolution is [...] that some of the similarity relations change. Objects which were grouped in the same set before are grouped in different sets afterwards and *vice versa*. Think of the sun, moon, Mars and earth before and after Copernicus; of free fall, pendular, and planetary motion before and after Galileo; or of the salts, alloys, and a sulphur–iron filing mix before and after Dalton" (Kuhn (1970b), p. 275).

[48] See Kuhn (1983a), pp. 51–53, and (1991a), pp. 92–94. A lexical structure plays a role very similar to that of Kant's mental categories (that are constituted by kind concepts, or kind terms), but unlike them it changes through time. See the example of the duck-billed platypus, evoked by Kuhn in his (1989a), p. 72, and (1991a), p. 92, and amply resumed by Umberto Eco in his (1997/1999), particularly chs. 2–4; see also Eco (1990), chs. 3–4, and (2003).

scientific kinds), some of which are antecedently available, in Hempel's sense. Despite the fact that they may be explicated, these concepts bear an essentially pre-linguistic nature.[49] Kinds, determined by kind terms, populate the world and divide it up into interconnected categories. Among the relationships that relate them to one another there is the "no-overlap principle", according to which terms referring to different species of the same kind cannot have referents in common.[50] Lexicons are necessary for the formulation of problems and for their solutions, and therefore for explaining nature; revolutions are described as changes in the lexicons of theories. The distinction between "normal science" and "extraordinary science" becomes a distinction between activities that require local taxonomic change and activities that do not.[51] Scientific communities are thus seen as linguistic communities that differentiate one from the other on the basis of the lexicon employed by their respective members: incommensurability is given by the difference of their respective lexical structures. As a consequence, since lexicons function in the constitution of the phenomenal world and the formulation of scientific problems, and since meanings are related to lexical structures, all previous characterization of incommensurability in terms of problems, meaning and world changes merge into the notion of lexical structure: they are different manifestations of such lexical changes.

Therefore, Kuhn highlighted the *local* feature of revolutions: instead of discussing too general entities, such as paradigms or disciplinary matrixes (that range from methodology to ontology), he focused his attention on a very limited class of objects, namely, kind terms. In this way, the meaning change of kind terms catches the revolutionary features of paradigm change. In particular, the meaning variance of kind terms involves the revision of the whole lexical taxonomy, and may give rise to incommensurability and communication difficulties between different scientific communities. On the other hand, since such meaning change only refers to a very limited class of terms, in a revolution several terms preserve their meaning and may provide the "common ground" required for a rational confrontation and choice.[52]

Natural kinds and the no-overlap principle

Kuhn's latest version of incommensurability is characterized by a growing emphasis on the semantics of natural kinds: it is impossible to translate terms of one taxonomic structure into another due to the restrictions that govern the relationships among natural kinds. Such restrictions derive from what Kuhn labelled the "no-overlap principle".[53]

---

[49] See Kuhn (1991a), p. 94, and (1993a).

[50] See Kuhn (1991a), p. 92; later, Kuhn reformulated the no-overlap principle by appealing to the notion of "contrast set", in order to reply to Hacking's objection in his (1993), pp. 286–287: see Kuhn (1993a), pp. 230–233. See also Buchwald (1989) and (1992), together with Kuhn (1993a), pp. 237–240.

[51] See Kuhn (1991a), p. 97.

[52] See Kuhn (1983c) (together with Hempel (1983)) and Sankey (1993a) and (1998).

[53] See Kuhn (1991a), p. 92.

As we have seen, lexicons are structured vocabularies of natural kinds.[54] Kuhn claimed that in order for scientists supporting different theories (i.e., different conceptual systems, or lexicons) to effectively communicate it is not required that speakers employ the same criteria for applying terms to the world. What is required is simply that they operate with "homologous lexical structures", with structured vocabularies that embody the same taxonomic system.[55] Analogously, in order for the translation from the lexicon of one theory to another to be possible, theories must share only a lexical taxonomy. If they do not share it, they are incommensurable.

Kuhn insisted on the fact that imposing purely extensional constraints on translation does not work: a translation that preserves reference can however be unable to preserve some crucial aspects of meaning.[56] The requisite of lexical homology aims at securing that terms belonging to inter-translatable languages refer to the same natural kind, and do not simply have the same extension. Since a single set of objects may belong to different natural kinds, the reference to a given natural kind requires that its members are determined as members of that kind, and not simply on an extensional basis. Reference to the same natural kind, therefore, is a stronger constraint than co-extensionality.

The constraint of referring to the same natural kind becomes, in Kuhn's later writings, the key ingredient of incommensurability. For Kuhn's argumentation in support of the translation difficulties between different lexical structures hinges on an aspect of the hierarchical nature of taxonomies of natural kinds: "no two kind terms, no two terms with the kind label, may overlap in their referents unless they are related as species to genus. There are no dogs that are also cats, no gold rings that are also silver rings, and so on: that's what makes dogs, cats, silver, and gold each a kind".[57] In other words, members of a given natural kind may belong to another natural kind only if one of the two kinds is itself a sub-class of the other, or belongs to it. For no natural kind can comprise members belonging to more than one category in a taxonomic structure, unless the kind in question is on a different hierarchical level and comprises members of subordinate kinds within the taxonomy. Since this constraint precludes the possibility of overlapping between memberships of kinds, Kuhn labelled it "no-overlap principle".

Such principle leads to untranslatability, as it can be seen from Kuhn's example of celestial taxonomy. Let us imagine attempting to translate the Ptolemaic term "planet" into the taxonomy of Copernican astronomy. Besides the planets classified as such in Copernican astronomy, the Ptolemaic category "planet" comprises also the Sun and the Moon. Translating the Ptolemaic term "planet" into the Copernican lexicon would therefore require the incorporation in the latter's taxonomy of a single

---

[54] See Kuhn (1983a), pp. 52–53, and (1991a), pp. 92–93. Kuhn describe lexicons as a "mental module" that stores the concepts and vocabulary: "the module in which each member of a speech community stores the kind term and kind concepts used by community members to describe and analyze the natural and social worlds" (Kuhn (1993a), pp. 238–239; see also pp. 229, 238 and 242).

[55] See Kuhn (1983a), pp. 51–53.

[56] See Kuhn (1983a), pp. 47–49.

[57] Kuhn (1991a), p. 92.

category that includes members of three different Copernican categories, namely, "planet", "star" and "satellite". However, such a category cannot be introduced in the Copernican taxonomy as a kind term, since the Ptolemaic category "planet" unites under a single natural kind what the Copernican scheme regards as members of distinct natural kinds. There arises the problem of what is the reason why it is not possible to integrate a kind belonging to a taxonomy within another taxonomic system. Kuhn did not outline his ideas with the details required to answer this question, but suggested that the reason has to do with the projectibility of natural terms: a property common to kind terms is that "They are projectible: to know any kind term at all is to know some generalizations satisfied by its referents and to be equipped to look for others".[58]

Given the relationship between the meaning of kind terms and the laws governing the members of one kind, untranslatability is due to the difference in the laws governing kinds of rival taxonomies. More particularly, a kind belonging to one taxonomy cannot be introduced into a rival taxonomy if the members of the kind, in the destination taxonomy, are classified as members of distinct kinds, and are therefore subject to distinct sets of natural laws. The Ptolemaic category of "planet", for instance, cannot be introduced as a unified kind within the Copernican scheme, since members of such a category would be subject to incompatible laws, that usually govern the behaviour of different kinds of celestial bodies. Some govern the behaviour usually exhibited by stars or satellites, others that of planets.[59] The impossibility of reducing one sort of taxonomy to another is reciprocal and therefore terms belonging to one lexical taxonomy cannot be translatable with terms indicating a rival taxonomy.

Incommensurability, comparability and rationality

As we saw in the previous chapter, the incommensurability thesis as expounded by Kuhn in *The Structure of Scientific Revolutions* won him the severest criticism, since it was understood as if he meant to say there are no theory-independent standards for theory choice: Kuhn was therefore charged of relativism and even of irrationalism.[60]

---

[58] Kuhn (1993a), p. 230; see also pp. 232–233. While projectibility of kind terms is the *justification* of the no-overlap principle, its *function* is to rule out the possibility of overcoming incommensurability too easily: see above, n. 24.

[59] But, we might object to Kuhn, are not all celestial bodies subject to one and the same physics, namely, Newton's celestial mechanics, or Einstein's relativity theory? What differentiates planets, stars and satellites within such physics is not so much the laws governing them, but the different role each object plays within the same picture – only boundary conditions set their respective "place" in the celestial context: as time goes by, conditions may change and stars may well turn into planets.

[60] In fact, some passages in *The Structure of Scientific Revolutions* openly lend themselves to such criticisms. Here is a typical example: "Like the choice between competing political institutions, that between competing paradigms proves to be a choice between incompatible modes of community life. Because it has that character, the choice is not and cannot merely by the evaluative procedures characteristic of normal science, for these depend in part upon a particular paradigm, and that particular paradigm is at issue. When paradigms enter, as they

However, already in the "Postscript – 1969" to the second edition of *The Structure of Scientific Revolutions* Kuhn attenuated his most extreme and radical assertions and acknowledges the existence of standards (namely, such "values" as quantitative precision, simplicity, consistency, and so on) for paradigm comparison.[61] He emphasized that scientists working within two rival paradigms may share the same standards but apply them differently in concrete situations. They learn how to proceed in practice not by following abstract rules, but through the exposure to previous and established exemplary problem solutions. For this very reason, it is a kind of knowledge that is almost impossible to separate from the cases observing which it is acquired. Therefore, when scientists face a new puzzle, they may disagree as to whether, for instance, the simpler solution is provided by theory *A* or *B*, or else whether different weights must be attributed to the several shared standards. It is a perfectly rational disagreement, and the only way to solve it is through the techniques of persuasion, by appealing and recurring to sensible arguments in order to show the opponents the preferability of one's choice.[62]

It is for these reasons that in "Objectivity, Value Judgement, and Theory Choice" (1973) Kuhn described the criteria, or standards, for theory choice as values, not rules. These values "provide *the* shared basis for theory choice"[63] and "function not as rules, which determine choice, but as values, which influence it".[64] Such a shared basis is not even capable of determining the choice of individual scientists, so that "every individual choice between competing theories depends on a mixture of objective and subjective factors, or of shared and individual criteria".[65]

There arises the question whether values do vary with time. In the positive case, Kuhn would lay himself open to the charge of relativism. In "Objectivity, Value Judgement, and Theory Choice" he offered a conditioned answer:

> Throughout this paper I have implicitly assumed that, whatever their initial source, the criteria or values deployed in theory choice are fixed once and for all [...]. Roughly speaking, but only very roughly, I take that to be the case. If the list of relevant values is kept short [...][66] and if their specification is left vague, then such values as accuracy, scope, and fruitfulness are permanent attributes of science.[67]

However, after the linguistic turn, Kuhn relinquished this condition: at the end of his last published work, "Afterwords" (1993), the five values listed twenty years before become permanent characteristics of science:

---

must, into a debate about paradigm choice, their role is necessarily circular. Each group uses its own paradigm to argue in that paradigm's defense" (Kuhn (1962a), p. 94).

61 See Kuhn (1970c), pp. 184–185.

62 See also Pera (1991) and Pera, Shea (eds) (1991).

63 Kuhn (1977c), p. 322.

64 Kuhn (1977c), p. 331.

65 Kuhn (1977c), p. 325.

66 In this paper Kuhn only mentions five values: accuracy, consistency, scope, simplicity and fruitfulness (see Kuhn (1977c), pp. 321–322; but see also pp. 323–325); on other occasions, the values he lists vary (see the references given above, in ch. 2, n. 212).

67 Kuhn (1977c), p. 335.

> As the developmental process continues, the examples from which practitioners learn to recognize accuracy, scope, simplicity, and so on, change both within and between fields. But the criteria that these examples illustrate are themselves necessarily permanent, for abandoning them would be abandoning science together with the knowledge which scientific development brings.[68]

As a historian,[69] Kuhn realized that some criteria or standards have actually changed, however very slowly. For instance, he noticed that in ancient times quantitative precision was required for astronomy, but not for physics, biology and so on. One of the major results of the scientific revolutions in the sixteenth and seventeenth centuries was the mathematization of several scientific disciplines. Only afterwards, that a theory must formulate quantitatively accurate predictions has become a routine requisite for nearly all disciplines. Kuhn seemed to believe that values (at least the ones he lists) have become permanent characteristics of science only after scientific revolutions. They operate both during normal science periods and revolutionary periods. New characteristics might certainly be added in the future, but Kuhn tends to think that it is unlikely that old ones disappear: "[their] rejection would be irrational".[70] In other words, Kuhn then thought that these characteristics are constitutive of science and scientific rationality. In the light of Kuhn's later views about theory choice, the charge of relativism and irrationalism simply drops.

Furthermore, it must be noticed that incommensurability, understood in terms of untranslatability, does not imply incomparability, due to its local nature. That is, only those propositions are untranslatable that contain a small subset of interrelated terms belonging to the lexicons of rival theories. Newtonian terms such as "force", "mass" and "weight" form a set of this kind for what concerns Ariostotelian physics. Nevertheless, two theories may share and in fact often actually do share a common vocabulary, made up by those terms that were antecedently available. In the case of Newtonian mechanics, in comparison with the Aristotelian one, they would comprise kinematics' terms such as "position" and "velocity", together with the qualitative sense of "mass", understood as portion of matter.[71] Taking them as a basis, it is possible to compare different theories by appealing to the above mentioned values. It is also possible to learn to understand what a theory says notwithstanding its incommensurability, in the sense of untranslatability. Understanding or interpreting

[68] Kuhn (1993a), p. 252.

[69] A physicist by training, but also historian and philosopher of science, Kuhn sees himself in the following terms: "Becoming a philosopher is, among other things, acquiring a particular mental set toward the evaluation both of problems and of techniques relevant to their solution. Learning to be a historian is also to acquire a special mental set, but the outcome of the two learning experiences is not at all the same. Nor, I think, is a compromise possible, for it presents problems of the same sort as a compromise between the duck and the rabbit of the well-known Gestalt diagram. Though most people can readily see the duck and the rabbit alternatively, no amount of ocular exercise and strain will educe a duck-rabbit" (Kuhn (1971c), pp. 5–6). And again: "I'm never a philosopher and a historian at the same time, but the two do interact. And that's the ideal arrangement, from my point of view" (Kuhn (I-1997a), p. 316).

[70] Kuhn (1993a), p. 252.

[71] See also Toulmin (1961), pp. 46–50.

a theory does not mean – *pace* Quine – to translate it in one's own language: on the contrary, it means acquiring a new language and becoming bilingual. It is bilingualism, not translation, that could bridge the communication gap separating members of different scientific communities.

Incommensurability and indeterminacy of translation

In his last papers, both published and unpublished, Kuhn focused on what he then regarded as the problem of incommensurability and, as we have seen, did so exclusively in terms of meaning alterations, thus leaving completely aside both the methodological and the ontological components of the incommensurability thesis. A major role in Kuhn's renewed interest in the linguistic aspect of scientific revolutions was certainly played and propelled by his close confrontation with Quine's ideas. Indeed, for many years Quine remained a critical point of reference that proved to be both very helpful for the articulation of Kuhn's own approach to the problems of translation and crucial for the development of his later views.

Kuhn established a link between his own incommensurability thesis and Quine's thesis of the indeterminacy of translation.[72] Briefly put, Quine's thesis is that "manuals for translating one language into another can be set up in divergent ways, all compatible with the totality of speech dispositions, yet incompatible with one another".[73] This thesis derives from the behaviourists' criticism of meaning: Quine held that verbal behaviour leaves meaning indeterminate and denied that there is more to meaning other than what is apparent in such behaviour. His famous example refers to occasional statements, such as the natives' utterance of the word "gavagai".[74] Quine claimed that the referent of "gavagai" is inscrutable: its ostension does not determine whether it refers to rabbits or stages, or brief temporal segments of rabbits,[75] and the translation of the natives' "apparatus of individuation" required in order to determine the translation also remains undetermined. The inscrutability of reference renders indeterminate the translation of sentences containing terms of this kind.[76]

---

[72] Kuhn highlights the parallel between incommensurability and the indeterminacy of translation on several occasions: see, for instance, his (1970b), pp. 268–269, (1970c), pp. 201–203, and (1976b), pp. 190–192. Later, as we shall see, he clearly distinguishes between these two notions: see Kuhn (1983a), 47–49, and (1989a), pp. 61–62.

[73] Quine (1960), p. 27.

[74] See Quine (1960), pp. 51–53. For a criticism, see Agassi (1988b): Quine's rejoinder is his (1988).

[75] See Quine (1960), pp. 51–52; see also pp. 29–30 and Quine (1969), pp. 30–35.

[76] In *Word and Object* Quine analyses how it is possible to translate expressions of an unknown language on the basis of verbal answers and perceptual stimuli. Quine claims that for the so-called "radical translations", that is, translations between different languages that evolved with unrelated cultures, it is possible to find different translations, at the same time logically incompatible and empirically equivalent. For example, when a radical translator *à la* Quine meets a stranger who, by pointing towards a rabbit, utters "gavagai", he might propose to translate such expression with "rabbit", or with "brief temporal stages of a rabbit", or else with "parts of a rabbit". According to Quine, these various proposals are logically incompatible and yet, if we judge them on the basis of the apparent behaviour in observable

Kuhn discussed the problems related to untranslatability and the thesis of the indeterminacy of translation, as expounded by Quine chiefly in his *Word and Object* (1960),[77] in several works.[78] At first, he adopted Quine's thesis in support of semantic incommensurability due to the meaning variance of the terms employed by alternative theories.[79] Theories expressed in different languages (that is, incommensurable theories) cannot be compared point-by-point on the basis of their empirical consequences, since according to Quine the statements of a theory can be translated into another's language in various and mutually incompatible ways, thus questioning the results of a point-by-point comparison. Advocates of different theories speak different languages and their ability to grasp their respective points of view are therefore necessarily limited by the imperfections of the processes of translation and of reference determination.[80]

Later, Kuhn sought to differentiate his own position from Quine's. In "Theory-Change as Structure-Change: Comments on the Sneed Formalism" (1976) he indirectly criticized Quine's idea of a radical translator and of a translation manual, arguing for the impossibility of a faithful and uniform translation:

---

situations, they are empirically equivalent. Quine's claim that it is always possible to find several empirically equivalent but logically incompatible translations has become known as the thesis of the "indeterminacy of translation".

[77] Quine (1960), but see also his (1969), (1970) and (1974).

[78] See Kuhn (1970b), pp. 267–269, (1970c), pp. 201–203, (1971a), p. 146, (1976b), pp. 190–192, (1977a), p. xxii, (1977c), pp. 338–339, (1979a), p. 126, (1983a), pp. 37–40 and 47–49, (1989a), pp. 58–65, (1990) and (1991a), pp. 90–94.

[79] "Proponents of different theories (or different paradigms, in the broader sense of the term) speak different languages – languages expressing different cognitive commitments, suitable for different worlds. Their abilities to grasp each other's viewpoints are therefore inevitably limited by the imperfections of the processes of translation and of reference determination" (Kuhn (1977a), pp. xxii–xxiii; see also Kuhn (1970b), pp. 267–270).

[80] By affirming that translation requires "the best available compromises between incompatible objects" (Kuhn (1970b), p. 267), Kuhn referred to Quine's discussion of indeterminacy to support the claim that "it is today a deep and open question what a perfect translation would be and how nearly an actual translation can approach the ideal" (*ibidem*, p. 268). He mentioned Quine's own example ("gavagai") to underline the epistemological difficulties involved in translating a language embodying different concepts: "Quine points out that, though the linguist engaged in radical translation can readily discover that his native informant utters 'Gavagai' because he has seen a rabbit, it is more difficult to discover how 'Gavagai' should be translated. […] Evidence relevant to a choice among these alternatives will emerge from further investigation, and the result will be a reasonable analytic hypothesis with implication for translation of other terms as well. But it will be only a hypothesis […]; the result of any error may be later difficulties in communication; when it occurs, it will be far from clear whether the problem is with translation and, if so, where the root difficulty lies" (*ibidem*). With these remarks Kuhn began to question Quine's view that it would not be possible to spot any differences between different translations. However, it must be noted that the empirical differences that it would be possible to point out among different translations seem to go beyond the mere behavioural reply to certain perceptual stimuli, and therefore trespass on what Quine would have regarded as empirical evidence.

> Unlike Quine, I do not believe that reference in natural or in scientific languages is ultimately inscrutable, only that it is very difficult to discover and that one may never be absolutely certain one has succeeded. But identifying reference in a foreign language is not equivalent to producing a systematic translation manual for that language. Reference and translation are two problems, not one, and the two will not be resolved together. Translation always and necessarily involves imperfection and compromise; the best promise for one purpose may not be the best for another; the able translator, moving through a single text, does not proceed fully systematically, but must repeatedly shift his choice of word and phrase, depending on which aspect of the original it seems most important to preserve.[81]

Unlike Quine, Kuhn held that even though it is possible to determine what the terms of another language or theory refer to, they can nevertheless turn out not to be uniformly or faithfully translatable.

By introducing a distinction between the process of translation from one language to another in case these are both known to the translator, and the process of interpretation of a language that is at first unknown to the interpreter, Kuhn claimed that Quine's radical translator is in all respects an interpreter learning a new language.[82] He highlighted that "Acquiring a new language is not the same as translating from it into one's own. Success with the first does not imply success with the second".[83] Kuhn employed this distinction to criticize Quine's thesis of the indeterminacy of translation and develop, in its stead, an even more radical untranslatability thesis. Indeed, Quine held that when facing different translations (or, to use his own term, different "analytical hypotheses") the point is not so much that we can never be certain about which is the correct one, rather, that there is not even an objective matter about which we can be right or wrong.[84]

The criticism of Quine became more explicit later, in "Commensurability, Comparability, Communicability" (1982). In this paper Kuhn sharply distinguished the process of learning a language from the process of interpreting it. This allowed him to conclude that translations are often actually impossible, contrary to what had been claimed by Quine, who had always defended the possibility of a plurality of translations, logically incompatible and empirically equivalent:

> There need be no English description coreferential with the native term 'gavagai'. In learning to recognize gavagais, the interpreter may have learned to recognize distinguishing features unknown to English speakers and for which English supplies no descriptive terminology. Perhaps, that is, the natives structure the animal world differently from the way English speakers do, using different discriminations in doing so. Under those circumstances, 'gavagai' remains an irreducibly native term, not translatable into English.[85]

---

[81] Kuhn (1976b), pp. 189–190.

[82] See Kuhn (1983a), pp. 37–40.

[83] Kuhn (1983a), p. 39; see also Kuhn (1989a), pp. 61–63, (1990), p. 300, and (1991a), pp. 92–93.

[84] See Quine (1960), pp. 73–79.

[85] Kuhn (1983a), pp. 39–40.

Kuhn concluded that, rather than a plurality of logically incompatible but empirically equivalent translations, *à la* Quine, translations are often not at all possible.[86]

At the same time, Kuhn said that the difference between the incommensurability thesis and that of the indeterminacy of translation is rooted in a fundamental difference in their respective basic assumptions: whereas Quine developed a theory of translation based on a purely extensional semantics, Kuhn always insisted on the necessity of taking the intensional aspects as well into account,[87] and on the fact that such intensional aspects are to be found not in individual concepts, but in the structural relations among them.[88] More particularly, he was interested in the structural relations between concepts that must be acquired through projection from individual instances: this was the idea at the basis of Kuhn's original notion of paradigm, and it explained Kuhn's insistence on the cognitive and learning processes that characterizes his subsequent works, since the late 1960s.[89] On the contrary, Quine did not think that there is much to learn from the analysis of the way in which structural relations are established between concepts acquired by projection from individual instances.

In the series of the three Shearman Memorial Lectures – that together with "Afterwords" and the first five chapters of Kuhn's last book[90] present Kuhn's most updated and defined version of his thought on the topic – Kuhn affirmed that Quine's radical translator is, in fact, a person learning a new language, and that most of the arguments Quine advanced about the indeterminacy of translation could very well and equally be read as arguments actually supporting its very impossibility.

As is particularly clear by reading some notes Kuhn drafted for a lecture on meaning,[91] he came to understand meaning variance exclusively in terms of change of extension and defined incommensurability entirely in terms of lack of correspondence between the structural relations that organize concepts, as it is exemplified in the particular case of taxonomies. Such focusing is probably affected by the confrontation with Quine's theses, but Kuhn's growing confidence of the fact that he had always been on the right track about incommensurability must also be considered: he was convinced that all was left to do was to develop in this direction the insights he had had from the very beginning.

---

[86] See Kuhn (1989a), pp. 61–63, and (1990), p. 300.

[87] See Kuhn (1983a), pp. 47–50: "What is it that translation must preserve? Not merely reference, I have argued, for reference-preserving translations may be incoherent, impossible to understand while the terms they employ are taken in their usual sense. That description of the difficulty suggests an obvious solution: translations must preserve not only reference but also sense or intension" (*ibidem*, p. 50).

[88] See Kuhn (1983a), pp. 50–53.

[89] At least since 1969, the year in which he writes what got eventually published as his (1974c).

[90] See, respectively, Kuhn (U-1987), (1993a) and (U-1982-). See also Kuhn (I-1997b), pp. 113–114.

[91] See Kuhn (U-1990a).

Translation and bilingualism

Until the 1970s Kuhn always maintained a connection between incommensurability and the inability to achieve a complete translation of one theory in another's language. In the early 1980s, however, he distinguished two aspects of the translation process: a *technical* aspect, consisting in the (not necessarily one-by-one) replacement of the words in one language into another's; and an *interpretative* aspect, during which language learners try to make sense of a significant portion of the foreign language by relating it to its linguistic context, but not to their native language. Incommensurability refers only to the former aspect: that two theories are incommensurable implies that their concepts cannot be mutually translated through word replacements, but the proponents of one can learn the theory of their rivals by interpretation.[92]

In the late 1980s Kuhn specified the meaning of such impossibility of translation by means of his theory of kinds. The meaning variance of kind terms inevitably leads to "a sort of untranslatability, localized to one or another area in which two lexical taxonomies differ".[93] This sort of untranslatability is the by-product of the difficulty of mapping a foreign taxonomy onto a native one. A "foreign" term is untranslatable not because we are unable to find its referents in its linguistic context, but because we are unable to find a corresponding term in the destination language whose referents do not overlap with those referred to by the term we wish to translate. Therefore, the impossibility to achieve a full translation does not provoke the total breakdown of communication between two different linguistic communities. We may be unable to translate a foreign term, but we can understand it through an interpretative process: for we can learn its meaning by directly identifying its referents in its very linguistic context, without appealing to our language.[94]

The actual possibility of cross-communication among proponents of two different scientific communities is secured not by translation (as Kuhn thought in the 1970s), but by bilingualism: "the process which permits understanding produces bilinguals, not translators".[95] Bilinguals can acquire a second language and express themselves through it without the mediation of the first: comprehension, then, is possible even without translation.[96] Just like bilinguals, historians and scientists can acquire

---

[92] Kuhn (1983a), especially pp. 37–40, is the key text for illustrating the turning point. For a detailed discussion of Kuhn's changing views on translation in the 1970s and the early 1980s see Hoyningen-Huene (1989a/1993), pp. 256–258); see also Chen (1997), pp. 261–264, and Sankey (1993a). Acquiring the language of one theory means learning something about the world: Kuhn's philosophical production after 1980 is nothing but an elaboration and a development of this insight, presented in Kuhn (1970b), pp. 270–271, and (1974c).

[93] Kuhn (1991a), p. 93.

[94] This is the case for the English word "mat", discussed in Kuhn (1991a), pp. 93–94.

[95] Kuhn (1991a), p. 93.

[96] But "bilingualism has a cost" (Kuhn (1991a), p. 93): unlike translators, bilinguals often say that there are things they can express in one language but not in the other. Once again, even in Kuhn's later reading of scientific revolutions, we encounter the "losses" that characterized his earlier reflections, ever since *The Structure of Scientific Revolutions*. The so-called "Kuhn loss" seems to be an essential and unavoidable component of scientific

terms not familiar to them without having to translate them into their contemporary vocabulary, or without having to associate them with concepts deployed within their scientific community.[97] According to Kuhn, this way secures rational confrontation: the choice among different, competing theories can be rational even if they are incommensurable. Meaning variance, and the relative impossibility of translation, does not entail impossibility of confrontation, nor does incommensurability involve relativism.

Towards a new evolutionary epistemology

Kuhn's last step consisted in a revision of the evaluation standards for the growth of science. In order to elaborate such standards he developed an evolutionary epistemology, recalling an idea he had already put forward, three decades earlier, in the closing pages of *The Structure of Scientific Revolutions*.[98] Then he employed the analogy with biological evolution to illustrate scientific revolutions: scientific progress can be compared with biological evolution since they are both the outcome of competition and selection. In both cases, it is a process *from* something, not a process *towards* anything.[99] There is "no coherent direction of ontological development",[100] no process zeroing in on the truth:

> I aim to deny all meanings to claim that successive scientific beliefs become more and more probable or better and better approximations to the truth and simultaneously to suggest that the subject of truth claims cannot be a relation between beliefs and a putatively mind-independent or 'external' world. [...] That claims to that effect are meaningless is a consequence of incommensurability.[101]

Kuhn further developed the analogy with biological evolution. First, he explained, the relationships among the individual members of a scientific community and the community itself are very similar to the relationships tying individual organisms to the species they belong to. In biological evolution organisms are characterized not only by their genotype, but also by the genic pool of the whole species they belong to. In this sense science is an intrinsically communitarian activity:

---

progress (see above, ch. 2, n. 204): it is the price historians and scientists must pay to achieve progress – and it is central ("particularly relevant", *ibidem*) in Kuhn's characterization of incommensurability throughout the 1980s and 1990s.

[97] "Faced with untranslatable statements, the historian becomes bilingual, first learning the lexicon required to frame the problematic statements and then, if it seems relevant, comparing the whole older system (a lexicon plus the science developed by it) to the system in current use" (Kuhn (1989a), pp. 76–77).

[98] See Kuhn (1962a), pp. 170–172.

[99] See Kuhn (1962a), pp. 170–171; see also the fourth section of Kuhn (1984). For a similar argument, see also Toulmin (1961), pp. 109–114.

[100] Kuhn (1970c), p. 206. "Theories do not 'dig' deeper and deeper into the constitution of the world, since the constitution of the world (that is, ontology) sometimes changes with them" (Barrotta (1998), p. 178): incommensurability between successive paradigms involves ontological discontinuity.

[101] Kuhn (1993a), pp. 244–245. See also Kuhn (1962a), pp. 170–173, (1970c), pp. 205–207, and (1979b), p. 206.

> Groups do not have experiences except insofar as all their members do. And there are no experiences, gestalt switches or other, that all the members of a scientific community must share in the course of a revolution. Revolutions should be described not in terms of group experience but in terms of the varied experiences of individual group members. Indeed, that variety itself turns out to play an essential role in the evolution of scientific knowledge.[102]

Secondly, scientific development and biological evolution share the same pattern of growth, namely, a tree-pattern:[103] Kuhn highlighted "the apparently inexorable (albeit ultimately self-limiting) growth in the number of distinct human practices or specialties over the course of human history".[104] *Proliferation* of distinct disciplines – a process akin to biological speciation[105] – is therefore the fundamental characteristic of scientific progress.[106]

Thirdly, both scientific development and biological evolution give rise, in the course of their growth, to isolated units. In the biological field they are populations isolated from the reproductive point of view, whose members face difficulties to interbreed with members of other populations. In the scientific field, they are communities of specialists, or experts, who communicate perfectly among themselves but face serious problems in communication with members of other communities.[107] Incommensurability is therefore inevitable for the growth of science.[108]

On the basis of this analogy with biological evolution Kuhn proposed an evolutionary epistemology in order to specify the evaluation standards for scientific development. The basic idea of traditional epistemology, a correspondence theory of truth that evaluates beliefs on the basis of their ability to reflect a mind-independent

---

[102] Kuhn (1989c), p. xiii. Kuhn relinquished the *Gestalt* switch image he introduced in *The Structure of Scientific Revolution*, for it applies to individual members of the community, not to the community itself.

[103] The structural conception of theories, elaborated by Joseph D. Sneed and, subsequently, by Wolfgang Stegmüller and Kuhn himself (see above, n. 1, for references), plays here a particularly relevant role.

[104] Kuhn (1992), p. 116.

[105] And not, as Kuhn thought for many years, to mutation: see Kuhn (1991a), p. 98. The analogy with biological speciation accounts for the proliferation of new specialties: it is a new consequence of scientific revolutions which was not present in Kuhn's early philosophical works.

[106] "Specialization and the narrowing of the range of expertise now look to me like the necessary price of increasingly powerful cognitive tools" (Kuhn (1991a), p. 98).

[107] See Kuhn (1991a), pp. 98–99.

[108] "[I feel] more strongly than ever that incommensurability has to be an essential component of any historical, developmental, or evolutionary view of scientific knowledge. Properly understood – something I've by no means always managed myself – incommensurability is far from being the threat to rational evaluation of truth claims that it has frequently seemed. Rather, it's what is needed, within a developmental perspective, to restore some badly needed bite to the whole notion of cognitive evaluation. It is needed, that is, to defend notions like truth and knowledge from, for example, the excesses of postmodernist movements like the strong programme" (Kuhn (1991a), p. 91): incommensurability, for Kuhn, is what is needed to avoid any relativism and irrationalism, and talk properly about the truth.

world, is not able to account for the change of beliefs themselves;[109] it therefore must be rejected and replaced by a weaker conception, internal to the lexicon itself.[110] For if an assertion can be rightly said to be true or false in the context of a given lexicon, the system of categories embedded in the lexicon can be, *per se*, neither true nor false. By rejecting the correspondence idea of truth Kuhn rejected the idea that the categorical structure of a theory can reflect the world in itself, independently of theory. We can talk about truth only within a given lexicon, that is, we can only evaluate assertions made within a given lexical context: "lexicons are not [...] the sorts of things that can be true or false"[111] – their logical status is that of meaning of the words in general, that is, of a convention that we can justify only in a pragmatic way. Furthermore, the lexicon marks the distance between the reality described and the theory describing it in various ways:

> Experience and description are possible only with the describer and described separated, and the lexical structure which marks that separation can do so in different ways, each resulting in a different, though never wholly different, form of life. Some ways are better suited for some purposes, some to others. But none is to be accepted as true or rejected as false; none gives privileged access to a real, as against an invented, world. The ways of being-in-the-world which a lexicon provides are not candidates for true/false.[112]

In other words, they are evaluated on the grounds of their ability to achieve an aim, not to reflect reality.

To account for the rationality of progressive change of our beliefs, said Kuhn, we have at our disposal only secondary standards (or criteria, or "values"), such as precision, consistency, scope, simplicity and fruitfulness. They are not fixed standards: they are context-dependent and reflect the constraints imposed by the historical moment and situations. The growth of knowledge is the outcome of the growth of different specializations, each of which is devoted to improving the current beliefs about a limited domain, so as to increase, in that area, its precision, consistency, fruitfulness, and so on.[113]

This evolutionary epistemology involves two significant changes in the meaning and scope of incommensurability. Scientific revolutions, Kuhn wrote in 1962, are episodes in the developmental process of individual sciences, or of an individual scientific specialty – but they also have, he now adds, "a second, closely related, and equally fundamental role: they are often, perhaps always, associated with an increase in the number of scientific specialties required for the continued

---

[109] In a continuous process of development there is no "fixed, rigid Archimedean platform [...] outside of history, outside of time and space" (Kuhn (1992), p. 115; see also p. 113 and Kuhn (1991a), pp. 95–96) that can provide the basis to measure the distance between present and "true" beliefs: there is no way to determine those that better reflect the external world, or the world-in-itself: see Kuhn (1992).

[110] See Kuhn (1991a), pp. 95–99, (1992), p. 115), and (1993a), pp. 244–245.

[111] Kuhn (1993a), p. 244.

[112] Kuhn (1991a), p. 104.

[113] See Kuhn (1992), p. 120.

acquisition of scientific knowledge".[114] In order to assess scientific development we may compare competing theories on the basis of the above mentioned secondary standards, or else we may measure the level of proliferation achieved in the process of knowledge production. Rational evaluation and shared standards are distinct, separated moments: in such a way, however incommensurability keeps playing a significant role in attributing different weights to the standards of appraisal adopted by rival scientific communities, the possibility of rational comparison, of rational assessment of the growth of knowledge is not *a priori* excluded.[115]

Finally, evolutionary epistemology allows for a redefinition of the functional role of incommensurability. For if according to traditional epistemology this notion bears exclusively negative effects, according to Kuhn it is nothing but a conceptual disparity separating different specialties: "Once the two specialties have grown apart, that disparity makes it impossible for the practitioners of one to communicate fully with the practitioners of the other. And those communication problems reduce, though they never altogether eliminate, the likelihood that the two will produce fertile offspring":[116] in Kuhn's eyes, incommensurability plays a positive function, that of provoking reciprocal isolation of communities of specialists, thus promoting their proliferation and , with it, the growth of the sciences.

Open Issues

In the last fifteen years of his life Kuhn repeatedly referred to "the more theoretical analyses on which I am currently engaged",[117] and to his "projected book".[118] In this large project, the book on which he was at work, he wanted to deal with the philosophical problems raised by *The Structure of Scientific Revolutions*, and particularly realism and truth, but also questions related to rationality and relativism: needless to say, incommensurability played a fundamental role in all these issues, and Kuhn aimed precisely at offering his ultimate word on it. Regrettably, at the time of his death, in June 1996, the book was largely unfinished.[119] However, a

[114] Kuhn (1993a), p. 250.

[115] See Chen (1997), pp. 265–268.

[116] Kuhn (1992), p. 120.

[117] Kuhn (1981), p. 14.

[118] Kuhn (1991a), p. 91. See also his (1989a), p. 58, (1992), p. 106, and (1993a), p. 228.

[119] Of the nine projected chapters, only the first five may be said to be fairly complete. The book, as it stands, is being edited by James Conant and John Haugeland. As far as I have been able to gather from the few people who have read the unfinished text, the first five chapters do not contain great novelties with respect of what we can find in Kuhn (1989a), (1991a), (1991b), (1992), (1993a), (U-1984), (U-1987), (U-1990a) and (U-1990b) – only, in the typescript we find more attention to uniformity and consistency in the argumentative discourse. In support of this, when Kuhn circulated the first five chapters of the book among an extremely restricted group of colleagues and friends, he used to add a copy of his (1989a), saying that in that paper he had actually managed to better express what he intended to say in those first chapters (my own sense is that he was unable to progress further than that). Kuhn's major concern in the unfinished typescript seems to be that of offering a defence of incommensurability not simply through scientific revolutions (as he had already done in

number of articles and papers, together with the completed part of the manuscript, indicate the direction in which he was moving.[120] In the years since the publication of *The Structure of Scientific Revolutions* Kuhn significantly changed his views on a number of issues[121] and particularly on the nature of the similarity and dissimilarity relations among concepts, on the role of incommensurability in the development of science and the relevance of history for the history and the philosophy of science.

One of the key themes of Kuhn's work after *The Structure of Scientific Revolutions* is the development of an account of the nature of "family resemblance" of concepts.[122] Since the "Postscript – 1969" to the second edition of *The Structure of*

The Structure of Scientific Revolutions), but also of the incommensurability affecting the continuous speciation process of new disciplines and sub-disciplines. Ultimately, Kuhn gives flesh and blood, enriching it with new examples from the history of science, to the skeleton he outlined in his (1992).

[120] See, in particular, Kuhn (1981), (1983a), (1986), (1989a), (1990), (1991a), (1992), (1993a), (I-1990), (I-1991), (I-1997a), (I-1997b), (U-1980), (U-1984), (U-1987), (U-1990a) and U-1990b). The unfinished manuscript of the book he was working on at the time of his death is Kuhn (U-1982-).

[121] Some may want to say that he laid different accents (or ascribed different weights) on the issues he raised in *The Structure of Scientific Revolutions* and on which he had been working on ever since. This is certainly correct. However, given the significant focusing on the linguistic aspects of the scientific enterprise that marks Kuhn's philosophical reflection in the 1980s and 1990s, which I take to be a radical impoverishment of his earlier insights, I prefer to adopt this phrasing.

[122] Traces of the discussion whether concepts are defined by necessary and sufficient conditions, valid in any circumstances, or a given concept can be explicated only by the ostension of typical examples, are to be found already in Plato (in *Euthyphro*, for instance, where Socrates asks Euthyphro to tell him what is piety and what is impiety, to which Euthyphro answers by making examples of pious and impious actions: Socrates replies that he did not ask for concrete instances of pious or impious actions, but for the nature of piety and impiety, that is, for the form that makes such actions pious or impious). In contemporary philosophy the dominant position, after Frege, has been that concepts can be defined by a set of characteristics each of which is individually necessary and the whole of which is altogether sufficient in order that and object is an instance of the defined concept. This conception dates back to the distinction, originally introduced in the School of Port Royal's logic textbook (better known as *Port-Royal Logic*), published in the seventeenth century, between "extension", that is, the class of objects referred to by a given concept, and "intension", that is, the list of the characteristics shared by all the objects referred to by a given concept: see Arnauld, Nicole (1685/1996), First Part, ch. 6: "I call the *comprehension* of an idea the attributes that it contains in itself, and that cannot be removed without destroying the idea. [...] I call the *extension* of an idea the subjects to which this idea applies. These are also called the inferiors of a general term, which is superior with respect to them" (*ibidem*, pp. 39–40). In his (1953) Wittgenstein criticizes this idea. By analysing the concept of "game" he shows that it might be impossible to find a definition of this concept that applies in each and every case. Therefore, an instance of one concept may happen to share some characteristics with another one, while the latter might happen to share others with a third instance of the same concept. Even if in this case it is true that all examples share some of their characteristics with each of the others, no individual characteristic must be necessarily shared by all instances. Not all games, for example, have various players, or end with a winner and a loser. But if no

Scientific Revolutions, Kuhn argued that all concepts, scientific concepts included, are based upon similarity and dissimilarity relations, rather than on definitions. Still, in the last significant paper he published[123] he introduced a distinction between what he calls "nomic" and "normic" concepts – a distinction that can be read as a reaction to the problems raised by an account of scientific concepts exclusively rooted on family resemblance relations.[124] For both kinds of concepts, however, Kuhn argued that, just as kind terms, they play a specific function, that can be traced back to the evolution of various neural mechanisms.

Normic concepts and nomic concepts

In his 1993 "Afterwords" Kuhn introduced the new distinction between normic and nomic concepts, based on the fact whether generalizations involving these concepts allow exceptions or not:

> Some of these generalizations are normic, admit exceptions. "Liquids expand when heated" is an example even though it sometimes fails, e.g. for water between 0 and 4 degrees centigrade. Other generalizations, though often only approximate, are nomic, exceptionless. In the sciences, where they mainly function, these generalizations are usually laws of nature: Boyle's law for gases or Kepler's laws for planetary motions are examples.[125]

This distinction resembles a previous distinction of Kuhn's between concepts "applied by direct inspection" and concepts in the determination of the reference of which "laws and theories also enter".[126] Concepts applied by direct inspection – or

---

individual characteristic (or set of characteristics) must be shared by all instances, it becomes impossible to define a concept by a list of its characteristics, each of which is individually necessary and the whole of which is altogether sufficient. Different instances of one concept, Wittgenstein argues, entertain only a "family" resemblance, that is, "a complicated network of similarities overlapping and criss-crossing" (Wittgenstein (1953), Part I, §66). On the basis of Wittgenstein's distinction, Kuhn advanced an account of concept acquisition rooted on perceived similarity rather than on rules, and subsequently developed it as years went by. According to his own account, a conceptual structure is constituted by the clustering of the objects referred to by concepts in classes of similarity. See Kuhn (1970a), (1970b), (1970c), (1974c), (1979b), (1983a), (1983b), (1989a), (1990), (1991a) and (1993a). For an analysis of Kuhn's argument, see Hoyningen-Huene (1989a/1993), ch. 3, and Andersen (2000) and (2001), pp. 41–46.

[123] Kuhn (1993a).

[124] See Wittgenstein (1953), Part I, §§65–67: concepts are webs of similarities that intertwine in many points: their constituting units depends on the set of decisions that have been taken by linguistic practice. According to Wittgenstein, it is possible to grasp an object not only for what it is *per se*, but also in the light of some similarity relation that allows us to see and identify it as an object of a certain kind, or as something that was once familiar to us (see also *ibidem*, Part II, section XI).

[125] Kuhn (1993a), p. 230. As to "normic generalizations", Kuhn refers to Scriven (1959). See Andersen (2001).

[126] Kuhn (1979b), p. 200. This distinction, in turn, is related to the logical positivists' distinction between theoretical and observational terms, even if it breaks from it entirely.

"basic terms", as Kuhn also calls them – are learned by ostension: they are terms such as "duck", goose" and "swan", that are learned in "contrast sets",[127] on the basis of similarity and dissimilarity relations among their instances. Analogously, concepts in the determination of the reference of which laws or theories also contribute – or, as Kuhn also called them, "theoretical terms" – are learned by showing concrete problem situations to which a certain law applies. For example, in order to teach students the concepts of "force" and "mass", the teacher may point to concrete situations in which Newton's second law applies such as, for instance, simple pendulum, free fall and harmonic oscillators. As in the case of the referents of basic terms, Kuhn held that what such situations

> have in common is not that they satisfy some explicit or even some fully discoverable set of rules and assumptions that gives the tradition its character and its hold upon the scientific mind. Instead, they may relate by resemblance and by modelling to one or another part of the scientific corpus which the community in question already recognizes as among its established achievements.[128]

Before "Afterwords", then, Kuhn highlighted the *resemblance* between two kinds of concepts, rather than their difference.

By introducing the distinction between normic and nomic concepts Kuhn emphasized the *difference* in the way in which the two kinds of concepts are acquired. Normic concepts are acquired in contrast sets, and Kuhn remarked that "The ability to pick out referents for any of these terms depends critically on the characteristics that differentiate its referents from those of the other terms in the set, which is why the terms involved must be learned together and why they collectively constitute a contrast set".[129] These characteristics of normic concepts are the same Kuhn held valid for concepts in general, on the basis of his account of the family resemblance among concepts. Furthermore, the core of this account was the assertion that it is not necessary that different exemplifications of one concept satisfy the same generalizations: the generalizations normic concepts belong to "are normic, admit exceptions" and "When terms are learned together in this way [that is, by ostension and in contrast sets], each comes with attached normic generalizations about the properties likely to be shared by its referents".[130]

Nomic concepts, on the contrary, are not learned in contrast sets. As Kuhn observed, nomic concepts such as "force" "are not normally in any contrast set at all".[131] However, Kuhn thought that it is not possible to acquire these concepts individually: we must acquire them together with other closely related ones (not by

---

Indeed, Kuhn clearly rejects it: "The distinction between a theoretical and a basic vocabulary will not do in its present form because many theoretical terms can be shown to attach to nature in the same way, whatever it may be, as basic terms" (Kuhn (1974c), p. 302, n. 11). See also Hoyningen-Huene (1989a/1993), pp. 90–93.

[127] See Kuhn (1993a), p. 230.

[128] Kuhn (1962a), pp. 45–46.

[129] Kuhn (1993a), p. 230.

[130] Kuhn (1993a), p. 230.

[131] Kuhn (1993a), p. 231.

contrast, but otherwise). Kuhn held that a concept such as "force" "must be learned with terms like 'mass' and 'weight'. And they are learned from situations in which they occur together, situations exemplifying laws of nature".[132]

Considering Kuhn's previous account of the concepts in whose extension also laws and theories are involved, it seems possible to interpret the distinction between normic and nomic concepts as a distinction not between concepts related by family resemblance and concepts that it is possible to define explicitly, but rather as a difference between the level at which family resemblance enters into play and the complexity of the ostensive acts on the basis of which concepts are introduced.

Whereas, in the case of normic concepts, several instances of each individual concept of the contrast set are shown in order to learn the concept in question, in the case of nomic concepts what gets highlighted is not instances of individual concepts, but complex problematic situations to which a given law applies and that involve the simultaneous employment of several concepts.[133] For example: in the case of normic concepts instances of individual contrasting concepts – like "duck", "swan" or "goose" – are shown, whereas in the case of nomic concepts what is shown is instances of application of a natural law – such as Newton's second law, $F = ma$, in which the concepts of "force", "mass" and "acceleration" are simultaneously employed.[134]

---

[132] Kuhn (1993a), p. 231.

[133] See Andersen (2001), together with Andersen, Nersessian (2000).

[134] Previously, Kuhn focused on the family resemblance between the problematic situations to which a given law could be applied, highlighting the similarity between normic and nomic concepts. However, although he later underlined their differences, Kuhn never said how the referents of individual normic terms like "force", "mass" and "acceleration", that are involved in non-contrasting relations such as Newton's second law, are to be identified. Kuhn's analysis is limited to the discussion of one single example, namely, the acquisition of the concepts of "force", "mass" and "weight". Kuhn claimed that before exposing students to Newtonian terminology, they must "already have a vocabulary adequate to refer to physical objects and to their locations in space and time. Onto this they must have grafted a mathematical vocabulary rich enough to permit the quantitative description of trajectories and the analyses of velocities and accelerations of bodies moving along them. Also, at least implicitly, they must command a notion of extensive magnitude, a quantity whose value for the whole of a body is the sum of its values for the body's parts" (Kuhn (1989a), p. 66). Later, however, some of the concepts that must be acquired can be available in a still qualitative form. The concept of "weight" is available in qualitative form prior to Newtonian "force" and is used during the latter's acquisition: it refers to the particular kind of force that causes a physical body to press onto its support when at rest, or to fall when it is unsupported. In the same way, the concept of "quantity of matter", that can be employed to introduce the Newtonian concept of "mass" is available in the quantifiable substratum underlying physical bodies, the stuff of which the quantity is conserved as the qualities of material bodies change (see Kuhn (1989a), pp. 68–69). In the case of "weight" and "mass", but unlike the case of "force", the qualitative features by which their referents are picked out are identical with those of pre-Newtonian usage: they can be determined on the basis of the same qualitative characteristics either in their Newtonian or pre-Newtonian context. According to Kuhn, Newtonian quantification is learned through exposure to problematic situations to which concepts are applied. He describes two standard methods: in the first, students are shown Newton's second law as a description of the way

The biological component of the lexicon

Kuhn's theory of concepts is an important part of his argument against realism. From his point of view, it is not language that needs to be adapted to the world, rather, it is the world that is an outcome of the mutual adaptation of language and experience. A lexicon allows the members of the community that employs it to have access to a world that is perceptually and conceptually subdivided in a certain way. Therefore, Kuhn claimed that "Like Kantian categories, the lexicon supplies preconditions for possible experience".[135] A lexicon should not be seen, then, as a set of beliefs, but as "a mental module prerequisite to having beliefs, a mode that at once supplies and bounds the set of beliefs it is possible to conceive".[136]

Kuhn expected that this mental module is biologically and culturally determined:

> Doubtless some aspects of that lexical structure are biologically determined, the products of a shared phylogeny. But, at least among advanced creatures (and not just those linguistically endowed), significant aspects are determined also by education, by the process of socialization, that is, which initiates neophytes into the community of their parents and peers.[137]

The biological component made Kuhn ponder the evolutionary origin of the module: "Presumably it evolved originally for the sensory, most obviously for the visual, system"; "it developed from a still more fundamental mechanism which enables individual living organisms to reidentify other substances by tracing their spatio-temporal trajectories".[138]

The evolutionary metaphor

Kuhn's evolutionary epistemology originates in *The Structure of Scientific Revolutions*, where he claims that the historical development of sciences is a development from something, not a teleological development towards any goal. This approach led Kuhn to compare the development of science with biological evolution: "scientific development is, like biological, a unidirectional and irreversible process".[139]

Kuhn originally saw the parallel between scientific and biological development in the roles played by theory choice and natural selection: "Verification is like natural

---

in which moving bodies behave; in the second, students are shown the law of universal gravitation. In both cases, students learn how to quantify the notion of force through a spring balance and by resorting to Hooke's law ($F = -kx$), and thanks to Newtonian concept of "force" they can quantify the notions of mass and weight. See Andersen, Nersessian (2000).

135 Kuhn (1991a), p. 104.

136 Kuhn (1991a), p. 94. "Some such taxonomic module", Kuhn continues, "I take to be prelinguistic and possessed by animals" (*ibidem*).

137 Kuhn (1991a), p. 101.

138 Kuhn (1991a), p. 94.

139 Kuhn (1970c), p. 206.

selection: it picks out the most viable among the actual alternatives in a particular historical situation".[140] From this point of view, what happens during the periods of crisis – when extraordinary science sees the proliferation of alternative theories that increasingly distance themselves from the previously accepted one – can be seen as a sort of multiple mutations of the theory, of which only the fitter, or more promising, will survive.

At the beginning of the 1990s Kuhn changed his mind about the parallel between the development of science and biological evolution, arguing that "revolutions, which produce new divisions between fields in scientific development, are much like episodes of speciation in biological evolution. The biological parallel to revolutionary change is not mutation, as I thought for many years, but speciation".[141] Kuhn introduced the metaphor of speciation to describe how in history the development of science has given birth to an increasing number of specialties. Such a proliferation can take place in various ways: a new specialty may rise by separating itself from the others, as in the case of computer science, that was born from mathematics; or it may form in the overlapping zone of already existing specialties, as in the case of physical chemistry or biochemistry, that grew as distinct research areas. It may look as if only in the former way has proliferation actually taken place, while in the latter it is nothing but a step forward in the ever increasing process of unification of the sciences. In any case, Kuhn underlined that the emergence of new specialties in the overlapping zone of already existing specialties does not lead to the unification of the latter. On the contrary: instead of becoming part of the disciplines that generated them, new specialties separate themselves more and more from their parents, developing their own specialized journals, their own associations and their own academic departments.[142] Therefore, Kuhn claimed that "Over time a diagram of the evolution of scientific fields, specialties, and subspecialties comes to look strikingly like a layman's diagram for a biological evolutionary tree".[143]

---

[140] Kuhn (1962a), p. 146; see also p. 173.

[141] Kuhn (1991a), p. 98. "And the problems presented by speciation (e.g. the difficulty in identifying an episode of speciation until some time after it has occurred, and the impossibility, even then, of dating the time of its occurrence) are very similar to those presented by revolutionary change and by the emergence and individuation of new scientific specialties" (*ibidem*). A second parallel between biological and scientific development concerns the unit which undergoes speciation: "In the biological case, it is a reproductively isolated population, a unit whose members collectively embody the gene pool, which ensures both the population's self-perpetuation and its continuing isolation. In the scientific case, the unit is a community of intercommunicating specialists, a unit whose members share a lexicon that provides the basis for both the conduct and the evaluation of their research and which simultaneously, by barring full communication with those outside the group, maintains their practitioners of other specialties" (*ibidem*). See also Kuhn (1992), pp. 119–120.

[142] See Kuhn (1991a), pp. 97–98, and (1992), pp. 116–118.

[143] Kuhn (1991a), pp. 97–98. "With much reluctance I have increasingly come to feel that this process of specialization, with its consequent limitation on communication and community, is inescapable, a consequence of first principles. Specialization and the narrowing of the range of expertise now look to me like the necessary price of increasingly powerful cognitive tools" (*ibidem*, p. 98). Proliferation of disciplines and sub-disciplines is not explainable in terms

Already in *The Structure of Scientific Revolutions* Kuhn had stated that "The net result of a sequence of such revolutionary selections, separated by periods of normal research, is the wonderfully adapted set of instruments we call modern scientific knowledge. Successive stages in that developmental process are marked by an increase in articulation and specialization".[144] However, in some respects the new metaphor for science in terms of speciation is quite different from the one introduced for revolutions at the end of *The Structure of Scientific Revolutions*.

In 1962 Kuhn introduced the notion of revolution to indicate those episodes in the history of science in which a theory is replaced by another one, incompatible with it. Later, in introducing the metaphor of speciation Kuhn highlighted that this destructive element[145] is "not nearly so directly present in biological speciation".[146] Even the historical examples to which he appeals to illustrate the new metaphor are rather different. If in his major book revolutions were instantiated by the Copernican revolution and the chemical revolution,[147] the process of speciation is now illustrated by examples drawn from the separation of the various components of mathematics during the seventeenth century,[148] or the development of molecular biology.[149] Although these examples are not worked out in detail,[150] it is clear that cases of this kind do not involve the same struggle for survival among old and new specialties that was in fact present in scientific revolutions.

In particular, there is a significant shift in the choice of examples from the history of science Kuhn considered in his as yet unpublished papers.[151] In *The Structure of Scientific Revolutions* one of his favourite examples to illustrate his philosophical views is that of the chemical revolution, related to the name of Antoine Laurent Lavoisier, the discovery of oxygen and the demise of the phlogiston theory. Another important example in that book drawn from the history of chemistry is Dalton's

---

of scientific revolutions, as Kuhn described them in the early 1960s. In his later works the scope of revolutions is considerably restricted: they occur on a much lesser scale and do not involve major changes of world-view. Rather, they occur within single specialties and have only "local" effects. Accordingly, as we have seen, is the notion of incommensurability significantly restricted. Interestingly, the historical case-studies Kuhn employs are also quite different from those considered in *The Structure of Scientific Revolutions* (the Copernican revolution, the chemical revolution, or the transition from Newton to Einstein).

[144] Kuhn (1962a), p. 172. "And the entire process may have occurred", Kuhn once again remarks, "without benefit of a set goal, a permanent fixed scientific truth, of which each stage in the development of scientific knowledge is a better exemplar" (*ibidem*, pp. 172–173).

[145] See Kuhn (1962a), p. 66, and (1961a), p. 208: "innovation is [...] necessarily destructive as well as constructive". On this issue, see above, ch. 2, n. 204.

[146] Kuhn (1992), p. 120.

[147] See Kuhn (1962a), pp. 116–117 and 118, respectively.

[148] See Kuhn (1992), pp. 116–117.

[149] See Kuhn (1991a), pp. 97–98.

[150] Apart from his more specific historical works, (1951b), (1952a), (1952b), (1952c), (1955a), (1955b), (1958a), (1958b), (1959b), (1960), (1961a), (1961b), (1971b), (1980b) and (1984) – and, most notably, his (1957) and (1978), together with Kuhn, Heilbron (1969) – the historical examples Kuhn draws upon in his philosophical works are quite sketchy and rarely worked out in sufficient detail.

[151] I am particularly referring to Kuhn (U-1984) and (U-1987).

law of definite proportions, of which Kuhn examines the conceptually revolutionary effects that led to the controversy between Joseph Louis Proust and Claude-Louis Berthollet.[152] In his later years, however, the historical examples he carefully considered are others: Aristotelian dynamics, the construction of Alessandro Volta's pile and Max Planck's discovery of the law of black-body radiation. Most likely, in Kuhn's view, these latter examples better supported the new conception of incommensurability Kuhn was struggling with.[153]

The most striking difference, however, is the role played by incommensurability in scientific revolutions and in the speciation of scientific disciplines, respectively. The process of speciation, that is, the mutual isolation of sub-disciplines, is generated by an increasing conceptual disparity among the tools developed: ever more specialized scientists, with their highly specific tools, designed on the basis of the necessities and aims of a given sub-specialty, inhabit a secluded niche, isolated from those of other sub-specialties.[154] Kuhn attributed such conceptual disparity to incommensurability:

> [...] what makes the specialties distinct, what keeps them apart and leaves the ground between them as apparently empty space [...] is incommensurability, a growing conceptual disparity between the tools deployed in the two specialties. Once the two specialties have grown apart, that disparity makes it impossible for the practitioners of one to communicate fully with the practitioners of the other.[155]

But the conceptual disparity between two specialties located on two different branches of the evolutionary tree of the sciences is very different from the conceptual disparity that separates two specializations, before and after a scientific revolutions. We might want to say that the lack of communication among the distinct niches – such as evolutionary biology and molecular biology – only means that they deal with different things: as in the case of astrophysics and biochemistry, they simply work and inquire on different phenomena. For theories like those of oxygen and phlogiston, for example, we would say that they deal with the same thing, and therefore are competing within the same niche to offer the best account for what happens in their common domain of application. In the terms of the speciation metaphor, that means that scientists inhabiting a given scientific niche peacefully coexist with scientists working in other niches. We have struggle and competition only in the case in which somebody tries to enter somebody else's niche or evade his own, changing the very nature of the niche and of the research practised by his inhabitants. Arguing

---

[152] It is to be noted that some of Kuhn's early works, as a historian of science, dealt exactly with seventeenth-century chemistry: see his (1951b), (1952a) and (1952b).

[153] However, it is also to be noted that in his (1989a) – the paper he deemed closest to his most recent and unpublished views – Kuhn goes back to inquire into the concepts implied by terms such as "mass" and "force", that is, the problems posed by the transition from Newtonian mechanics to the special theory of relativity, one of the most significant historical examples Kuhn employs in *The Structure of Scientific Revolutions* to argue for his theses (and particularly to support incommensurability).

[154] See Kuhn (1991a), pp. 102–103, and (1992), p. 120.

[155] Kuhn (1992), p. 120.

for incommensurability both inter- and intra-niches – that is, both among different sub-specialties and a new speciality and an old one, whose problems we expect the former resolves – Kuhn seems to be shading off the difference between the invaders into a given niche and the inhabitants of another.

History and philosophy of science

The "new philosophy of science" is also known as the "historical philosophy of science" as a consequence of the new and important role it attributes to history in the study of the dynamics of scientific knowledge. As Mary Hesse remarked in her review of *The Structure of Scientific Revolutions*,

> it cannot be disputed that this is the first attempt for a long time to bring historical insights to bear on the philosophers' account of science, and whatever the puzzles that remain to be solved, Kuhn has at least outlined a new epistemological paradigm which promises to resolve some of the crises currently troubling empiricist philosophy of science. Its consequences will be far reaching.[156]

Three decades later, during a conference in Kuhn's honour at the Massachusetts Institute of Technology, Michael Friedman claimed that

> *The Structure of Scientific Revolutions* (1962) forever changed our appreciation of the philosophical importance of the history of science. [The picture of science Kuhn drew] has itself sparked a revolution of sorts in the philosophy of science [...] Whatever the fate of this new philosophy may be, it is clear beyond the shadow of a doubt, I think, that careful and sensitive attention to the history of science must remain absolutely central in any serious philosophical consideration of science.[157]

Indeed, *The Structure of Scientific Revolutions* opens with a strong and explicit admonition: "History, if viewed as a repository for more than anecdote or chronology, could produce a decisive transformation in the image of science by which we are now possessed".[158] Thirty years later, in some of his last published writings Kuhn touched again on this point, that had been the *leitmotiv* of his entire philosophical production:

> When I first got involved, a generation ago, with the enterprise now often called historical philosophy of science, I and most of my coworkers thought history functioned as a source of empirical evidence. That evidence we found in historical case studies, which forced us to pay close attention to science as it really was. Now I think we overemphasized the empirical aspect of our enterprise (an evolutionary epistemology need not be a naturalized one). What has for me emerged as essential is not so much the details of historical cases as the perspective or the ideology that attention to historical cases brings with it.[159]

---

[156] Hesse (1963), p. 287. See also Andersen (2001).

[157] Friedman (1993), p. 37.

[158] Kuhn (1962a), p. 1.

[159] Kuhn (1991a), p. 95; see also his (1992), pp. 106–107. On Kuhn's idea of history, see Gattei (2000b), pp. 294–296.

The perspective Kuhn now deems essential is that according to which science is not seen as a static body of knowledge, but as an ever-developing enterprise.[160] Interesting questions for the philosopher of science do not refer to the rational motivations to regard a particular belief as true, but those that make that belief change: when questions about rationality, objectivity or evidence arise, they are addressed not to the beliefs that were current either before or after the change, but rather to the change itself – philosophy should address not the rationality of belief, but the rationality of incremental change of belief. Kuhn thought that this view may lead to setting aside several of the problems that have troubled philosophers of science for centuries. The theoretical character of observations does not constitute an obstacle for rationality any more: for the philosopher who adopts the historical perspective

> the *rationality* of the conclusions requires only that the observations invoked be neutral for, or shared by, the members of the group making the decision, and for them only at the time the decision is being made. By the same token, the observations involved need no longer be independent of all prior beliefs, but only of those that would be modified as a result of the change.[161]

Furthermore, the evaluation of beliefs in order to weight their truth usually involves a set of secondary criteria such as accuracy, consistency, simplicity, scope or fruitfulness, that all too often have proven to be highly controversial: "To ask which of two bodies of belief is *more* accurate, displays *fewer* inconsistencies, has a *wider* range of applications, or achieves these goals with the *simpler* machinery does not eliminate all ground for disagreement, but the comparative judgement is clearly far more tractable than the traditional one from which it derives".[162]

The price to pay for this perspective to become acceptable is that of relinquishing the realist assumption of the existence of a stable domain of mind-independent entities and the correspondence theory of truth. From Kuhn's point of view, however, by adopting this perspective the conclusions about theory choice, truth and realism that were previously drawn from historical inquiry may now be derived

> from first principles. Approaching them in that way reduces their apparent contingency, making them harder to dismiss as a product of muckraking investigation by those hostile to science. And the approach from principle yields, in addition, a very different view of what's at stake in the evaluative processes that have been taken to epitomize such concepts as reason, evidence, and truth. Both these changes are clear gains.[163]

What the nature of these "first principles" might be, however, remains to be explained.

---

160 See Kuhn (1992), pp. 111–112. "The historian [...] always picks up a process already under way, its beginnings lost in earlier time. Beliefs are already in place; they provide the basis for the ongoing research whose results will in some cases change them; research in their absence in unimaginable, though there has nevertheless been a long tradition of imagining it" (Kuhn (1991a), p. 95).

161 Kuhn (1992), p. 113.

162 Kuhn (1992), p. 114.

163 Kuhn (1992), p. 112.

The deep affinity between Kuhn's theory of scientific revolutions – as it is expressed first in *The Structure of Scientific Revolutions* and more explicitly in the 1980s and 1990s – and Carnap's theory of linguistic frameworks is by no means accidental. Indeed, it reflects the intellectual situation that characterizes the philosophical reflection on the history and philosophy of science at the beginning of the twentieth century.

In the "Preface" to *The Structure of Scientific Revolutions* Kuhn tells how he switched from his studies in the physics of materials to the history of science and describes his own apprenticeship as a historian: "I continued to study the writings of Alexandre Koyré and first encountered those of Emile Meyerson, Hélène Metzger and Anneliese Maier. More clearly than most other recent scholars, this group has shown what it was like to think scientifically in a period when the canons of scientific thought were very different from those current today".[164] Then, in the first, introductory chapter, significantly titled "A Role for History", Kuhn explains the reasons that lie at the roots of his rejection of the model of scientific growth by accumulation:

> [...] historians of science have begun to ask new sorts of questions and to trace different, and often less than cumulative, developmental lines for the sciences. Rather than seeking the permanent contributions of an older science to our present vantage, they attempt to display the historical integrity of that science in its own time. They ask, for example, not about the relation of Galileo's views with those of modern science, but rather about the relationship between his views and those of his group, i.e. his teachers, contemporaries, and immediate successors in the sciences. Furthermore, they insist upon studying the opinions of that group and other similar ones from the viewpoint – usually very different from that of modern science – that gives those opinions the maximum internal coherence and the closest possible fit to nature. Seen through the works that result, works perhaps best exemplified in the writings of Alexandre Koyré, science does not seem altogether the same enterprise as the one discussed by writers in the older historiographic tradition.[165]

It is not surprising that Kuhn sees himself decidedly within the historiographic tradition initiated by Koyré with his *Galileo Studies* in the late 1930s, a tradition that imposed the history of science as an autonomous and independent discipline in the immediate post-war years.

In a 1968 survey of the developments of the history of science Kuhn resumed the idea of the rejection of the cumulative model begun, he said, with "the influence, beginning in the late nineteenth century, of the history of philosophy":[166] a new

---

[164] Kuhn (1962a), pp. vii–viii. Kuhn is here referring to Koyré (1939–1940), Meyerson (1908, 1926), Metzger (1923) and (1930) and Maier (1949). See also Koyré (1931).

[165] Kuhn (1962a), p. 3 (on the new approach to the study of Galileo, compare Kuhn's words with Segre (1991)). The paragraph closes with these words: "By implication, at least, these historical studies suggest the possibility of a new image of science. This essay aims to delineate that image by making explicit some of the new historiography's implications" (*ibidem*).

[166] Kuhn (1968a), p. 107.

> attitude towards past thinkers came to the history of science from philosophy. Partly it was learned from men like Lange and Cassirer who dealt historically with people or ideas that were also important for scientific development. [...] And partly it was learned from a small group of neo-Kantian epistemologists, particularly Brunschvicg and Meyerson, whose search for quasi-absolute categories of thought in older scientific ideas produced brilliant genetic analyses of concepts which the main tradition in the history of science had misunderstood or dismissed.[167]

Finally, in a historiographic/philosophical addendum to a 1984 essay on Planck, Kuhn added some interesting remarks. Replying to some questions about the relationship between some of his works on Planck and the theory of scientific revolutions expounded in *The Structure of Scientific Revolutions*, Kuhn explained that "The concept of historical reconstruction that underlies it has from the start been fundamental in both my historical and my philosophical work. It is by no means original: I owe it primarily to Alexandre Koyré; its ultimate sources lie in neo-Kantian philosophy".[168]

All the above mentioned scholars agree, on general Kantian grounds, in rejecting an ingenuously empiricist epistemology in favour of highlighting the constraints posed by the mind itself. And almost all of them (with the possible exception of Emile Meyerson) distance themselves from Kant by acknowledging that conceptual frameworks significantly evolve through the history of the sciences, and are therefore relative to the various stages of theoretical progress.[169] From this point of view the logical positivists are closer to the Marburg School's neo-Kantianism, as articulated by Cassirer, according to which conceptual frameworks are exemplified, in their purest form, by mathematical physics and mathematical logic. The logical positivists were a step ahead of Cassirer, since they aimed at formulating philosophy too as a branch of exact mathematical sciences – that is, of *Wissenschaftslogik*. In so doing, they eliminated the history of science from their idea of philosophy. It is to Kuhn's great merit, in this context, that he reintroduced the history of science as perhaps the most important reference point of the philosophy of science itself. This process of "historicization", however, inevitably raised the problem of social and cultural relativism that dominates post-Kuhn debates nowadays. There arises the question whether it is possible to deal with this problem in a more satisfying way, continuing to emphasize the relevance of the developments in mathematics, mathematical physics and mathematical logic (as both the neo-Kantian members of the Marburg

---

[167] Kuhn (1968a), p. 108. In these same pages Kuhn mentions the works of Burtt (1925) and Lovejoy (1936), and refers to "the modern historiography of science" founded by "E.J. Dijksterhuis, Anneliese Maier, and especially Alexandre Koyré" (*ibidem*).

[168] Kuhn (1984), p. 361; this essay was later republished as the "Afterword" to the second edition (1987) of Kuhn's dense study *Black-Body Theory and Quantum Discontinuity 1894–1912* (Kuhn (1978)), the culmination of several years of inquiry and research into one of the revolutions in physics of the twentieth century, that comprises Kuhn, Heilbron, Forman, Allen (1967), Kuhn, Heilbron (1969), Kuhn (1975b), (1975c), (1980b), (1980c) and (1984); see also other related pieces, such as Kuhn (1967a), (1967b), (1968b), (1972), (1976c) and (1983d), together with his (V-1980).

[169] See Friedman (2000), (2001) and (2002a).

School and the logical positivists did) and at the same time acknowledge the relevance of the actual historical evolution of the sciences (as did both the Marburg School and the historiographic tradition up to Kuhn).

Chapter 5

The Shadow of Positivism

> The natural result of any investigation is that the investigators either discover the object of search, or deny that it is discoverable and confess it to be inapprehensible, or persist in their search. So, too, with regard to the objects investigated by philosophy, this is probably why some have claimed to have discovered the truth, others have asserted that it cannot be apprehended, while others again go on inquiring.
>
> *Sextus Empiricus*

The received reading of twentieth-century philosophy of science presents Kuhn's position as diametrically opposed to that of Logical Positivism, in whose coffin *The Structure of Scientific Revolutions* is supposed to have hammered the last nail. The neo-empiricist movement has often been described as strenuously advocating a sort of foundationalism of an empirical nature, according to which all knowledge must be traced back to an empirically certain basis of indubitable observation reports. From such a view there would follow the impossibility of genuine scientific revolutions *à la* Kuhn, since scientific progress must follow an eminently cumulative pattern of growth,[1] something Kuhn explicitly rejects from the very beginning.[2] Otherwise,

---

[1] This, as I said, is the received reading. Strictly speaking, however, the logical positivists' efforts to trace knowledge back to indubitable observation reports are not inconsistent with revolutions. To be sure, logical positivists would regard revolutions as rare and unusual events that, on the whole, careful gathering of data and abstracting from them would and should minimize. But new observations can lead to revisions of even the most established theories and no logical positivist would deny that the Copernican revolution or the replacement of Newton's theory of gravitation by Einstein's general theory of relativity were authentic revolutions. They would not have thought revolutions involved things like incommensurability and "world change", and they would have regarded radical discontinuity as unlikely – but while on the whole cumulative, in unusual and rare circumstances there would be non cumulative changes. Logical Positivism, then, is compatible with but *not committed* to a cumulative model.

[2] *The Structure of Scientific Revolutions* opens by rejecting such a pattern off-hand, in its very first introductory chapter, "A Role for History"; Kuhn, however, does not relate it directly to Logical Positivism here. Later on, in the ninth chapter of the book, "The Nature and Necessity of Scientific Revolutions", he rejects the conception, "closely associated with the early logical positivism and not categorically rejected by its successors", that "would restrict the range and meaning of an accepted theory so that it could not possibly conflict with any later theory that made predictions about some of the same natural phenomena" (Kuhn (1962a), p. 98). By claiming that the meaning of a theory is exhausted by its logical implications within a class of theoretically neutral observation sentences, logical positivists would indeed defend such a view.

if we accepted Kuhn's own model, the progress of science would be scattered by radical discontinuities, totally incompatible with the naïve form of empiricism advocated by logical positivists. It is not surprising, then, that Kuhn's theory of scientific revolutions is commonly viewed as the determining factor of positivism's demise.[3]

However, at a closer look things happen to be quite different. Indeed, in the past fifteen years an ever-increasing volume of research has been devoted to a thorough and careful study of the birth, development and decline of the neo-empiricist movement.[4] These studies have shown that the traditional picture of the relationship between Kuhn and Logical Empiricism is to be regarded as a myth – a gross, generic and from several fundamental respects substantially misleading simplification.

Carnap and Kuhn

Emblematic, among all others, is the case of Rudolf Carnap, one of the leading and most influential members of the Vienna Circle, a point of reference for the logical positivistic movement in the United States – certainly one of the most remarkable exponents of the logical-empiricist approach to the philosophy of science from which Kuhn was allegedly so determined to distance himself.

To many, the name of Carnap evokes ideas such as the search for an empirical foundation in epistemology, verifiability theory of meanings in semantics, aversion for any form of metaphysics, confirmation theory in methodology, the cumulative model for the growth of knowledge, a dry, formalistic philosophical style, concerned with logical analysis and the rational reconstruction of language in science. On the contrary, the name of Kuhn is commonly associated with practically opposed concepts: an informal, evolutionary style that pays great attention to the actual practice of science, the sharp distinction between normal, extraordinary and revolutionary periods in the history of science – respectively characterized by a scientific activity wholly committed to puzzle solutions under the guide of paradigms (without any references to forms of confirmations or falsifications), the emergence of crisis and breakdown of the commitment, and paradigm shifts characterized by *Gestalt* switches, actual conversions akin to religious and political ones, incommensurability and semantic holism, all fields in which little space seems to be left for experience and reason.

---

[3] See, for example, Suppe (1974), where Logical Empiricism is presented as "the Received View", with which more recent views are contrasted (among them Kuhn's). See also Rorty (1979), pp. 332–333 and 335–342, and Giere (1988), p. 32.

[4] See the essays by Coffa (1991), Uebel (ed.) (1991), Friedman (1992a), (1999), (2000) and (2001), Stadler (ed.) (1993) and (2003), Galison (1993), (1995) and (1996), Seiler, Stadler (1994), Cartwright, Cat, Fleck, Uebel (1996), Stadler (1997), Richardson (1998), Parrini (2002) and Parrini, Salmon, Salmon (eds) (2003).

In the light of some recent studies,[5] this picture appears at the very least superficial. In particular, among the undeniable divergences,[6] emerge deep and unexpected similarities. For at a closer look their respective philosophies are revealed to be more complex and articulated than is usually thought. And if we focus on the successive phases of their reflections – the period following *The Logical Structure of the World*, beginning in the early 1930s, and Kuhn's "linguistic turn", in the 1980s and 1990s – we notice subtle but significant modifications that, instead of widening the gap separating their views, strikingly seem to be bridging it.

In fact, in Kuhn's reflection on the nature and character of scientific revolutions it is possible to find the informal counterpart of the reltivized conception of *a priori* constitutive principles developed by logical positivists. The Kuhnian distinction between paradigm change during extraordinary science periods and the activity peculiar to normal science periods reflects Carnap's distinction between change of language or linguistic framework and the rule-governed activity operated within such frameworks.

During normal science periods scientists operate within a generally shared conceptual framework that defines rules and evaluative criteria – that is, the laws of the linguistic game – of a given research area. Such criteria are not questioned during the periods of normal science. On the contrary: these very criteria and standards are what makes the puzzle-solving activity peculiar to normal science possible. In Carnap's own terms, they constitute the rules governing a given linguistic framework and defining the set of internal questions. Just as for Carnap the logical rules of a linguistic framework define or constitute the notion of correctness or validity with reference to a framework, so a particular paradigm, governing a given stage, or phase, of normal science, generally involves some shared rules (even if only tacitly shared, perhaps) that define or constitute what is regarded as a valid or correct solution to a problem in that particular period of normal science.

During the periods of revolutionary theoretical change those very criteria, previously shared and agreed upon, become something researchers start casting their doubts upon, depriving scientists of the references on the basis of which they could motivate and support the transition to a new paradigm. In Carnap's terms, scientists face an external question, concerning the replacement of the rules governing a given linguistic framework with new rules, often radically different from them.[7]

---

[5] See, in particular, Reisch (1991), Irzik, Grünberg (1995), Friedman (1991), (1992a) and (1993), Earman (1993), Axtell (1993) and Irzik (2002).

[6] See particularly Earman (1993), pp. 21–24.

[7] The distinction between internal and external questions is explicit in Carnap (1950b), but the core idea is already clearly presented in his (1934): here it is principally applied to the debate over the foundations of mathematics among logicism, formalism and intuitionism, each of which is reinterpreted as a proposal to reformulate the overall language of science in accordance with the respective systems of logical rules (with or without the principle of excluded middle, for instance). Besides constituting a central point in Carnap's philosophy of formal languages (or linguistic frameworks), such a distinction can be seen to be particularly relevant to highlight the close resemblance between Kuhn's and Carnap's positions as regards truth: see below, n. 55.

Two letters from Carnap to Kuhn

It is not amazing, then, that in 1960 Carnap, in his capacity as editor (with Charles Morris) of the *International Encyclopedia of Unified Science*,[8] writes to Kuhn to express his personal approval for the inclusion of *The Structure of Scientific Revolutions* in the grand neo-positivistic editorial project:

> I believe that the planned monograph will be a valuable contribution to the Encyclopedia. I am myself very much interested in the problems which you intend to deal with, even though my knowledge of the history of science is rather fragmentary. Among many other items I liked your emphasis on the new conceptual frameworks which are proposed in revolutions in science, and, on their basis, the posing of new questions, not only answers to old problems.[9]

Two years later, in a second letter, Carnap shares with Kuhn his impressions after reading the final version of the manuscript of *The Structure of Scientific Revolutions*:

> I am convinced that your ideas will be very stimulating for all those who are interested in the nature of scientific theories and especially the causes and forms of their changes. I found very illuminating the parallel you draw with Darwinian evolution: just as Darwin gave up the earlier idea that the evolution was directed towards a predetermined goal, men as the perfect organisms, and saw it as a process of improvement by natural selection, you emphasize that the development of theories is not directed toward the perfect true theory, but is a process of improvement of an instrument.[10] In my own work on inductive logic in recent years I have come to a similar idea: that my work and that of a few friends in the step for step solution of problems should not be regarded as leading to "the ideal system", but rather as a step for step improvement of an instrument. Before I read your manuscript I would not have put it in just those words. But your formulations and clarifications by

---

[8] As we saw in the first chapter, this ambitious project – started at the end of the 1930s by one of the core exponents of the Vienna Circle, Otto Neurath – was directed by Rudolf Carnap and Charles Morris; its board of scientific advisers comprised, besides some of the most significant names of the international philosophical and scientific community (Niels Bohr, John Dewey, Federigo Enriques, Jan Łukasiewicz, Richard von Mises, Bertrand Russell, Alfred Tarski, Edward C. Tolman) also the most eminent exponents of the neo-positivistic movement: apart from Carnap and Morris, also Herbert Feigl, Philipp Frank, Jørgen Jørgensen, Arne Naess, Ernest Nagel, Hans Reichenbach, Louis Rougier and Joseph H. Woodger (among them, at least in the early stages of the movement, also Lizzie Susan Stebbing, untimely passed away in 1943). Furthermore, it is to be noted that Kuhn's doctoral adviser at Harvard was Percy W. Bridgman (a Nobel laureate for Physics, in 1946), who had deep philosophical interests and became famous for promoting "operationism", a variant of positivistic empiricism: see Bridgman (1927).

[9] From Carnap's letter to Kuhn, 12 April 1960, in Reisch (1991), p. 266. There Carnap also expresses his interest for other manuscripts Kuhn had sent to him. For the analysis of the similarities between Carnap's and Kuhn's philosophical stances, I will particularly follow Reisch (1991), Irzik, Grünberg (1995) and Irzik (2002).

[10] See Kuhn (1962a), pp. 170–173.

> example and also your analogy with Darwin's theory helped me to see clearer what I had in mind.[11]

Carnap says he found the analogy with Darwin's evolutionary theory "illuminating" not on the basis of his own knowledge of the history of science, which Carnap himself describes as "fragmentary" in his first letter, but rather for shedding light on his own and others' work in the "step for step solution of problems". Just as biological evolution is not "directed towards a predetermined goal", so Carnap's principle of tolerance[12] secures that there is no ideal philosophical model for a scientific theory, no "ideal system" on which philosophical analysis will eventually converge. On the contrary: various philosophical goals will eventually generate "species" of philosophical "instruments" intended to clarify and reconstruct the scientific reasoning on the grounds of a particular advantage or in view of a given aim. Just as biological species may become more adapted within their respective ecological niches, so these various instruments may become more "effective" and "fruitful" in achieving their goal.[13] On the other hand, these instruments may turn out to be useless and become obsolete: it is a point Carnap raises in "Empiricism, Semantics, and Ontology" (1950), written perhaps in those "recent years" during which he had come "to a similar idea" of the evolution of philosophical instruments: "Let us grant to those who work in any special field of investigation the freedom to use any form of expression which seems useful to them; the work in the field will sooner or later lead to the elimination of those forms which have no useful function".[14]

Carnap says he came to this formulation thanks indeed to the formulations, the examples and the analogy to Darwin's theory advanced by Kuhn in the manuscript of *The Structure of Scientific Revolutions*. Not only, then, has the analogy with Darwinian evolutionism evoked "a similar idea" of progress in Carnap's work, but it also concurred with the wide historical framework provided by Kuhn – this is what Carnap probably refers to with "formulations and clarifications by example". This part of the analogy between progress in Logical Empiricism (or at least in

---

[11] From Carnap's letter to Kuhn, 28 April 1962, in Reisch (1991), pp. 266–267. In a note written the day before this letter Carnap directly refers to ch. XIII of *The Structure of Scientific Revolutions* ("Progress through Revolutions"): "The development of scientific theories", he writes, "is to be understood analogously [by analogy, that is, with the evolutionary perspective described by Kuhn in the final chapter of his book]: that is, not as directed towards an ideal, true theory, the *one and only* true theory about the world, but as a development understood as a step towards a better form, through the selection of one form out of many competing ones. The selection is made on the grounds of the preference expressed within the community of scientists, keeping in mind that many sociological, cultural, etc. factors play also a role. Not: 'we come closer to the truth'; rather: 'we improve an instrument'" (RC 082-03-01). See also Irzik (2003).

[12] See Carnap (1963a), p. 55: "every one is free to choose the rules of his language and thereby his logic in any way he wishes".

[13] See Carnap (1963a), pp. 66–67, and Kuhn (1991a), pp. 102–103 and 104, (1992), p. 120, and (1993a), p. 250.

[14] Carnap (1950b), p. 221.

inductive logic) and the history of science, understood as a succession of paradigms and conceptual frameworks, turns out to be fundamental to understanding the depth of the relationship between Carnap and Kuhn: just as the latter thinks that different paradigms are incommensurable, since each comprises its own criteria for theory appraisal and choice, so for the former different philosophical languages need to be evaluated not on the basis of a single canon of adequacy but, rather, on the grounds of the different aims in view of which they are introduced.

Normal science and extraordinary science

According to Kuhn, history of science roughly witnesses two different kinds of periods: those of "normal science", during which a particular scientific community works without asking too many questions on the paradigm (or lexicon) it supports and is committed to, which dictates the same rules and criteria for everyday scientific practice to all its members; and those of "extraordinary science", during which is questioned the consensus shared by the community members and the very commitment that unites scientists during normal science periods. Along similar lines, according to Carnap there are two substantially different kinds of activity associated with the linguistic frameworks within which our theories are formulated: the possibility of judging "internal" questions on the grounds of the logical rules shared within an individual linguistic framework, and the possibility of judging "external" questions that, by hypothesis, do not and cannot presuppose such logical rules.[15]

Just as for Carnap the logical rules of a linguistic framework are constitutive of the notion of propriety or validity relative to that framework, so for Kuhn the particular paradigm governing a given period of normal science involves generally accepted (albeit frequently tacit) rules, constituting what is to be regarded as a valid or proper solution to a given problem. Just as for Carnap external questions, concerning which linguistic framework is to be adopted, are not governed by logical rules but rather require appeal to less rigorously defined considerations of pragmatic and conventional nature, so for Kuhn the revolutionary paradigm changes in which periods of extraordinary science culminate do not follow generally accepted rules, as during the periods of normal science, but rather require something resembling a conversion.[16]

Since linguistic forms are not right or wrong *per se*, but are better or worse to the extent that they succeed in reaching the aim in view of which they had been

---

[15] According to Carnap, the standard procedure to deductively test a scientific theory on the basis of its observational predictions ends up by not being entirely governed by pre-established rules. For, he explains, when facing a clash between theoretical predictions and observational results we confront three possibilities: either we reject the theoretical assertions from which unsuccessful predictions have been derived, or we reject the observation statements clashing with the theory in question, or else we modify the logical rules of the language so as to eliminate the inconsistency between observation and theory. Carnap relates this point of view to the epistemological holism he associates with the works of Duhem and Poincaré, thus making particularly explicit the relationship between his own philosophy and traditional empiricist foundationalism (see Carnap (1934), §84).

[16] See Reisch (1991), Friedman (1992a), (1993) and (2001), and Earman (1993).

constructed, for Carnap a sort of planning of language is required. And if the principle of linguistic planning in philosophy is extended so as to include also the choices between radically different theories – as indeed Carnap does – the image of the scientific revolution that results from it is very similar to Kuhn's. Normal science seems then to coincide with the scientific activity practiced within a theoretical framework, that is, within a given scientific language. In this context language remains stable and scientific activity consists in assigning truth values to theoretical sentences. On the contrary, revolutionary periods occur, according to Carnap, when the adequacy and the value of a language are questioned. At this moment the terms of the debate change and it is not easy to establish what the new form and structure of the scientific language must be. Since the value of linguistic frameworks lies in their usefulness in view of a certain aim, revolutionary scientific debate will hinge on values, principles and beliefs concerning what language would better serve that aim and how it could be constructed.

Kuhn underlines this very same point while defending his own historiography against the charge of attributing an irrational nature to science: "Debates over theory-choice cannot be cast in a form that fully resembles logical or mathematical proof",[17] since debates over the different merits of incommensurable theories concern the very premises that include the canonical and generally accepted criteria on the grounds of which scientists can evaluate them. Until these premises are clarified, debate will still hinge on values and *desiderata* of a practical nature.

As the two letters referred to above highlight, Carnap welcomes Kuhn's monograph in the *International Encyclopedia of Unified Science* without expressing any reservation on the implications for scientific rationality it is usually assumed the book contains. Maybe he sensed a feature that escaped later critics' analyses: scientific activity somehow depends on practical decisions that, since they regard first principles, cannot be justified by appealing to those very first principles. In Carnap's eyes, that does not entail any implications of irrationality.

Most importantly, perhaps, Carnap (as opposed to many other critics) sensed that the radical challenge of *The Structure of Scientific Revolutions* was not at rationality, but at realism.

A later attempt of clarification

In the light of some recent literature on the similarities of their respective views, in one of his later papers Kuhn takes the opportunity to speak of his reaction to Carnap's letter: at the time he was writing *The Structure of Scientific Revolutions*, he says,

> I was almost totally innocent of the post-*Aufbau* Carnap, and discovering him has distressed me acutely. Part of my embarrassment results from my sense that responsibility required that I know my target better, but there is more. When I received the kind letter in which Carnap told me of his pleasure in my manuscript, I interpreted it as mere politeness, not as an indication that he and I might usefully talk.[18]

---

[17] Kuhn (1970c), p. 199.

[18] Kuhn (1993a), p. 227.

Before that, however, he adds a sort of disclaimer: "Whatever role the problems encountered by positivism may have played in the background for *The Structure of Scientific Revolutions*, my knowledge of the literature that attempted to deal with those problems was decidedly sketchy when the book was written".[19] And he further qualifies his own words by immediately highlighting the "deep difference" separating his views from Carnap's, which Kuhn thinks persists despite the "deep parallels"[20] between them. Contrary to what might be reasonably expected, such a difference does not lie in the fact that Carnap's linguistic rules should always be explicitly stated, while Kuhn's criteria and scientific standards are largely tacit and are therefore reinforced by consensus and implicit conventions rather than by an explicit set of formal rules. Rather, Kuhn highlights how he, as opposed to Carnap, is concerned from the very start with the historical *development* of knowledge: "I have seen each stage in the evolution of a given field as built – not quite squarely – upon its predecessor, the earlier stage providing the problems, the data, and most of the concepts prerequisite to the emergence of the stage that followed".[21] In addition, "while the cognitive significance of language change was for [Carnap] merely pragmatic",[22] Kuhn says he has always insisted that

> some changes in conceptual vocabulary are required for the assimilation and development of observations, laws, and theories deployed in the later stage (whence the phrase "not quite squarely" above). Given those beliefs, the process of transition from old state to new becomes an integral part of science, a process that must be understood by the methodologist concerned to analyze the cognitive basis for scientific beliefs. Language change is *cognitively* significant for me as it was not for Carnap.[23]

Kuhn's point seems to be this: for Carnap, language change involves "external" questions and therefore has a purely pragmatic character, not cognitive or epistemic in the single sense in which Carnap understands the term "epistemology". For although Carnap links his idea of language change to scientific revolutions, he never provides a detailed discussion of such revolutions. That would require a historical analysis that does not belong to the proper scope Carnap assigns to epistemology, or "logic of science" (*Wissenschaftslogik*) – namely the formulation and analysis of a variety of possible linguistic frameworks within which the results of individual sciences may be properly fitted. What is fundamental, for Carnap, is that the only extant and genuinely philosophical problems are merely formal, that is, they pertain to the application of the logic of language to the particular sciences. Although several empirical problems may arise in the analysis of historical transitions from one theory to another through a revolution (and in his letters to Kuhn Carnap expresses his interest in such questions), the only genuinely philosophical questions concern the (merely a-temporal) articulation of the logical structures of the two different

---

19 Kuhn (1993a), p. 227; see also Kuhn (I-1997a), pp. 305–306.

20 Both expressions are in Kuhn (1993a), p. 227.

21 Kuhn (1993a), p. 227.

22 Kuhn (1993a), p. 227.

23 Kuhn (1993a), pp. 227–228.

languages considered each time.[24] For Kuhn, on the contrary, as is evident from the first chapter of *The Structure of Scientific Revolutions*, the point is precisely that historical analysis of scientific change may have, in the first place, a genuinely philosophical character.

However, in his later years Kuhn explicitly acknowledges how his philosophy is in several respects very similar to the relativized view of *a priori* previously developed by Reichenbach and Carnap:

> Though it is a more articulated source of constitutive categories, my structured lexicon resembles Kant's a priori when the latter is taken in its second, relativized sense. Both are constitutive of *possible experience* of the world, but neither dictates what that experience must be. Rather they are constitutive of the infinite range of possible experiences that might conceivably occur in the actual world to which they give access. Which of these conceivable experiences occurs in that actual world is something that must be learned, both from everyday experience and from the more systematic and refined experience that characterizes scientific practice. They are both stern teachers, firmly resisting the promulgation of beliefs unsuited to the form of life the lexicon permits. What results from respectful attention to them is knowledge of nature, and the criteria that serve to evaluate contributions to that knowledge are, correspondingly, epistemic. The fact that experience within another form of life – another time, place, or culture – might have constituted knowledge differently is irrelevant to its status as knowledge.[25]

As years went by, then, the similarities between Carnap's and Kuhn's philosophies deepened and radicalized, rather than diminished.

The linguistic conception of scientific theories

The treatment of scientific theories in terms of languages is characteristic of the whole of Carnap's philosophical output. They are, generally speaking, linguistic structures involving observational and theoretical vocabularies, and rules for the formation of propositions. Such a way of conceiving theories – clearly inspired by formal mathematical systems – spans over the whole of Carnap's philosophy of science, but his remarks on the scientific practice are also expressed in these terms. In his eyes, scientists do not refine, or change, a theory, but alter or modify the truth values of some of its propositions. As he himself admits in the first of the above quoted letters to Kuhn, his knowledge of the history of science is "rather fragmentary": his historical remarks on concrete scientific practice, then, betray personal intuitions rather than historical inquiries into which was or still is actual scientific practice. In spite of all this, Carnap's idea of scientific revolutions is strikingly similar to the picture Kuhn offers of them, drawing upon his substantial historical knowledge of revolutionary science.

---

[24] See Carnap (1934), §72 ("Philosophy Replaced by the Logic of Science"), and (1936a).

[25] Kuhn (1993a), p. 245.

Linguistic frameworks and lexicons

Carnap's linguistic turn takes place in *The Logical Syntax of Language* (1934) and culminates in "Empiricism, Semantics, and Ontology" (1950).[26] Its key notion is that of a linguistic framework. The linguistic framework *L* of a theory comprises a set of syntactical and semantic rules:[27] among other things, these include meaning postulates expressing the logical meaning dependencies that hold between the meanings of the primitive descriptive terms of *L*, and rules for assigning the degree of confirmation to a hypothesis relative to a body of evidence. The most important syntactic rules are formation rules, that establish what a proposition of *L* can be, and transformation rules, that define what a consequence of a proposition in *L* is; semantic rules comprise rules for the truth conditions of the propositions of *L*, rules for designation, that specify the relationships between the terms of *L* and what they designate, and meaning postulates, that express the logical subordination of meaning in force among the meanings of the primitive descriptive terms of *L*.

Carnap thinks that every scientific theory is embedded within a linguistic framework. A scientific theory consists of the conjunction of theoretical postulates and correspondence rules. Theoretical postulates serve to introduce primitive theoretical terms into the theory and express the fundamental laws about a certain domain of phenomena. In conjunction with theoretical postulates, correspondence rules provide a partial interpretation of theoretical terms, facilitating their application to phenomena by permitting the derivation of observation statements.

Carnap's position with respect to linguistic frameworks encapsulates a fundamental distinction and two philosophical principles. He works with a sharp distinction between internal and external questions of reality and existence.[28] On the one hand, internal questions of existence have theoretical answers that can be provided, completely or in part, on the basis of the rules governing the framework. External questions, on the other, raise the problem of the reality of entities prior to the endorsement of the linguistic framework. Thus the answers to external questions either lack cognitive sense because they are pseudo-questions, or they need to be reconstructed as proposals, justifiable only on the grounds of considerations of practical utility, relative to the adoption or rejection of a linguistic framework as a whole, and as such they fall under the scope of Carnap's principle of tolerance, according to which each one is free to choose the logical rules governing his own language.

His second principle is the doctrine of meaning holism, according to which the rules of a language contribute to the meanings of primitive terms occurring in them, and a change of the rules results in a change in their meaning. Such semantic holism pertains both to the logical-mathematical terms and to the empirical primitive ones. As to the former, Carnap offers this formulation:

---

[26] Carnap (1934) and (1950b), respectively. The revised edition of this latter essay, published in 1953, contains an important qualification of the term "framework", that is now "used only for the system of linguistic expressions, and not for the system of the entities in question" (Carnap (1950b, 1953) p. 509, footnote).

[27] See Carnap (1950b) and (1963b), pp. 889–905.

[28] See Carnap (1950b).

> Up to now, in constructing a language, the procedure has usually been, first to assign a meaning to the fundamental mathematico-logical symbols, and then to consider what sentences and inferences are seen to be logically correct in accordance with this meaning. [...] no conclusion arrived at in this way can very well be otherwise than inexact and ambiguous. The connection will only become clear when approached from the opposite direction: let any postulates and any rules of inference be chosen arbitrarily; then this choice, whatever it may be, will determine what meaning is to be assigned to the fundamental logical symbols.[29]

On the other hand, Carnap's meaning holism with respect to the primitive terms introduced by theoretical and meaning postulates, while it can be found in his published writings,[30] is perhaps best expressed in a note he wrote in 1955 for one of Herbert Feigl's papers:

> I would not say that the *postulates* of a theory, e.g. the fundamental laws of physics, are *"true in virtue of the meanings of the terms"*. I am inclined at the present moment (quite tentatively, pending further clarification) to distinguish between two kinds of postulates. The *M-postulates* determine only meanings and hence are analytic (e.g. "eagles are birds", "no bachelor is married"). The *P-postulates* (e.g. physical laws) contribute to the determination of the meanings, but also to the description of the world and are hence synthetic. Thus their truth is not entirely due to their meanings.[31]

This passage makes clear Carnap's adherence to semantic holism in a sense very similar to Kuhn's. Carnap's P-postulates function very much like Kuhn's symbolic generalizations in the disciplinary matrix he described in 1969.[32] They both have a double function, semantic and factual. It is precisely for this reason that they resist piecemeal refutation by experiment.

Indeed, there is a distinct similarity between the activity during normal science periods, *à la* Kuhn, and the activity within Carnap's linguistic frameworks, and consequently a similarity also between their respective conceptions of scientific revolutions. According to Carnap, scientific revolutions occur in two ways: by a change in the rules of the linguistic framework of the theory or by a change in the theoretical postulates. The most succinct formulation of this is found in Carnap's 1963 reply to Quine:

> I should make a distinction between two kinds of readjustment in the case of a conflict with experience, namely, between a change in the language, and a mere change in or addition of, a truth-value ascribed to an indeterminate statement (i.e. a statement whose truth value is not fixed by the rules of language, say by the postulates of logic, mathematics and physics). A change of the first kind constitutes a radical alteration, sometimes a revolution,

---

29 Carnap (1934), p. xv. A certain form of holistic conception of meaning emerges also from other logical positivists' works in the 1950s: see Earman (1993), pp. 11–12, and Friedman (1987) and (1992b); see also Richardson (1998).

30 See Carnap (1963b), pp. 958–960.

31 Carnap, "Comments on Feigl's 'Philosophy of Science'", 18 January 1955 (RC 090-63-24); this note refers to the paper Feigl delivered during a 1954 Congress in Zürich. See also Irzik (2003).

32 See Kuhn (1970c), pp. 182–184.

> and it occurs only at certain historically decisive points in the development of science. On the other hand, changes of the second kind occur every minute. A change in the first kind constitutes, strictly speaking, a transition from a language L_n to a new language L_n+1.[33]

Scientific revolutions

A revolution is the shift from one linguistic framework to another. A linguistic framework is defined by its rules:[34] altering them would alter the scientific language and result in a revolution. Particularly interesting would be the adoption of different meaning postulates and rules of confirmation. The former would give us new terms or new meanings for old terms, and the latter would give us new assignments for degrees of confirmation.

Alternatively, a revolution can also occur if the theoretical postulates of the theory change. Since one of the functions of these postulates (that express the fundamental laws of the theory) is to introduce new theoretical terms, the addition of such terms is tantamount to the addition of new postulates for them, and that is precisely what happens during scientific revolutions: the class of theoretical terms "will generally be changed only when a radical revolution in the system of science is made, especially by the introduction of a new primitive theoretical term and the addition of postulates for that term".[35]

It is because the theoretical postulates introduce theoretical terms and partially contribute to their meanings that they are recalcitrant to refutation by experience, like analytic sentences. Indeed, the above quoted passage from Carnap's reply to Quine ends by noting just that: "To be sure, this status [the analyticity of language] has certain consequences in case of changes of the second kind [that is, non-revolutionary changes], namely, that analytic sentences cannot change their truth-value. But this characteristic is not restricted to analytic sentences; it holds also for certain synthetic sentences, e.g. physical postulates and their logical consequences".[36]

---

33 Carnap (1963b), p. 921. Carnap distinguishes between two kinds of modifications the scientists generally introduce to theory when it conflicts with experience. The first kind involves the evaluation or re-evaluation of individual propositions (or perhaps even sub-theories) within a wider and more stable linguistic framework, or language: it is the typical activity of normal science periods, as described by Kuhn. The second kind considers the possibility of completely re-elaborating the very structure of the linguistic framework in which the previous activity fitted in, till transforming it, in fact, into another language: it is the typical activity of extraordinary science periods, that result in a revolution.

34 The same holds for Wittgenstein's linguistic games: see his (1953), Part I, §§185–242: where arguments no longer work, all that is left is a common behaviour, a custom we must share in order not to cancel the intersubjective dimension of the agreement that guides us in using a given rule. No interpretation (that is, no formulation of the rule) can encapsulate its singularities, because it cannot determine the way in which it will be applied. Training is necessary in order to form agreement, and it is only because it is the condition for agreement that a rule achieves its end and fulfils its prescriptive character. Only since it is shared is a rule in fact a rule: only when it belongs to a consolidated practice does it acquire its meaning, otherwise it is only an empty form, that can be applied in various and different ways. See Kripke (1981).

35 Carnap (1956), p. 51.

36 Carnap (1963b), p. 921.

That not every theoretical change constitutes a radical revolution is also evident from Carnap's treatment of correspondence rules. He writes:

> I am not thinking now of a revolution in physics, in which an entirely new theory is developed, but of less radical changes that modify existing theories. Nineteenth-century physics provides a good example, because classical mechanics and electromagnetics had been established, and, for many decades, there was relatively little change in fundamental laws. The basic theories of physics remained unchanged. There was, however, a steady addition of new correspondence rules, because new procedures were continually being developed for measuring this or that magnitude.[37]

This passage makes clear also that Carnap views the formulation of new correspondence rules as an activity akin to Kuhn's notion of "normal" paradigm articulation, which in no way involves revolutionary changes.

Given the similarities between, on the one hand, the way in which Kuhn's symbolic generalizations and Carnap's theoretical postulates function and, on the other, between what Kuhn says about lexicons and what Carnap says about linguistic frameworks, it is not at all surprising that the two philosophers have nearly identical views about how scientific revolutions occur. Both think that there are mainly two ways in which revolutions in science occur: either by change of the paradigm (or theory, or lexicon), or by lexical change, that is, change of the linguistic framework.

To sum up, then, both Carnap and Kuhn consider a scientific theory as embedded in a linguistic structure that restricts scientific activity. As a consequence, they both sharply separate between scientific activity within a theory and its lexical (or linguistic) framework, and scientific activity outside it – or, in other words, between normal and revolutionary science.[38] Finally, they both hold a form of local meaning holism.

Incommensurability

Once the thesis of meaning holism is endorsed, the thesis of semantic incommensurability necessarily follows for both philosophers. Carnap did not articulate this consequence nor develop its consequences, but already in 1936 he wrote: "In translating one language into another the factual content of an empirical statement cannot always be preserved unchanged. Such changes are inevitable if the structures of the two languages differ in essential points".[39] In this passage we

---

37 Carnap (1966), pp. 237–238.

38 See also above, ch. 4, n. 143.

39 Carnap (1936b, 1949), p. 126. This is one of the papers Carnap gave at the *First International Congress for The Unity of Science*, held in Paris on 15–23 September 1935. The passage I have just quoted continues along a line that very closely resembles some passages from *The Structure of Scientific Revolutions* and other works of Kuhn. Take for example the following: "[…] while many statements of modern physics are completely translatable into statements of classical physics, this is not so or only incompletely so with other statements. The latter situation arises when the statement in question contains concepts (like, e.g. 'wave function' of 'quantization') which simply do not occur in classical physics; the essential point being that these concepts cannot be subsequently included since they presuppose a different

find one of the characteristic theses of post-positivistic philosophy of science: the non-existence of neutral facts, independent from the theoretical context in which they are revealed, and semantic incommensurability, that is, the difficulty of mutual translation between different paradigms or lexicons.[40]

The idea that one language can comprise propositions that cannot be translated into another language is also present in a paper published several years later: "It seems to me obvious that, if two men wish to find out whether or not their views on certain objects agree, they must first of all use a common language to make sure that they are talking about the same objects.[41] It may be the case that one of them can express in his own language certain convictions which he cannot translate into the common language; in this case he cannot communicate these convictions to the other man. For example, a classical mathematician is in this situation with respect to an intuitionist or, to a still higher degree, with respect to a nominalist".[42]

---

form of language. [...] some statements of classical physics could not be translated into the new language, and others only incompletely. (This means not only that previously accepted statements would have to be rejected; but also that to certain statements – regardless of whether they were held true or false – there is no corresponding statement at all in the new language)" (*ibidem*). See also Carnap (1946), pp. 598–602.

[40] See also Wittgenstein (1969), §65.

[41] For others it is not obvious at all – quite the opposite: "frameworks, like languages, may be barriers. They may even be prisons. But a strange conceptual framework, just like a foreign language, is no absolute barrier: we can break into it, just as we can break out of our own framework, our own prison. And just as breaking through a language barrier is difficult but very much worth our while, and likely to repay our efforts not only by widening our intellectual horizon but also by offering us much enjoyment, so it is with breaking through the barrier of a framework. A breakthrough of this kind is a discovery for us. It has often led to a breakthrough in science, and it may do so again" (Popper (1976), p. 61). See also Popper (1970), p. 56, and (1974d), pp. 1152–1153. This attitude, in Popper's eyes, has a fundamental ethical value: "While we *may* interpret our views [according to the myth of the framework], we do not *have* to do so. We can choose to pursue an aim or goal – such as the aim of understanding better the universe in which we live, and ourselves as part of it – which is autonomous of the particular theories or frameworks that we construct to try to meet this aim. And we can choose to set ourselves standards of explanation, and methodological rules, which will help us achieve our goal and which it is *not* easy for any theory or framework to satisfy. Of course, we may choose not to do this: we may decide to make our ideas self-reinforcing. We may set ourselves no task other than one we know our present ideas can fulfil. We certainly can choose to do all this. But if we do choose to do this, not only will we be turning our backs on the possibility of learning that we are wrong, we will also be turning our backs upon that tradition of critical thought (stemming from the Greeks and from culture clash) which has made us what we are, and which offers us the hope of further self-emancipation through knowledge)" (Popper (1976), pp. 60–61).

[42] Carnap (1963b), pp. 929–930. Interestingly, Carnap seems to refer to a sort of incommensurability thesis in mathematics – an extension Kuhn never attempted (and that was suggested by Eugenio Frola some fifteen years before Kuhn and Feyerabend: see above, ch. 3, pp. 91–96) in his (1992) Giulio Giorello remarks that in a revolution new problems and new conjectures, new methods "to discover the truth", unexpected connections and new ontologies do not produce, in mathematics – as opposed as to natural sciences – something resembling the consequences of a scientific revolution as Kuhn understands it. In particular,

I think these passages capture the gist of what Kuhn, in his later writings, describes as "local incommensurability".[43]

Truth

Now I wish to focus my attention on what I take to be a central issue to understand both the close relationship between Kuhn and Logical Positivism, and the nature of the divide separating Popper and the philosophical tradition Kuhn belongs to: the notion of truth.

In the first edition of *The Structure of Scientific Revolutions* (1962) Kuhn hardly refers to the concept of truth: he has no need of it, not even in order to characterize and explain progress:[44] "The developmental process described in this essay has been a process of evolution *from* primitive beginnings – a process whose successive stages are characterized by an increasingly detailed and refined understanding of

---

there would be no trace, in mathematics, of the characteristic "Kuhn losses": loss in the ability to explain certain phenomena, loss of some (authentic) scientific problems, due the narrowing of a certain scientific field, and, finally, (partial) loss of communicability among practitioners of different scientific disciplines (see above, ch. 2, n. 204). If, in physics as well as in other natural sciences, progress seems to be possible only at the price of some regress, in mathematics, according to Giorello's analysis, such losses in explanatory power seem not to take place: after a revolution (in a broader, not strictly Kuhnian, sense) we would have a substantial enrichment of mathematical content. To the issue of revolutions in mathematics is dedicated the whole of Gillies (ed.) (1992): examples are the discovery of irrational numbers, the introduction of the infinitesimal calculus, the discovery of non-Euclidean geometries or the transition from geometry to algebra. In a letter to Gillies of 2 April 1990 (declining the offer to write a preface for that book), while admitting that he had never been concerned with the problem of revolutions in mathematics, Kuhn declares himself "persuaded that there are subtle but important differences between mathematics and the natural sciences with respect to the developmental pattern outlined in *Structure*, and that work on them (the differences) would clarify the nature of both". He also draws a most interesting parallel between himself and Lakatos, highlighting how their respective different training (Kuhn as a theoretical physicists, Lakatos as a mathematician) probably constituted the major source of their disagreements. Such a different background informed the very terminology they adopted. Take, for example, their different view of the prelude to a revolution: for Kuhn this is a crisis, while for Lakatos it is the degenerative phase of a research programme. Or, again, the quite distinctive way in which mathematicians and physicists confront the works of their predecessors: indeed, Lakatos cannot accept the incommensurability thesis as advanced by Kuhn and Feyerabend. "These are only two examples of cases in which an apparent disagreement can", Kuhn says, "be traced to genuine differences in the fields on which our methodologies were, in the first instance, based". An exception, notes Kuhn, is the choice of "paradigm" instead of "research programme": indeed, Kuhn regrets he did not employ the latter for the first, wider sense of "paradigm" (I thank Donald Gillies for pointing this letter out to me and for providing me with a copy of it). See also Kuhn (1970b), (1971a) and (1980a).

[43] See Kuhn (1983a), pp. 35–37, and (1991a), p. 97.

[44] See Kuhn (1962a), pp. 170–173. In fact, in the first edition of *The Structure of Scientific Revolutions* the term "truth" appears only in a quotation from Francis Bacon (on p. 18).

nature. But nothing that has been or will be said makes it a process of evolution *toward* anything".[45]

In the 1969 "Postscript" to the second edition of the book (1970) he introduces two arguments against the notion of truth implicit in the traditional view of progress as increasing verisimilitude.[46] Let me quote Kuhn's own words at some length:

> A scientific theory is usually felt to be better than its predecessors not only in the sense that it is a better instrument for discovering and solving puzzles but also because it is somehow a better representation of what nature is really like. One often hears that successive theories grow ever closer to, or approximate more and more closely to, the truth. Apparently generalizations like that refer not to the puzzle-solutions and the concrete predictions derived from a theory but rather to its ontology, to the match, that is, between the entities with which the theory populates nature and what is 'really there'.
>
> Perhaps there is some other way of salvaging the notion of 'truth' for application to whole theories, but this one will not do. There is, I think, no theory-independent way to reconstruct phrases like 'really there'; the notion of a match between the ontology of a theory and its 'real' counterpart in nature now seems to me illusive in principle. Besides, as a historian, I am impressed by the implausibility of the view. I do not doubt, for example, that Newton's mechanics improves on Aristotle's and that Einstein's improves on Newton's *as instruments for puzzle-solving*. But I can see in their succession *no coherent direction of ontological development*. On the contrary, in some important respects, though by no means in all, Einstein's general theory of relativity is closer to Aristotle's than either of them is to Newton's.[47]

Kuhn's arguments against a progressive approach to the truth are therefore of two kinds: an epistemological argument and a historical one. However, the latter seems to

[45] Kuhn (1962a), pp. 170–171. He then urges us to give up the concept itself in order to get rid of some of the problems which have afflicted the history of Western thought: "We are all deeply accustomed to seeing science as the one enterprise that draws constantly nearer to some goal set by nature in advance. But need there be any such goal? Can we not account for both science's existence and its success in terms of evolution from the community's state of knowledge at any given time? Does it really help to imagine that there is some one full, objective true account of nature and that the proper measure of scientific achievement is the extent to which it brings us closer to that ultimate goal? If we can learn to substitute evolution-from-what-we-know for evolution-toward-what-we-wish-to-know, a number of vexing problems may vanish in the process. Somewhere in this maze, for example, must lie the problem of induction" (*ibidem*, p. 171). In his (2003) Brendan Larvor argues that in *The Structure of Scientific Revolutions* Kuhn allowed some features of his procedure and experience as a historian of science to pass over into his model for the growth of scientific knowledge: the fact that science is partly directed by extra-scientific factors, incommensurability, the absence of any ahistorical standard of rationality and, most particularly, that science cannot be shown to be heading towards the truth – all these appear as methodological commitments rather than historical-philosophical theses. This reading of Larvor's fits very well with Feyerabend's 1960–1962 charge to Kuhn of deliberately blending descriptive and prescriptive elements in his book (see above, ch. 3, pp. 126 and 128–130).

[46] See Kuhn (1970a), pp. 205–207.

[47] Kuhn (1970c), pp. 206–207, emphasis added. Once again, regrettably, Kuhn does not substantiate his latter claim: see above, ch. 2, n. 204, for further references.

be contradicting the former: for if the notion of truth is inconsistent, how can history tell us that successive theories do not succeed in approaching closer and closer to the truth? And furthermore: how does Kuhn explain the affinity between Einstein's theory and Aristotle's, given the incommensurability that separates them?

But the first argument is unclear, too. Hoyningen-Huene interprets it in the following way:

> The [...] argument is epistemological;[48] it procceds from the assumption that it's essenially meaningless to talk of what there really is, beyond (or ouside) of all theory. If this insight is correct, it's impossible to see how talk of a "match" between theories and absolute, or theory-free, purely object-sided reality could have any discernable meaning. How could the (qualitative) assertion of a match, or the (comparative) assertion of a better match, be assessed? The two pieces asserted to match each other more or less would have to be accessible independently of one another, where one of the pieces is absolute reality. But if we had access to absolute reality [...] what interest would we have in theories about it?.[49]

But if this is the sense in which the above quoted passage from Kuhn's "Postscript – 1969" is to be understood, then his argument is quite a weak one.[50] Why does the fact that we know that there is a correspondence between a theory and reality require independent access to each of them? Take, for example, as Alexander Bird suggests, the correspondence between a key and a lock: I know there is a correspondence between the thread form of the key and the gears of the lock not because I have independent access to those gears, but because I know that that key opens that lock. Secondly, what Hoyningen-Huene calls "insight" is clearly false. For

> we have an intuitive notion of the possibility of error and of ignorance. And Kuhn must share this, since the only satisfactory explanation of the origin of anomalies is that the world is not exactly as our theories say it is. If error and ignorance can be shared by all of us, then there must be a way things are that is "beyond" theory. Kuhn is conflating metaphysical, semantic and epistemological questions here. Even if it were impossible to assess the assertion of a match, that would not make that assertion meaningless, unless one had some sort of verificationist view about meaning (another positivist trait).[51]

That is, we can speak of truth even in the absence of a criterion for truth.

---

48 See also Kuhn (1970b), pp. 265–266, and (1979b), pp. 205–206.

49 Hoyningen-Huene (1989a/1993), pp. 263–264. Scientific progress must therefore be interpreted, according to Kuhn, not in terms of an increasing approximation to the truth, but only as an instrumental improvement of scientific knowledge: "Conceived as a set of instruments for solving technical puzzles in selected areas, science clearly gains in precision and scope with the passage of time. As an instrument, science undoubtedly does progress" (Kuhn (1979b), p. 206. See also Kuhn (1962a), pp. 172–173, and (1970c), p. 206.

50 The issue is raised also by Alexander Bird in his (2000), pp. 226–228.

51 Bird (2000), pp. 227–228.

Lexicons and realism

The basic idea of traditional epistemology, a correspondence theory of truth that assesses beliefs on the grounds of their ability to reflect the world, independently of the mind, cannot account for the change of the very beliefs, according to Kuhn. Therefore, it must be rejected and replaced with a weaker conception, internal to the lexicon itself.[52] For if a statement can be properly said to be true or false within the context of a given lexicon, the system of categories embedded in the lexicon cannot be, *per se*, true or false.[53] By relinquishing the correspondence theory of truth Kuhn rejects the idea that the system of categories of a theory may reflect the world-in-itself, independently of theory. We may speak of truth only within the context of a given lexicon, that is, we may only assess the assertions stated within a given lexical context: "lexicons are not [...] the sorts of things that can be true or false":[54] their logical status is that of words' meaning in general, that is, of a convention we can only justify in a pragmatic way.[55] Truth is internal to lexicon in the sense that its use

[52] The notion of necessary truth may be replaced by "something like a redundancy theory of truth": Kuhn (1991a), p. 99; see also *ibidem*, pp. 95–99, (1992), p. 115, and (1993a), pp. 244–245.

[53] See Wittgenstein (1969), §205: "If the true is what is grounded, then the round is not *true*, nor yet false"; and "[...] why should the language-game rest on some kind of knowledge?" (§477; see also §559). According to Wittgenstein, a language game presents no gaps, since together with its possible moves it also defines the space which makes those very moves possible: just as the rules of the game define which moves belong to it, so the grammar of the language circumscribes what is meaningful. Nothing meaningful can therefore remain outside its boundaries and establish itself as a mark of the incompleteness of the language game (*incommensurability*). A game to which new rules are added is not a richer game, but simply a new game (*paradigm shift*). Therefore, a language game is criterion to itself – like the sample standard metre unit preserved at The International Bureau of Weights and Measures of Sèvres, near Paris, it is not itself measurable, since it is not possible to measure what is to be the unit of measurement: its having a length cannot be ascertained, but it is a feature which displays itself in the way we use it when measuring (see Wittgenstein (1953), part I, §50, and §241). "If you tried to doubt everything you would not get as far as doubting anything. The game of doubting itself presupposes certainty" (Wittgenstein (1969), §115; see also §160, §450 and §625. Doubting the paradigm means, on the one hand, condemning oneself to silence; on the other, it means extending the practice of doubt beyond what is reasonable (i.e. meaningful) to doubt. Stephen Toulmin, a pupil of Wittgenstein's thinks along the same line: "There must always be some point in a scientist's explanations where he comes to a stop: beyond this point, if he is pressed to explain further the fundamental basis of his explanation, he can say only that he has reached rock-bottom" (Toulmin (1961), p. 42). The difference with Popper's perspective (see the passages from *The Logic of Scientific Discovery* and *Die beiden Grundprobleme der Erkenntnistheorie* quoted above, pp. 19 and 97) is substantial.

[54] Kuhn (1993a), p. 244. Kuhn made this concept quite explicitly already in *The Structure of Scientific Revolutions*: "there is no standard higher than the assent of the relevant community" (Kuhn (1962a), p. 94).

[55] More than ever, the striking similarity with Carnap is evident. According to Carnap, internal questions can be answered by referring to the logical rules of a given linguistic framework. In this case we have genuine theoretical questions, to which the notions of "correct" or "incorrect", "true" or "false" clearly and unproblematically apply. Researchers

is restricted to assessing claims made within the context of the lexicon: truth claims in one lexicon are not relevant for those made in another, nor can truth be applied to a lexicon itself.[56]

In other words, Kuhn decidedly rejects the idea that the structure which constitutes the theory might reflect the way the world is, independently of theory. The lexicon embodies a linguistic convention that marks the distance between the reality described by a theory and the theory describing it in different ways:

> Experience and description are possible only with the described and describer separated, and the lexical structure which marks the separation can do so in different ways, each resulting in a different, though never wholly different, form of life. Some ways are better suited to some purposes, some to others. But none is to be accepted as true or rejected as false; none gives privileged access to a real, as against an invented, world. The ways of being-in-the-world which a lexicon provides are not candidate for true/false.[57]

Lexicons are assessed on the basis of their ability to serve a particular function, not to reflect reality.[58] To quote again Kuhn's own words:

> [...] what replaces the one big mind-independent world about which scientists were once said to discover the truth is the variety of niches within which the practitioners of these various specialties practice their trade. Those niches, which both create and are created by the conceptual and instrumental tools with which their inhabitants practice upon them, are as solid, real, resistant to arbitrary change as the external world was once said to be. But, unlike the so-called external world, they are not independent of mind and culture,

---

sharing a given linguistic framework can engage in theoretically genuine disputes about such internal questions. On the contrary, external questions, essentially involving a choice among different linguistic frameworks, are not genuinely rational in this sense. For, in the latter case, we are confronted with questions of a purely pragmatic or instrumental character about the adequacy or appropriateness of a given framework, designed in view of a given aim. This means, in the first place, that answers to external questions cannot be assessed by appealing to dichotomies like "correct" or "incorrect", "true" or "false", but nearly always involve problems of degrees. Secondly, such a distinction implies that answers to external questions are necessarily relative to the goals individual researchers aim at – more cautious researchers, fearing to contradict themselves, could, for example, prefer the weaker rules of intuitionist logic, while those interested in a wider applicability of physics may opt for the more binding rules of classical logic. See Carnap (1928a), (1934), (1935), (1936–1937), (1956) and (1963a); for a discussion, see Reisch (1991), pp. 270–274.

56 Putnam's "internal realism" closely resembles this stance: see, in particular, Putnam (1978), Part 4, (1981), chs. 5–7, and especially ch. 3, (1983), chs. 2, 11 and 13, (1987) and (1990), Part I. In his introduction to the latter collection, James Conant remarks: "Having originally stood for the dream of realizing our natural human aspirations to knowledge and objectivity, 'philosophical realism' now names an intellectual current that ultimately serves only to corrode our conviction in the possibility of attaining either" (p. xv). For a somewhat revised position, see Putnam (1994).

57 Kuhn (1991a), p. 104.

58 Although Kuhn assumes the existence of an independent reality throughout his work, his position involves idealistic leanings: see Hoyningen-Huene (1989a/1993), pp. 267–271, and (1989b); see also Nola (1980a), Sankey (1997a), chs. 2–4, and Ghins (1998).

> and they do not sum to a single coherent whole of which we and the practitioners of all the individual scientific specialties are inhabitants.[59]

The idea that lexicons (or paradigms) are not and cannot be true or false *per se* is but a variant of Logical Positivism's justificationism: it is the idea that truth is grounded on the solidarity of beliefs within a given scientific community, an immediate consequence of Kuhn's highlighting of the communitarian character of science. Positivists as well placed particular emphasis on community: they regarded communal collaboration as important for the production and justification of scientific knowledge, which they in turn regarded as important for the unity of science. It is this very emphasis that fuels Kuhn's conception of science as a social institution and his attempt to define scientific knowledge, if not truth itself, in terms of the consensus of belief that is forged among its members.[60]

The strength of a community and the solidarity of belief sound attractive because they offer ease and disengagement.[61] But it is a two-edged sword, for when we add to Kuhn's ideas that scientists must commit themselves uncritically to a paradigm and that it is appropriate for the profession to ostracize those who disagree, then it becomes difficult to see how science can uphold and encourage freedom of thought.

Coherence theory and correspondence theory

Kuhn's arguments against the correspondence theory of truth have distinguished precedents: we can find something similar in Kant and also in James.[62] However, particularly relevant here are his precursors among logical positivists, chiefly Neurath and Carnap. In a 1935 article (his very first publication[63]) Hempel describes the progressive shift, in some of the major exponents of Logical Positivism, from a correspondence theory of truth to a (restrained) coherence theory: such a shift, that goes hand in hand with some shifts in their conceptions of the nature of perceptive

---

59 Kuhn (1992), p. 120.

60 Most interestingly, in his comments on the typescript of *The Structure of Scientific Revolutions*, Feyerabend spots this point and highlights its root in Wittgenstein's philosophy: "[…] advance of knowledge, so I would have thought, has nothing to do with membership in communities (Wittgenstein notwithstanding)" (Feyerabend (1995a), p. 356).

61 Wrote Kant: "Laziness and cowardice are the reasons why such a large portion of men, even when nature has long emancipated them from alien guidance (*naturaliter maiorennes*), nevertheless gladly remain immature for life. For the same reasons, it is all too easy for others to set themselves up as their guardians. It is so convenient to be immature! If I have a book to have understanding in place of me, a spiritual adviser to have a conscience for me, a doctor to judge a diet for me, and so on, I need not make any efforts at all. I need not think, so long as I can pay; other will soon enough take the tiresome job for me" (Kant (1784/1970), p. 54).

62 See Kant (1781, 1787), B234–236, and James (1904).

63 Hempel (1935, 2000): it was Richard Jeffrey, a few years ago, who first pointed out this article to me. I wish to acknowledge his help with it and dedicate this section to him, *in memoriam*.

knowledge and observation, presents a striking anticipation of Kuhn's reflection on these issues.

In his article Hempel briefly refers to Wittgenstein's *Tractatus Logico-Philosophicus*, "the logical and historical starting point of the Vienna Circle's researches",[64] characterized by a correspondence theory of truth: "a statement is to be called true if the fact or state of affairs expressed by it exists; otherwise the statement is to be called false".[65] Wittgenstein's ideas concerning truth were rather generally adopted by the members of the early Vienna Circle. The first to raise doubts – which soon developed into a vigorous opposition – was Otto Neurath. And the first to recognize the importance of Neurath's ideas was Carnap, who joined some of Neurath's theses and gave them a more precise form.

Hempel offers a "crude, but typical formulation"[66] of Neurath's main theses:

> Science is a system of statements which are of one kind. Each statement may be combined or compared with each other statement (e.g. in order to draw conclusions from the combined statements or to see if they are compatible with each other or not). But statements are never compared with a 'reality', with 'facts'. None of those who support a cleavage between statements and reality is able to give a precise account of how a comparison between statements and facts may be accomplished – nor how we may possible ascertain the structure of facts. Therefore, the cleavage is nothing but the result of a redoubling metaphysics, and all the problems connected with it are mere pseudoproblems.[67]

As we can see, Neurath's doubts about the possibility of a correspondence between facts and propositions – a central theme of Wittgenstein's *Tractatus Logico-Philosophicus* – and access to reality, are the very same as Kuhn's, as read and understood by Hoyningen-Huene.[68]

Neurath's ideas involve a coherence theory of truth. As Hempel explains:

> Carnap developed, at first, a certain form of a suitable coherence theory, the basic idea of which may be elucidated by the following reflection: If it is possible to cut off the relation of sentences to 'facts' from Wittgenstein's theory and to characterize a certain class of statements as true atomic statements, one might perhaps maintain Wittgenstein's important ideas concerning statements and their connections without further depending

[64] Hempel (1935, 2000), p. 10. Towards the end of the 1920s some members of the Vienna Circle, particularly Friedrich Waismann and Moritz Schlick, undertook a close confrontation with Wittgenstein and the ideas advanced in his *Tractatus Logico-Philosophicus*: see Waismann (1967) and Wittgenstein, Waismann (2003). On Wittgenstein's views of truth, see his (1953), Part I, §§71, 77 and 133, and (1969), §§105, 370, 403, 457–458 and 519.

[65] Hempel (1935, 2000), p. 10.

[66] Hempel (1935, 2000), p. 10. For a more detailed exposition, see Neurath (1931a), (1931b), (1931c), (1932a), (1932b) and (1933).

[67] Hempel (1935, 2000), p. 11.

[68] See above, p. 212. Hoyningen-Huene's interpretation, it is worth noticing, is endorsed by Kuhn himself: "No one, myself included, speaks with as much authority about the nature and development of my ideas" (Kuhn (1989c), p. xi).

upon the fatal confrontation of statements and facts – and upon all the embarrassing consequences connected with it.[69]

Hempel takes this to be the first step in the logical positivists' progressive abandonment of Wittgenstein's theory of truth towards that of Carnap and Neurath: by replacing the concept of atomic facts by that of protocol statements, the problematic correspondence with "external reality" is substituted by a comparison with the basic elements of experience.

The second step involves a change of view concerning the formal structure of the system of scientific statements. It consists in loosening the verificationist conception of meaning typical of Wittgenstein's thought:[70] in so doing universal statements, such as scientific hypotheses, can be regarded as meaningful even if they do not receive a logically conclusive verification by singular statements. Furthermore, Hempel remarks, also several propositions that appear to be singular in form possess a logical hypothetic form. The singular statements we adopt depend upon which formal system we choose.[71] Thus, also a second fundamental principle of the *Tractatus Logico-Philosophicus* must be abandoned: it is no longer possible to define the truth or falsehood of certain basic statements, whether or not they may be atomic statements or protocol statements, or other kinds of singular statements. "So", Hempel writes, "the refined analysis of the formal structure of the systems of statements involves an essential loosening or softening of the concept of truth; […] In science a statement is adopted as true if it is sufficiently supported by protocol statements".[72]

However, the principle of reducing the test of each statement to a certain kind of comparison between the statement in question and a certain class of basic statements which are allegedly deemed to be ultimate and admit no doubt is still a leftover from Wittgenstein's view. The third and last phase of the step-by-step evolution from a correspondence theory into a restrained coherence theory of truth may be characterized, in Hempel's outline, as the process of eliminating even this characteristic. The idea is then to regard protocol statements not as absolutely reliable, but as akin to the other scientific statements for what concerns their revisability. Though we do appeal to protocol statements when a theory needs to be tested, protocol statements themselves can no longer be conceived as constituting an unalterable basis for the whole system of scientific statements.[73] The chain of testing

---

[69] Hempel (1935, 2000), p. 11. "The desired class of propositions", Hempel continues, "presented itself in the class of those statements which express the result of a pure immediate experience without any theoretical addition. They were called protocol statements, and they were originally thought to need no further proof" (*ibidem*).

[70] "According to Wittgenstein, a proposition that cannot ultimately be verified has no meaning: in other words, a statement has a meaning when and only when it is a truth-function of the atomic propositions" (Hempel (1935, 2000), p. 12).

[71] Our choice, as particularly Neurath emphasized, however logically arbitrary, is practically restricted by psychological and sociological factors.

[72] Hempel (1935, 2000), p. 13.

[73] Neurath says that we do not renounce a judge who says whether a statement in question is to be adopted or rejected – only, the judge (that is, the system of protocol statements) is

steps has no absolute last link: it depends upon our decision when to break off the testing process. Science is not a pyramid rising on a solid basis – rather, Neurath presents us with an image of science as a boat that must be constantly repaired while at sea: there is no dry dock that allows for restoring it from the keel up.[74]

Carnap and Neurath are no idealists, though Hempel urges us to understand: by no means do they intend to say that there are no facts, only propositions. What they actually mean to say[75] is that each non-metaphysical consideration of philosophy belongs to the domain of the logic of science, unless it concerns an empirical question, and therefore is proper to empirical science. And it is possible to formulate each statement of the logic of science as an assertion concerning certain properties and relations to scientific propositions only. So the concept of truth may be characterized "as a sufficient agreement between the system of acknowledged protocol statements and the logical consequences which may be deduced from the statement and other statements which are already adopted".[76]

Hempel's outline of the development of the logical positivists' coherence theory of truth leads to a position very close to Kuhn's own.[77] Not only in Kuhn's philosophy statements describing their observations play the same role as protocol statements in the positivists' philosophy of science portrayed by Hempel, but the third step in the progressive dismissal of the early Wittgenstein's ideas, rejecting the foundational reliability of protocol statements, goes hand in hand with Kuhn's idea of the theory-ladenness of observations. However different their starting points can be, the picture that results is nearly identical: although observation is the basis for scientific beliefs, not even it is free from revision in the light of theoretical change.

Once again it is clear how Kuhn is not the anti-positivist thinker he is generally taken to be. Quite the contrary: the best way to understand his thought seems to be that of framing it within the tradition it in fact belongs to, that is, the Logical Positivism or Empiricism of Neurath and Carnap. Just like them, he rejects the characteristic assumptions of a certain kind of positivism, typical of the followers of Wittgenstein's early philosophy, such as Moritz Schlick. The latter's reply to Carnap's and Neurath's progressive shift away from Wittgenstein is that their positions lead to relativism about truth: for, to the coherence theory of truth it may be objected that there might be several different and incompatible systems presenting a satisfactory internal coherence. A rejoinder may be to accept it and therefore make truth relative to the various coherent systems. It is Kuhn's move: if we regard the beliefs shared within the tradition of normal science as one of these coherent systems, then the

---

removable.

[74] See Neurath (1932a), p. 92: "There is no way to establish fully secured, neat protocol statements as starting points of the sciences. There is no *tabula rasa*. We are like sailors who have to rebuild their ship on the open sea, without ever being able to dismantle it in dry-dock and reconstruct it from the best components". Carnap supports the same view: there are no absolutely first statements for establishing science, each empirical statement (including protocol statements) may require further justification.

[75] See especially Carnap (1934) and (1935).

[76] Hempel (1935, 2000), p. 15.

[77] On the friendship and mutual influence of Kuhn and Hempel, see Kuhn (1983c), (1993a) and (I-1997a), p. 309, Hempel (1983) and (1993) and Wolters (2003).

relitivized "truth" of Carnap's and Neurath's coherence theory ends up coinciding with the idea of "truth" as relative to the various paradigms.[78] And the coincidence becomes even more striking if we consider the close resemblance between Carnap's formal linguistic frameworks and Kuhn's lexicons, or structured vocabularies.

In "Truth and Confirmation" (1936) Carnap underlines that he prefers to speak of the confrontation between propositions and facts, rather than their comparison:

> There has been a good deal of dispute as to whether in the procedure of scientific testing *statements must be compared with facts* or as whether such comparison be unnecessary, if not impossible. If 'comparison of statement with fact' means the procedure which we called the first operation [that is, the confrontation of a statement with observation] then it must be admitted that this procedure is not only possible, but even indispensable for scientific testing. Yet it must be remarked that the formulation 'comparison' is not quite appropriate here. Two objects can be compared in regard to a property which may characterize them in various ways [...]. We therefore prefer to speak of 'confrontation' rather than 'comparison'. Confrontation is understood to consist in finding out as to whether one object (the statement in this case) properly fits the other (the fact); i.e. as to whether the fact is such as it is described in the statement, or, to express it differently, as to whether the statement is true to fact.[79]

"Furthermore", Carnap continues, "the formulation in terms of 'comparison', in speaking of 'facts' or 'realities', easily tempts one into the absolutistic view according to which we are said to search for an absolute reality whose nature is assumed as fixed independently of the language chosen for its description. The answer to a question concerning reality however depends not only upon that 'reality', or upon the facts but also upon the structure (and the set of concepts) of the language used for that description".[80]

A particularly telling parallel between the logical positivists and Kuhn becomes evident from the conclusion of the above mentioned 1935 article by Hempel:

> what characteristics are there according to Carnap and Neurath's views, by which to distinguish the true protocol statements of our science from the false ones of a fairy tale? As Carnap and Neurath emphasize, there is indeed no formal, no logical difference between the two compared systems, but there is an *empirical* one. The system of protocol statements, which we call true and to which we refer in everyday life and science, may only be characterized by the historical fact that it is the system which is actually adopted by mankind, and especially by the scientists of our culture circle; and the 'true' statements in general may be characterized as those which are sufficiently supported by that system of actually adopted protocol statements.[81]

---

78 See Hempel (1935, 2000), pp. 15–17.

79 Carnap (1936b), p. 125.

80 Carnap (1936b), pp. 125–126.

81 Hempel (1935, 2000), pp. 17–18. Truth is not reduced without qualification to the formal properties of a system of statements. Therefore, Carnap and Neurath do not support a pure coherence theory, but what Hempel calls a "restrained" coherence theory of truth. Not so Kuhn: see his (1992), p. 120.

But "How do we learn to produce 'true' protocol statements?",[82] asks Hempel.

> Obviously by being conditioned. Just as we accustom a child to spit out cherry-stones by giving it a good example or by grasping its mouth, we condition it also to produce, under certain circumstances, definite spoken or written utterances (e.g. to say, 'I am hungry' or 'This is a red ball'). And we may say that young scientists are conditioned in the same way if they are taught in their university courses to produce, under certain conditions, such utterances as 'The pointer is now coinciding with scale-mark number 5' or 'This word is Old-High-German' or 'This historical document dates from the 17th century'. Perhaps the fact of the general and rather congruous conditioning of scientists may explain to a certain degree the fact of a unique system of science.[83]

The logical positivists' departure from the correspondence theory of truth is grounded on the very same concerns at the basis of Kuhn's perplexities about the problematic correspondence of a theory with reality. Two decades after the "Postscript – 1969" to the second edition of *The Structure of Scientific Revolutions* Kuhn writes:

> what is fundamentally at stake is rather the correspondence theory of truth, the notion that the goal, when evaluating scientific laws or theories, is to determine whether or not they correspond to an external, mind-independent world. It is that notion, whether in an absolute or probabilistic form, that I'm persuaded must vanish together with foundationalism. What replaces it will still require a strong conception of truth, but not, except in the most trivial sense, correspondence truth.[84]

And he continues: "[…] we must learn to get along without anything at all like a correspondence theory of truth. Buts something like a redundancy theory of truth is badly needed to replace it".[85] Both for Kuhn and the logical positivists the rejection of the correspondence theory goes hand in hand with their respective anti-realism.

Finally, it must be noted that Carnap subsequently abandoned coherence theory – both, presumably, for the inconveniencies involved in that approach, and for the appeal of Tarski's correspondence theory of truth, developed in the early 1930s.[86] The fact that Kuhn remains attached to that approach testifies that the roots of his reflection might plunge deep in the early phase of the neo-positivistic movement, rejecting one of its most radical developments.

At the roots of Kuhn's position: justificationism and "language games"

Kuhn's position is rooted both in justificationism and in a particular way of posing problems which is typical of Wittgenstein and his followers.[87] Taken together, these

---

[82] Hempel (1935, 2000), p. 18.

[83] Hempel (1935, 2000), pp. 18–19.

[84] Kuhn (1991a), p. 95.

[85] Kuhn (1991a), p. 99.

[86] See Tarski (1932) and (1935).

[87] Following William Bartley's suggestion, we could call the latter "Wittgensteinian problematic": see Bartley (1990), chs. 14–15. See also Wisdom (1974b) and Irzik, Grünberg

two closely interwoven aspects work together and reinforce one another, forcing the compartmentalization of knowledge and the limitation of rationality.

One single problem lies at the roots of both of them: the problem of induction. For their development hinges on the assumption that the problem of induction has not been and cannot be resolved.[88] However, if we suppose it is possible to solve it and inquire what the consequences of its solution are, both from the methodological and the philosophical point of view, it will be possible to see things from an entirely different perspective.

From David Hume (1711–1776) onwards, it has been asserted that there are two kinds of inference: deductive inference, which defines logic; and inductive inference, which defines the natural sciences.[89] The two apply, so to say, to different fields and must not be confused: in Hume's view (at least as I understand him) the problem of induction is simply dissolved once we learn not to apply the standards of deductive logic to judge inductive inference: once we realize that the two principles cannot be unified, the task of the philosopher is simply that of describing and clarifying the standards of deductive and of inductive reasoning. Most logical positivists, while maintaining the unity of the sciences, accepted this "methodological" division. Wittgenstein extended this approach: each discipline, or field, or "language game", or "form of life" is alleged to have its own standards, or principles, or "logic", which need not conform to or be reducible to any other standards or (external) principle and which, again, is the special task of the philosopher to *describe* and *clarify* – not in the least to judge, defend or criticize.[90] There is no arguing or judging among disciplines: criticism, evaluation and explanation would no longer be proper philosophical aims. *Knowledge is essentially divided, and description is all that remains to the philosopher*. All he can do is to describe the logics, grammars or first principles of the various kinds of discourse, and the many sorts of language games and forms of life in which they are embedded. *Philosophical critique is no longer of content, but of criteria application*: as Paul Feyerabend put it, all that is left are "consolations for the specialists".[91]

(1995), particularly sections 2 and 6.

88 By no accident in the closing pages of *The Structure of Scientific Revolutions* Kuhn speaks of "dissolution" rather than "solution" of the problem of induction: "If we can learn to substitute evolution-from-what-we-do-know for evolution-toward-what-we-wish-to-know, a number of vexing problems may vanish in the process. Somewhere in this maze, for example, must lie the problem of induction" (Kuhn (1962a), p. 171); see also Kuhn (1974d), p. 508, (1984), pp. 361–362, and (1989a), pp. 76–77 and 85–86.

89 "Instead of being a faulty sort of deduction, induction is fundamental, defining science – just as deduction is fundamental, defining logic" (Bartley (1990), p. 219).

90 See Wittgenstein (1953), Part I, §66: "don't think, but look!"; "[...] we may not advance any kind of theory. There must not be anything hypothetical in our considerations. We must do away with all *explanation*, and description alone must take its place" (*ibidem*, §109); see also *ibidem*, §89, and his (1969), §189. It is no accident that Wittgenstein had already used the term "paradigm" to refer to what rules in an activity very similar to Kuhnian "normal science": see his (1953), Part I, §§50 and 54.

91 This is the title of Feyerabend (1970a); for a discussion of this paper and the related Feyerabend (1978b), see above, ch. 2, pp. 73–82. The affinities between Kuhn's position

At the heart of this view lies a dangerous form of imperialism, according to which disciplines and their practitioners must *conform*: they must not judge one another, and they must not try to describe a common world in collaboration with other disciplines, since each one has its own. Kuhn's philosophy, regardless of its author's intentions, tries to convince us that no fundamental decision or responsibility on our part is required, that we really should not waste our time on criticism or on trying to understand, and that everything will and must go well if we only fall into step behind the institutionalized scientific dogma of the day and its experts. The risk is to replace philosophical and scientific values of truth, rationality, and the freedom of thought with political power, solidarity and (blind, dogmatic) commitment to belief.[92] If truth is nothing more than solidarity, then all of our questions are political questions – and the only question that matters is: which side are you on?

---

and Wittgenstein's are substantial: see, in particular, Wittgenstein (1921), 4.112, (1953), Part I, §109, and (1969), §189. From Lakatos' point of view, the author of the *Philosophical Investigations* is an intellectual defender of the *status quo*: for him, Wittgenstein's followers set themselves the task to discourage every incursion from outside and attempt to overthrow from inside a "linguistic game" or "form of life" (see Lakatos (1976)). For Wittgenstein philosophy has no cognitive function – rather, it has a "therapeutic" function (see his (1953), Part I, §§109, 133 and 255). The descriptive task which characterizes philosophy concerns the rules governing the use of our language, that is, the grammar of the terms that constitute it: "description" refers to the description of language games, and it aims at showing the rules of those games and hence the structures which characterize them. Concerning rules, and not facts, description has an exemplary value. There is a close parallelism between the role of Kuhn's "exemplars" ("or exemplary problem solutions") and Wittgenstein's "examples": see Kuhn (1970a) and (1974c) (both written in 1969, a crucial year for Kuhn's philosophical development), and Wittgenstein (1953), Part I, §§71, 77 and 133. For the similarities between Kuhn and Wittgenstein see also Sharrock, Read (2002), Read (2003a), (2003b) and (2004).

[92] Since it is impossible to justify our beliefs, only a critical quest for truth can prevent the scientific community from degenerating into a closed society, that allows room for research only if it conforms to its parameters and abides by its rules, and that ostracizes those critically minded who strike too deep. For, as Feyerabend argued (see his (1970a)) it is not the plurality of beliefs that constitutes the major threat to science, but their institutionalization and regimentation into pre-established frames. Progress is made by devising new interesting ideas and testing them critically, not incorporating them in the standard scientific dogma of the day. Recall Kant's words: "*Enlightenment is man's emergence from his self-incurred immaturity*. Immaturity is the inability to use one's own understanding without the guidance of another. The immaturity is *self-incurred* if its cause is not lack of understanding, but lack of resolution and courage to use it without the guidance of another. The motto of enlightenment is therefore: *Sapere aude!* Have courage to use your own understanding!" (Kant (1784/1970), p. 54). "*Sapere aude!*" was the motto of the Bund der Aletophilen (Society of Truth-Lovers) founded in 1736 by Count Ernst Christoph von Manteuffel (1676–1749) with the aim of spreading the philosophy of Leibniz and Wolff. The words are taken from one of Horace's epistles: "dimidium facti qui coepit habet: sapere aude" ("Ad Lollium", in *Epistulae*, Book I, 2, 40). In the eighteenth century this motto, from an exhortation to moral courage, as it was in Horace, assumed a decidedly more definite meaning as an exhortation to the boldness of knowledge. It is employed also by Schiller in his (1795/1967), p. 51. See also Venturi (1959): in this article Venturi traces the vicissitudes through which the quotation from Horace passed on its way to becoming an international slogan.

The dangers of the scientific institution

The fundamental epistemological fact of the twentieth century was the crisis of foundationalism. Its fundamental epistemological problem has been how to respond to it. Some philosophers have concluded that scientific knowledge is unjustified and hence irrational after all. Others – indeed, the majority – have opted for retaining the idea that scientific knowledge is justified, but for weakening either the idea that truth is correspondence with reality or the idea that justification shows that a statement is true. Each of these responses retains the foundationalist theory of rationality, according to which it is irrational to accept a belief that has not been justified and obligatory to accept one that has. Wittgenstein's idea that science is grounded in a form of life; Carnap's idea that it is grounded in the external questions of a linguistic framework; Kuhn's (similar) idea that it is grounded in the acceptance of a scientific paradigm; Rorty's idea that it is grounded in the solidarity of community – each of these is a return to Hume, since the fundamental idea of each of them is that our knowledge is grounded, and must be grounded if we are to regard it as rational.

But the crisis of foundationalism has no implication whatsoever for truth. It does not show that truth does not exist, it does not belittle it and it does not reduce it to solidarity and consensus. It only shows that – as Socrates beautifully said – we are living in the twilight zone between knowledge and ignorance, where the views that we hold may be true, but where we are unable to know that they are.[93] The failure of foundationalism as a concept and as a research programme does not involve the failure of epistemology: however difficult to reach, the twilight zone between knowledge and ignorance is an ideal well worth holding on to.

Kuhn's relativism gives rise to a sort of conservative defence of whatever belief system is construed as rational according to the established scientific community. Although revolutionary science is acknowledged, a critical attitude is systematically discouraged: instead, normal science is regarded as the essence of the scientific enterprise, and dogmatic commitment to a paradigm (or a lexicon) is upheld as a necessary prerequisite for rational knowledge and social harmony.

Institutionalism might be a well-intentioned[94] response to the crisis of foundationalism. But it is an inadequate and perhaps even dangerous response. It

---

[93] See Plato, *Symposium*, 202a.

[94] Steve Fuller may disagree: in his controversial *Thomas Kuhn. A Philosophical History of Our Times* (2000) he urges that *The Structure of Scientific Revolutions* be read "as an exemplary document of Cold War era" (Fuller (2000), p. 5). In such a context, Kuhn plays the role as a normal scientist in the Cold War political paradigm constructed by Harvard President James B. Conant, director of the National Defense Research Committee during the Second World War (which supervised the construction of the first atomic bomb). And *The Structure of Scientific Revolutions* appears as a work whose elegance and systematicity testify how Kuhn simply took Conant's politics of science as uncontroversial, as a taken-for-granted worldview. To put it bluntly, *The Structure of Scientific Revolutions* expresses Cold War mentality in a more abstract – and hence more convenient – form, shaping both academic and public perceptions of science: "Kuhn's sheltered institutional setting enabled him to articulate the Conant paradigm without having to register opponents or even the events of the day. Arguments that would be understood as contestable in a political setting, because agents see them as potentially affecting the course of

is well-intentioned because it is meant to oppose irrationalism, but it is inadequate because it fails to rid itself of the demand that we justify our beliefs. And it is dangerous because such a demand, in the absence of our traditional ideal of truth, can easily lead to an even more authoritarian approach to science than the infallibilist one that it was supposed to replace. And authority is authority – regardless of whether it is the authority of the Pope, the authority of experience, or the authority of the scientific institution.[95] What is worse, Kuhn's institutionalism allows for and even invites the parochial policies of making outsiders of those who criticize the insiders too sharply, and of rejecting alternative theories as meaningless instead of critically engaging with them.[96] All this fits very well with the policy of never admitting that you are wrong and that your opponents are right.[97] As a consequence, in Kuhn's (as well as

---

events, can easily acquire the status of fact when transferred to the depoliticized environment of the academy, where agents are removed from the levers of change. In other words, we may have a situation in which abstraction implies less, not more, critical reflection on the material conditions of thought" (*ibidem*, p. 6). For a multi-faceted discussion of Fuller's book, see Gattei (ed.) (2003); Fuller's replies are in his (2004).

95 See Kuhn (1962a), p. 94: "[…] there is no standard higher than the assent of the relevant community". Imre Lakatos understood this aspect very well: "The clash between Popper and Kuhn is not about a mere technical point in epistemology. It concerns our central intellectual values, and has implications not only for theoretical physics but also for the underdeveloped social sciences and even for moral and political philosophy. If even in science there is no other way of judging a theory but by assessing the number, faith and vocal energy of its supporters, then this must be even more so in the social sciences: truth lies in power" (Lakatos (1970), p. 93). Used to the severe and intense critical confrontation of Popper' seminars, Lakatos immediately grasped the underlying theme of Kuhn's proposal. In his eyes – as well as in those of other pupils of Popper's, like Joseph Agassi, William Bartley, Paul Feyerabend and John Watkins – Kuhn embodies a dangerous approach, authoritarian and dogmatic, to philosophy and science. That is why, while working towards the realization of the 1965 Bedford Colloquium, he invited Kuhn to speak on "Dogma versus Criticism" and asked Feyerabend to reply with a paper on "Criticism versus Dogma" (see above, ch. 2, p. 52). See also Notturno (1985), chs. 1, 6, 8 and 11.

96 See Kuhn (1962a), p. 149. If not actually meaningless, theories cannot be rationally discussed due to the incommensurability that allegedly separates them. More than ever, incommensurability betrays its real nature: however usually referred to as a problem, it proves to be an easy *way out* of problems, for it allows for and invites critical disengagement. Instead of confronting problems, scientists may always appeal to incommensurability to avoid them.

97 This is an element that often characterizes Kuhn's contributions to public debates and that reflects in his writings: see, for example, his (1963b), (1970b), (1974a), (1974d), (1980c), (1983b), (1984), (1989b) and (1993a). Kuhn's history of science does not conceal controversy and error but, regrettably, his philosophy of science plays them down. Controversy is a vital and regular factor in the scientific tradition: by saying that for most of the time leading scientists rightly shield from criticism the ruling scientific idea of the day he does not do it justice. In fact, Kuhn viewed dissent as mere verbal variance. This had a cost: the more he managed to defend it, the more he came to view all dissent as verbal. Had he rewritten *The Structure of Scientific Revolutions*, he says, he "would emphasize language change more and the normal/revolutionary distinction less" (Kuhn (1983a), p. 57). However, contrary to Kuhn's own intentions (I guess), this renders all revolutions merely verbal. Kuhn did not always conceal his dissent: he expressed dissent from Popper, Carnap and Reichenbach (Kuhn

in the logical positivists') view community, politics and power all become far more important than truth.[98]

Kuhn and Popper: Clashing Metaphysics[99]

The received view of the history of the philosophy of science in the twentieth century, some recent and important studies notwithstanding, presents Kuhn as the philosopher chiefly responsible for the demise of Logical Positivism. The analysis conducted so far suggests that this picture is mistaken from several points of view.

(1970b), p. 234), and later on from Quine, Putnam and van Fraassen (Kuhn (1983a), (1989a) and (1993b)). However, he always deemed general assent essential and voiced as much accord as he could. He voiced accord with Feyerabend (Kuhn (1970b), pp. 237 and 246), with Lakatos (*ibidem*, p. 240), with Agassi (Kuhn (1966)), with Popper (Kuhn (1970a), pp. 2–3, and (1970b), p. 241), with Watkins (Kuhn (1970b), p. 241), with Masterman (*ibidem*, p. 272, n. 1, and (I-1997a), p. 300). The case of Hempel is particularly telling. They enjoyed a "very considerable rapprochement" (Kuhn (1993a), p. 247) and Kuhn found that that their views "were perhaps not quite so different as we both then thought" (Kuhn (1993a), p. 225; see also his (1983c), p. 208, and (I-1997a), p. 309). Hempel was a friend (Kuhn (1983c), pp. 209–210, and (1993a), pp. 224–226) and so Kuhn had to set up an agreement with him anyway: Carnap had deemed every descriptive concept "purely" observational or else "purely" theoretical; Hempel agreed, Kuhn disagreed. "A few years later" (Kuhn (1993a), p. 226) Hempel moved to Kuhn's view: according to Kuhn, he replaced Carnap's distinction with one between old and new concepts, or between "antecedently available terms" and those learned together with a new theory (Hempel (1966), pp. 74–75). This way he "implicitly adopted a developmental or historical stance" (Kuhn (1993a), p. 226) – *implicitly*; he put things "in a sort of historical developmental perspective" (Kuhn (I-1997a), p. 309) – *sort of*. What is worse, Kuhn linked assent with approval. Carnap was an inductivist to the last (see his (1963b), p. 998), and so was Hempel. Kuhn was an anti-inductivist: he should have respected inductivism without giving it his consent. On these issues, see Agassi's remarks in his (1997) and (2002); the latter paper, in particular, deepens and substantiate the considerations I am here only outlining. See also Popper (1940), especially pp. 316–317.

[98] Also according to Wittgenstein we must once and for all take leave of Plato's image of knowledge as the outcome of a relentlessly critical attitude, of a reason that does not accept anything without doubting or questioning it. Accordingly, we should replace dialogue and its reciprocal and symmetrical structure with the asymmetric notion of teaching, that presupposes the learner's disposition to follow docilely, with the docility encapsulated already in the root of the Latin verb "doceo" (to teach). The learner must be docile because this attitude is part and parcel of the process of learning – he must not question the foundation that renders the very act of learning possible (see Wittgenstein (1953), Part I, §§ 143, 160, 449 and 457–458). As opposed to Popper (see, for example, his (1976)), for Wittgenstein questions about foundations must be banned because they cast doubt on the possibility of agreement and thereby cancel the room for rational discussion. The keystone of Wittgenstein's *Philosophical Investigations* is the reasons that make us think of rules as institutions, as codified and repeatable customs. Along the same lines, Kuhn portrays science as an institution and draws a picture of knowledge, if not of truth itself, as founded on the consensus and solidarity of belief established among the members of the scientific community.

[99] The topic of this section is further developed in my (2002b), (2003), (2005b), (2005c).

For sure, Kuhn played a major role in the "historical turn" that marked philosophy of science in the last third of the past century, thus contributing to the radical shift of focus from logic and language analysis to a more historically-informed approach, concerned with the dynamics of theory change and conceptual-change. However, Kuhn did not manage to break entirely with the preceding philosophical tradition: his works, especially in the light of the "linguistic turn" of the 1980s and 1990s, are soaked with the very empiricist philosophy he was determined to discard and saw himself to be rejecting.

Far from spelling the demise of Logical Empiricism, then, Kuhn's philosophy is its living legacy. Some historians – most notably Michael Friedman – tend to see his as a sort of unfinished revolution and attempt to resuscitate Logical Positivism from its ashes by a Kuhn-informed rehabilitation of its main tenets.[100] According to this view, Kuhn's could not have been but an unfinished revolution: however great his contribution, he only triggered a transformation and left his (and the positivists') heirs to complete it, adjusting its scope and drawing all its consequences. Kuhn, in other words, would play the role of the Copernicus he himself so well depicted in *The Copernican Revolution*, or that of the reluctant revolutionary Max Planck. A child of his times, he would find himself at the midpoint of a bend on a road: on the one hand, he could be seen as the last of positivists, while on the other as the first of their successors.

I think this is wrong. As the previous chapters show, I agree with this analysis but I do not accept the project underlying it and therefore its conclusions. However important in the history of the philosophy of science, Logical Positivism proved to be a mistaken approach. Kuhn's philosophy partly corrected it and certainly improved on it considerably, suggesting new directions for the debate. But its roots in the positivist tradition doomed it to the very same failure Logical Positivism had to come to terms with. Kuhn's later substantial revision and narrowing of his original theses, as expressed in *The Structure of Scientific Revolutions*, together with the difficulties that prevented him from completing his last book, betray the collapse of his research programme. Such a collapse reveals the wider crisis of foundationalism, an approach that spans the whole Western philosophical tradition.

Knowledge without foundations

Alone in the context of the philosophy of science of the past century, Karl Popper's critical rationalism provides a middle way between two, opposite authoritarian approaches to science and society: dogmatism and relativism.[101] It offers an account of how scientific knowledge can be objective and rational without being certain, appealing to induction and grounding itself upon expert opinion, consensus and authority of any kind. As soon as we give up the idea that we can show that our theories are true, the epistemic problem becomes the problem not of what to believe, but of how to question what we believe. By regarding criticism, not description, as

---

[100] See, for example, Friedman (1991), (1993), (2001), (2002a) and (2002b).

[101] See Gattei (2002c), (2006) and (forthcoming), as well as Gattei (ed.) (2002).

the alternative to justification, it offers an account in which truth, and not power, is the key element.

Popper defends the possibility of the growth of knowledge without attempting to found it: the task of the philosophy of science is not that of justifying propositions but, rather, that of inquiring into methods and criticizing procedures by highlighting their contradictions. Philosophy of science seeks to clarify, criticize and improve scientific knowledge and practice, not to found them.

As opposed to many philosophers who, confronting the "Fries' trilemma" (infinite regress, dogmatism or psychologism) have opted for psychologism or for some form of dogmatism, thus weakening their ideas of justification and truth, Popper gives up the idea that justification is a requirement for scientific knowledge: this cannot and need not be justified.[102] The process of testing of our theories does not produce incontrovertible results, since these very results are nothing but hypotheses that need to be tested in their turn. It is a process without a natural end and that may, in principle, go on *ad infinitum*. Any decision to cut it short and to accept a statement is conventional: any statement always lacks a definitive verification, or a foundation, and may always be revised in the future. Such an acceptance is dogmatic, in a sense, but it is an innocuous form of dogmatism: when new doubts emerge, scientists renew testing. Subjective convictions contribute to the consensus that ends the testing process – but, again, this is harmless psychologism: scientists do not ground statements and though they accept them in accordance with an objective methodological rule, stipulating admission of confirmed test statements into science, these rules are not so strict as to impose unique moves in all situations and allow for moves that lead to the upholding of errors. Dogmatism, infinite regress and psychologism all play a role in scientific practice, but are rendered innocuous by the hypothetical character of science.[103]

Science progresses not by discovering unshakable truths, but by eliminating its own errors: change is its hallmark and intersubjectivity (that is, Kant's surrogate for objectivity) is the sole basic rule. And though this is no guarantor of progress, it is what has kept the scientific enterprise continuing to progress thus far. Convention and experience modify, rather than determine, each other. Corroboration is an assessment of how well a theory stands up to tests: it represents a rough estimate that has no implication for truth-value or probability. Our experiences are theory laden: theory informs all actions and decisions, and these cannot be justified. Experience and access to reality remain problematic, but it is possible to learn from it all the same.[104] Metaphysical realism is a necessary working hypothesis, though there is no way of knowing for sure what the facts are and whether a statement actually corresponds to them: truth (i.e. correspondence) remains an ideal, a regulative idea, always sought and never sure to have obtained (indeed, we may well obtain truth,

[102] See Gattei (2007), ch. II.

[103] See Popper (1935, 1959), pp. 93–94 and 104–105; see also Hacohen (2000), p. 230.

[104] See Popper (1945, 1966), vol. II, p. 383.

but not justified certainty that we have obtained it).[105] The "linguistic turn" no longer threatens realist metaphysics and objective science.

Popper's critical rationalism is closely linked with the search for truth. As human beings, we should be aware of our fallibility and critical of our theories – but we can move from the awareness of our fallibility to the criticism of our theories only if we deliberately aim at the truth.[106] This is why for Popper – as opposed to Kuhn – truth plays the role of the regulative idea of scientific research and rational discussion. Rationality requires no foundation, only critical dialogue: it is the end of foundationist philosophy.[107]

The role of reason

Whereas foundationalist philosophies equate the rationality of scientific knowledge with its justification and this, in turn, with logic, that is, with deductive or inductive argumentation, Popper establishes an equation among rationality, criticism and logic, taking the latter to mean exclusively deductive argumentation. Indeed, valid deductive arguments are the only ones that allow us to preserve the truth from the premises to the conclusion: for it is impossible – that is, inconsistent or contradictory – that a deductively valid argument has true premises and false conclusion.[108]

All logical arguments can do is to test the possible inconsistency between the premises and the conclusion of a valid inference: they cannot force us to choose, or to accept the truth of the former rather than the latter, or vice versa. However, a valid (deductive) logical argument can clarify the choice in front of us. Not so an invalid argument, such as an inductive one: since its conclusion does not follow from its premises, there is no contradiction whatsoever in accepting the latter and rejecting the former. Inductive arguments, in other words, cannot force us to choose. If deductive arguments allow us a critical control over our scientific debates, the invalidity of inductive arguments makes them entirely worthless for critical thinking: the problem with them is not so much that they are unable to justify their conclusions – rather, that they offer us no reason to question our assumptions.

We criticize a statement by trying to show that some of its logical consequences are false. Criticism, that is, tries to show the falsity of a theory by showing its

---

[105] For a contrasting view see Devitt (1984, 1991): while defending them both, Devitt insists on a sharp distinction between a correspondence theory of truth and realism, arguing that no doctrine of truth is in any way constitutive of realism.

[106] See Notturno (2000), p. 109.

[107] As Jorge L. Borges beautifully said: "Two Greeks are talking: Socrates, perhaps, and Parmenides. / We better never know their names; the story, thus, will be more mysterious and calm. / The subject of the dialogue is abstract. They sometimes allude to myths, which they both disbelieve. / The reasons they advance may abound in fallacies and have no aim. / They do not quarrel. They do not want to persuade nor to be persuaded, they do not think of winning or losing. / They agree on one single thing: they know that discussion is the not impossible path to reach a truth. / Free from myth and metaphor, they think or try to think. / We shall never know their names. / This conversation of two strangers somewhere in Greece is the capital fact of History. / They have forgotten prayer and magic" (in Borges (1984)).

[108] For what follows, see Notturno (2000), ch. 5.

inconsistency, either with itself or with other statements that we deem true. With the sole exception of contradictions, however, logic alone cannot show the falsity of a statement.[109] For if two statements are contradictory with one another, logic tells us (at most) that one is true and the other is false but it does not tell us – and, indeed, cannot tell us – which is true and which is false. And since no statement can be justified (or proven true), it follows that the acceptance or rejection of a criticism always entails a judgement. In other words, for Popper *rationality is not so much a property of knowledge, as a task for humans*.[110] It is not the content of a theory, or a belief, that is true, but rather the way we hold it (that is, the way we defend or attack it). Appealing to reason means nothing but taking a decision.[111] And we take different decisions depending on what is the goal we are determined to achieve, or we have set ourselves to achieve. This is why truth, for Popper (as opposed to Kuhn, Wittgenstein and others) remains the regulative idea for science.[112]

Truth as the regulative idea

There are many similarities between Popper and Kuhn: both are interested in progress and the choice among competing scientific theories; both reject the inductivist conception of progress by accumulation and generalization from "facts"; both highlight the priority of theory over observation; both reject the idea that science can achieve a final, justified theory. Many were their reciprocal misunderstandings, too – but more significant are their differences. And I think that the fundamental difference between Popper and Kuhn is not about the possibility of falsification, the existence, role and nature of normal science, or the alleged incommensurability between different scientific theories. It is about the role of truth, the value of criticism and the nature of the bond that unites scientists into a community.[113]

Popper and Kuhn agree that there is no such thing as an objective criterion for truth, but Kuhn takes this to mean that truth plays no role at all in theory appraisal and theory choice, while Popper maintains that truth plays the role of a regulative idea. As a consequence, Kuhn characterizes the bond uniting scientists in terms of shared beliefs: since it is not possible to prove the truth of such beliefs, scientists

---

[109] Galileo too viewed logic as an instrument of criticism, not of discovery: see his (1632/1967), p. 35, and (1638/1914), pp. 137–138: "Logic, it appears to me, teaches us how to test the conclusiveness of any argument or demonstration already discovered and completed; but I do not believe that it teaches us to discover correct arguments and demonstrations". See also Frege (1919/1977), p. 1: "Just as 'beautiful' points the way for aesthetics and 'good' for ethics, so do words like 'true' for logic. All sciences have truth as their goal; but logic is also concerned with it in a quite different way: logic has much the same relation to truth as physics has to weight or heat. To discover truths is the task of all sciences; it falls to logic to discern the laws of truth".

[110] See Popper (1994b), p. 134.

[111] See Popper (1945, 1966), vol. II, pp. 380–381. See also Gattei (2002a) and (2002d).

[112] Thus we can better understand why the problem of the actual correspondence between an axiomatic system and reality proved to be a key passage point in Popper's early intellectual development.

[113] See Notturno (2000), pp. 230 and 239.

cannot help but uncritically commit themselves to them. Popper, on the other hand, characterizes this bond in terms of the search for truth, believing that only truth and the critical attitude enable a scientific community to be an open society.[114]

In hindsight, we can see how the radical challenge of *The Structure of Scientific Revolutions* was not to rationality, but to realism: Kuhn's thrust was actually directed not so much against the rationality of theory appraisal and theory choice, as against the epistemic, or truthlike, character of the theories so chosen, since it is not possible to say that they are better approximations to the truth, i.e. to reality. However, in so doing, Kuhn[115] confused and conflated the concept of truth with a criterion for truth, "by considering it nonsense to speak of truth in the absence of a decision procedure for determining whether or not a statement is true".[116] In Popper's own words:

> It is decisive to realize that knowing what truth means, or under what conditions a statement is called true, is not the same as, and must be clearly distinguished from, possessing a means of deciding – a *criterion* for deciding – whether a given statement is true or false.[117]

And, even more clearly:

---

[114] See Notturno (2000), ch. 10, especially pp. 238–239.

[115] And also some recent interpreters of Popper's thought, such as Geoffrey Stokes. In ch. 8 of his *Popper. Philosophy, Politics and Scientific Method*, Stokes attempts to reconcile Popper's critical rationalism with the critical theory of the Frankfurt School: "despite their initial antagonism, there are areas of 'convergence' or 'reconciliation' between critical rationalism and critical theory" (Stokes (1998), p. 144). Crucial to this attempted reconciliation is Stokes' discussion of truth. He describes Theodor Adorno's and Jürgen Habermas' subjectivist theories of truth, and then wrongly assigns to Popper a similar view: Habermas' consensus theory of truth, he maintains, "has affinities with Popper's arguments on the subject" (*ibidem*, p. 155; see also the discussion on pp. 155–158). "The Popperian concept of objectivity as an intersubjectivity leading to consensus – Stokes continues – denotes a theory of truth that is proceduralist and that shares a great deal with Habermas's theory" (*ibidem*, p. 156); "In practice, both Popper and Habermas advocate a proceduralist method that may be called a 'consensus theory of truth'" (*ibidem*, p. 168). Not so. Stokes confuses the concept of truth with that of corroboration, thus misplacing him alongside Habermas: intersubjectivity leads to improvement; and the corroboration of a theory (or its rational, tentative acceptance) does *not* imply that it is true: even the most well-corroborated and explanatorily powerful theories (such as Newton's, for example) can be false – and actually proved to be so. On Stokes' account, Popper's and others' failure to provide a formal definition of verisimilitude leaves us with but a consensus theory of truth: "Unless we have a satisfactory objective *criterion* of truth (and it is doubtful whether any such criterion can be formulated), then we remain with a consensus theory which utilizes a variety of criteria for theory choice. This does not entail a total relativism in which no secure rational judgement can be made. It merely claims that such judgements will derive from critical discourse upon the best available reasons and that the conclusions reached will change as the interests, ethics, evidence and reasons change" (*ibidem*, p. 141; but see pp. 139–141). Stokes' error is of the same of Kuhn's: see Gattei (2002a), especially pp. 249–250.

[116] Notturno (2000), p. 240.

[117] Popper (1945, 1966), vol. II, p. 371.

> [...] *the absence of a criterion of truth does not render the notion of truth non-significant any more than the absence of a criterion of health renders the notion of health non-significant. A sick man may seek health even though he has no criterion for it. An erring man may seek the truth even though he has no criterion for it.*[118]

Popper's critical attitude finds in the pursuit of truth its very rationale: for it is an attitude whose relevance lies precisely in the role that truth plays if we are to improve our knowledge of the real world. Falsifiability is the advice that, as rational human beings, we should be aware of our fallibility and critical of our theories. But we are able to move from the awareness of our fallibility to criticism of our theories only if we are consciously aiming at the truth.[119] That is why truth, on Popper's account, is the regulative idea of scientific inquiry and rational discussion:

> With the idol of certainty (including that of degrees of imperfect certainty or probability) there falls one of the defences of obscurantism which bar the way of scientific advance. For the worship of this idol hampers not only the boldness of our questions, but also the rigour and the integrity of our tests. The wrong view of science betrays itself in the craving to be right; for it is not his *possession* of knowledge, of irrefutable truth, that makes the man of science, but his persistent and recklessly critical *quest* for truth.[120]

Kuhn and the Legacy of Logical Positivism

Notwithstanding the enormous impact Kuhn had and still has on the history and the philosophy of science, the huge influence he exerted and still exerts on the most various fields, the perception one has of his thought today, after less than a decade from his death, has grown considerably dim: while his legacy on the institutionalization and compartmentalization of the philosophical and scientific knowledge is quite considerable, Kuhn's philosophy has left only some light traces on contemporary ways of doing philosophy of science. Furthermore, his influence on both the history and the sociology of science is rather questionable, beset with misunderstandings and confusions as it is.[121]

---

[118] Popper (1945, 1966), vol. II, p. 373; see also the whole discussion, on pp. 369–396.

[119] "The main philosophical malady of our time is an intellectual and moral relativism, the latter being at least in part based upon the former. By relativism – or, if you like, scepticism – I mean here, briefly, the theory that the choice between competing theories is arbitrary; since either, there is no such thing as objective truth; or, if there is, no such thing as a theory which is true or at any rate (though perhaps not true) nearer to the truth than another theory; or, if there are two or more theories, no ways or means of deciding whether one of them is better than another": Popper (1945, 1966), vol. II, p. 369.

[120] Popper (1935, 1959), pp. 280–281.

[121] For what follows, see Bird (2002), pp. 443–445; see also Preston (2004) and Bird (2004).

The Linguistic Turn: A Wrong Turn

All this is largely due to Kuhn's failure to provide an adequate explanation of what is perhaps his major contribution to twentieth-century philosophy of science, namely, his identification of paradigms, in their sense of exemplary problem solutions,[122] as the propelling engine of the development of a scientific field. In a sense, he was ahead of his time: at the time of writing *The Structure of Scientific Revolutions* he did not have the tools at his disposal to develop this idea, that is, the results on neural nets in cognitive science, that naturally complement the psychological researches Kuhn drew upon in his early works.

However, these results became available during Kuhn's lifetime and he might have employed them – but he never linked such researches to his own concept of paradigm. Possibly, the reason is that by that time Kuhn's researches were already heading in another direction.[123] Whereas *The Structure of Scientific Revolutions* draws upon several examples of scientific discoveries from the history of science, his later writings are "much more philosophical in style and *a priori* in method".[124] Whereas, for example, in 1962 his explanation of the relationship between observation, theory and reality was informed by *Gestalt* psychology and by the results of his Harvard colleagues, the experimental psychologists Jerome Bruner and Leo Postman, in his later (and partly still unpublished) writings Kuhn draws upon a very limited and focused set of historical examples, preferring to support his views with considerations closely resembling Wittgenstein's philosophy of language, and characterizing his own position in terms of "a sort of post-Darwinian Kantianism".[125]

This was no mere change of style or terminology:

> The move from a naturalistic to an *a priori* approach was a move in the direction opposite to the prevailing movement of philosophy itself. Naturalised epistemologists, led by Quine, came to deny that questions concerning the nature and possibility of knowledge are part of *a priori* philosophy; rather, they are the concern of empirical cognitive psychology. More specifically, Kuhn's philosophical approach betrayed commitments characteristic of the positivists and logical empiricists he intended to be rejecting. Although Kuhn had been instrumental in bringing philosophy to repudiate logical empiricism, by the late 1970s the vanguard of philosophy had overtaken Kuhn in this direction. From the perspective of the new causal theory of reference and causal or reliabilist theories of knowledge, *Kuhn's thesis of incommensurability and his rejection of truth-related progress appear to be conservative and positivist.*[126]

Maybe the ideas advanced by Kuhn at the beginning of the 1960s could have given rise to interesting developments, had he continued pursuing the study of the

---

[122] See Hoyningen-Huene (1989a/1993), pp. 154–162, and Gattei (2000b), pp. 302–308.

[123] See above, ch. 3, p. 120 and n. 167.

[124] Bird (2002), p. 444.

[125] Kuhn (1991a), p. 104; see also Kuhn (1993a), p. 245, and (I-1997a), p. 264. For an extensive discussion of the Kantian themes in Kuhn's later philosophy, see Hoyningen-Huene (1989a/1993) and Irzik, Grünberg (1998).

[126] Bird (2002), pp. 444–445, emphasis added.

mutual relationship between the history and the philosophy of science.[127] This, of course, is mere speculation. On the contrary, as a matter of fact, the philosophical approach that characterizes the last phase of Kuhn's intellectual development, the "linguistic turn" that marks his production from the beginning of the 1908s on, is a failure for just the same reasons that Logical Positivism was failure also. Kuhn's philosophical turn was a dead wrong turning.[128]

An unfinished revolution

Was Copernicus himself a Copernican? Interestingly enough, this question is often answered in the negative. In fact, he triggered a revolution the conclusion of which he would have been unable to recognize. But the Copernican revolution (the Scientific Revolution *par excellence*) came to an end only with Newton, well over one hundred years after Copernicus' death. The revolution against Logical Empiricism and Logical Positivism was not only well under way at the beginning of the 1980s, but started half a century earlier, even before Kuhn wrote *The Structure of Scientific Revolutions* – and yet Kuhn rejected it.

The reasons are probably various.[129] First, as mentioned, he was not properly trained as a philosopher and it is therefore unlikely he was fully aware of the details of the philosophical tradition within which or against which (allegedly, at least) he was working. In this respect, *The Structure of Scientific Revolutions* is quite telling: it contains very few footnotes and references, mostly to the works of historians of science. The philosophers mentioned are few and often share Kuhn's particular

---

[127] That is Agassi's own wish at the end of his review of *The Structure of Scientific Revolutions*: see Agassi (1966a), p. 122. Later, in his review of Kuhn (2000b), that takes the form of a "personal obituary" (Agassi (2002), p. 394), he cannot but notice how the book did not come up to his expectations. Recently, Alexander Bird has advanced an interesting (and questionable) hypothesis of the way in which Kuhn's philosophical position might have evolved had he not turned his back on the empirical element in his thinking and had instead developed it, to give a thoroughly naturalistic account of theory-change, world-change and incommensurability: see Bird (2002), pp. 445–451.

[128] A cause partially explaining the wrong turn is probably the fact that Kuhn was a "A Physicist who became a Historian for Philosophical Purposes" (see Kuhn (I-1997a)): lacking a proper philosophical training, he was not aware of the historical and dialectical provenance of the ideas he was dealing and working with (for a conspicuous example that caused a great fuss and was the source of repeated confusion and misunderstanding, see above, ch. 3, pp. 118–121). "He was able to identify certain ideas as being characteristic of positivism or empiricism, such as the thesis that observation and perception are pre-theoretical. These he attacked and thereby helped to undermine positivism. But at the same time he was unaware that other (related) theses, which he happily adopted, were also central to positivism, such as the theoretical-context account of the meaning of theoretical terms, or the conviction that truth-as-correspondence is inaccessible. It is the partial rejection and partial retention of positivism that causes Kuhn to expound apparently radical theses such as the thesis of incommensurability" (Bird (2002), p. 445). Whatever the reason, Kuhn was never able to achieve the far more radical and complete rejection of Positivism that was in fact achieved by other philosophers.

[129] See Bird (2002), pp. 459–463.

philosophical concerns and aims, such as Norwood R. Hanson. We find no reference to the major exponents of the positivistic movement, such as Schlick, Carnap, Feigl, Nagel or Hempel. Wittgenstein is very briefly and critically mentioned.[130] In fact, the idea Kuhn had of Positivism was rather stereotypical and oversimplified: in his eyes, it takes observations to be foundational both epistemologically and semantically; and it disregards history of science, deeming scientific progress a mere accumulation of knowledge aimed at approximating the truth by stages. The former thesis is indeed characteristic of positivism, while the view of progress as increasing verisimilitude is a thesis acceptable to many non-positivists (most notably Popper).

Furthermore, the Positivism Kuhn thought he was rejecting embraced rather more than these two claims: he was wrong to think that rejecting these two claims would amount to a root-and-branch rejection of Positivism (and, more generally, empiricism). It is certainly to Kuhn's (albeit, and quite significantly, not exclusively to his) merit that philosophy has repudiated some centuries-old tenets and has been able to reconcile itself with the lessons from the history of science. But, in fact, Kuhn's revolution is unfinished, for too many aspects of his thought contain a significant residue of that very Positivism he thought he was distancing himself from. Just like Copernicus who, while dealing the first fatal blow to the Aristotelian–Ptolemaic worldview, was also irrevocably soaked in that very same way of thinking, so Kuhn can be regarded the last exponent of the philosophical tradition he was determined to reject. He inaugurated the historical revolution in the philosophy of science – a revolution whose scope and significance goes much beyond what Kuhn himself was able to foresee.

---

[130] See Kuhn (1962a), p. 45.

Bibliography

ABBAGNANO, Nicola

1947 "Il problema filosofico della scienza", in Abbagnano, Buzano, Buzzati-Traverso, Frola, Geymonat, Persico (1947), pp. 139–161.

ABBAGNANO, Nicola, BUZANO, Piero, BUZZATI-TRAVERSO, Adriano, FROLA, Eugenio, GEYMONAT, Ludovico, PERSICO, Enrico

1947 *Fondamenti logici della scienza*, Turin: De Silva, 1947.

ACHINSTEIN, Peter, BARKER, Stephen F. (eds)

1969 *The Legacy of Logical Positivism. Studies in the Philosophy of Science*, Baltimore: The Johns Hopkins Press, 1969.

AGASSI, Joseph

1963 *Towards an Historiography of Science*, *History and Theory*, Beiheft 2, The Hague 1963.

1964a "The Nature of Scientific Problems and Their Roots in Metaphysics", in Bunge (ed.) (1964), pp. 189–211; reprinted in Agassi (1975), pp. 208–233.

1964b "The Confusion Between Physics and Metaphysics in Standard Histories of Science", in Henry Guerlac (ed.), *Ithaca: Proceedings of the Xth International Congress for the History of Science*, Paris: Hermann, 1964, pp. 231–238, 249–250; reprinted in Agassi (1975), pp. 270–281.

1966a Review of Kuhn (1962a), *Journal of the History of Philosophy*, IV, 1966, pp. 351–354; reprinted as "Kuhn on Revolutions", in Agassi (1988a), pp. 117–122.

1966b "Revolutions in Science, Occasional or Permanent", *Organon*, 3, 1966, pp. 47–61; reprinted in Agassi (1981a), pp. 104–118.

1968 "Science in Flux. Footnotes to Popper", in Robert S. Cohen, Marx W. Wartofsky (eds), *In Memory of Norwood Russell Hanson*, Dordrecht: D. Reidel Publishing Company, 1968, pp. 293–323; reprinted in Agassi (1975), pp. 9–39.

1971 "Tristram Shandy, Pierre Menard, and All That. Comments on Criticism and the Growth of Knowledge", *Inquiry*, 14, 1971, pp. 152–163; reprinted as "Kuhn and His Critics. Rational Reconstruction of the Ant Heap", in Agassi (1988a), pp. 315–328.

1973 "Continuity and Discontinuity in the History of Science", *Journal of History of Ideas*, 34, 1973, pp. 609–626; reprinted in Agassi (1981a), pp. 283–299.

1975 *Science in Flux*, Dordrecht–Boston: D. Reidel Publishing Company, 1975.

1981a *Science and Society. Studies in the Sociology of Science*, Dordrecht–Boston: D. Reidel Publishing Company, 1981.

1981b "Scientific Schools and Their Success", in Agassi (1981a), pp. 164–191.

1983 "The Structure of Quantum Revolution", *Philosophy of the Social Sciences*, 13, 1983, pp. 367–381; reprinted in Agassi (1993b), pp. 139–155.

1984 "Training to Survive the Hazard Called Education", *Interchange*, 15, 1984, pp. 1–14.

1986 "God Saves us From Our Friends: Enemies We Have No More", *Philosophia*, 16, 1986, pp. 209–238; reprinted as "After Lakatos. The End of an Era", in Agassi (1988a), pp. 329–352.

1987 "The Autonomous Student", *Interchange*, 18, 1987, pp. 14–20.

1988a *The Gentle Art of Philosophical Polemics: Selected Reviews and Comments*, La Salle, Illinois: Open Court Publishing Company, 1988.

1988b "Ixmann and the Gavagai", *Zeitschrift für allgemeine Wissenschaftstheorie*, 19, 1988, pp. 104–116.

1993a *A Philosopher's Apprentice: In Karl Popper's Workshop*, Amsterdam–Atlanta: Rodopi, 1993.

1993b *Radiation Theory and the Quantum Revolution*, Basel–Boston–Berlin: Birkhäuser, 1993.

1997 "Let a Thousand Flowers Bloom: Popper's Popular Critics", *Anuar*, VII, 1997, pp. 5–24.

2002 "Kuhn's Way", *Philosophy of the Social Sciences*, 32, 2002, pp. 394–430.

AGASSI, Joseph, JARVIE, Ian C.

1979 "The Rationality of Dogmatism", in Theodore F. Geraets (ed.), *Rationality To-Day*, Ottawa: University of Ottawa Press, 1979, pp. 353–362; reprinted in Agassi, Jarvie (eds) (1987), pp. 431–443.

AGASSI, Joseph, JARVIE, Ian C. (eds)

1987 *Rationality: The Critical View*, Dordrecht: Martinus Nijhoff Publishers, 1987.

AGAZZI, Evandro (ed.)

1986 *La filosofia della scienza in Italia nel '900*, Milan: Franco Angeli, 1986.

AJDUKIEWICZ, Kazimierz

1949 *Zagadnienia i kierunki filozofii: Teoria poznania. Metafizyka*, Warsaw: Czytelnyk, 1949; English translation by Henryk Skolimowski and Anthony Quinton, *Problems and Theories of Philosophy*, London–New York: Cambridge University Press, 1973.

1978 *The Scientific World-Perspective and Other Essays 1931–1963*, edited by Jerzy Giedymin, Dordrecht: D. Reidel Publishing Company, 1978.

ALLÉN, Sture (ed.)

1989 *Possible Worlds in Humanities, Arts and Sciences*, Proceedings of Nobel Symposium 65, Berlin: Walter de Gruyter, 1989.

ANDERSEN, Hanne

2000 "Kuhn's Account of Family Resemblance. A Solution to the Problem of Wide-Open Texture", *Erkenntnis*, 52, 2000, pp. 313–337.

2001 *On Kuhn*, Belmont, California: Wadsworth, 2001.

ANDERSEN, Hanne, NERSESSIAN, Nancy

2000 "Nomic Concepts, Frames, and Conceptual Change", *Philosophy of Science*, 67, 2000, pp. S224–S241.

ANDERSSON, Gunnar (ed.)

1985 *Rationality in Science and Politics*, Dordrecht–Boston: D. Reidel Publishing Company, 1985.

ANDRESEN, Jensine

1999 "Crisis and Kuhn", in Margaret W. Rossiter (ed.), *Catching Up with the Vision. Essays on the Occasion of the 75th Anniversary of the Founding of the History of Science Society*, a Supplement to *Isis*, 90, 1999, pp. S43–S67.

ANTISERI, Dario

1982 "Il ruolo della metafisica nella scoperta scientifica e nella storia della scienza", *Rivista di filosofia neo-scolastica*, 74, 1982, pp. 68–108.

ARNAULD, Antoine, NICOLE, Pierre

1685 *La logique ou l'Art de penser, contenant, outre les regles communes, plusieurs observations nouvelles, propres a former le jugement*, Amsterdam: Abraham Wolfgang, 1685[6]; English translation by Jill V. Buroker, *Logic, or, The Art of Thinking. Containing, besides Common Rules, Several New Observations Appropriate for Forming Judgement*, Cambridge–New York: Cambridge University Press, 1996.

ASQUITH, Peter D., NICKLES, Thomas (eds)

1983 *PSA 1982. Proceedings of the 1982 Biennial Meeting of the Philosophy of Science Association*, 2 vols., East Lansing, Michigan: Philosophy of Science Association, 1983.

AXTELL, Guy S.

1993 "In the Tracks of the Historicist Movement. Re-assessing the Carnap–Kuhn Connection", *Studies in History and Philosophy of Science*, 24, 1993, pp. 119–146.

AYER, Alfred J.

1936 *Language, Truth and Logic*, London: Victor Gollancz, 1936, 1946[2].

BAKER, George P., HACKER, Peter M.S.

1985 "Wittgenstein and the Vienna Circle: The Exaltation and Deposition of Ostensive Definition", in Brian F. McGuinness, Aldo G. Gargani (eds),

Wittgenstein and Contemporary Philosophy, Pisa: ETS Editrice, 1985, pp. 5–32.

BALDAMUS, Wilhelm

1972 *The Role of Discoveries in Social Science*, in Teodor Shanin (ed.), *The Rules of the Game. Cross-disciplinary Essays on Models in Scholarly Thought*, London: Tavistock, 1972, pp. 276–302.

1977 *Ludwik Fleck and the Development of the Sociology of Science*, in Peter R. Gleichmann, Johan Goudsblom, Hermann Korte (eds), *Human Figurations. Essays for Norbert Elias*, Amsterdam: Amsterdams Sociologisch Tijdschrift, 1977.

1979 "Das exoterische Paradox der Wissenschaftsforschung. Ein Beitrag zur Wissenschaftstheorie Ludwik Flecks", *Zeitschrift für allgemeine Wissenschaftstheorie*, 10, 1979, pp. 213–233.

BARONE, Francesco

1953 *Il neopositivismo logico*, Turin: Edizioni di Filosofia, 1953, 1986[3].

BARROTTA, Pierluigi

1998 *La dialettica scientifica. Per un nuovo razionalismo critico*, Turin: UTET Libreria, 1998.

BARTLEY, William W., III

1968a "Theories of Demarcation Between Science and Metaphysics", in Lakatos, Musgrave (eds) (1968), pp. 40–64.

1968b "Reply", in Lakatos, Musgrave (eds) (1968), pp. 102–119.

1969 "Sprach- und Wissenschaftstheorie als Werkzeuge einer Schulreform: Wittgenstein und Popper als österreichische Schullehrer", *Conceptus*, 3, 1969, pp. 6–22.

1970 "Die österreichische Schulreform als die Wiege der modernen Philosophie", *Club Voltaire. Jahrbuch für kritische Aufklärung*, IV, 1970, pp. 349–366.

1974 "Theory of Language and Philosophy of Science as Instruments of Educational Reform: Wittgenstein and Popper as Austrian Schoolteachers", in Robert S. Cohen, Marx W. Wartofsky (eds), *Methodological and Historical Essays in the Natural and Social Sciences*, Dordrecht: D. Reidel Publishing Company, 1974, pp. 307–337.

1976 "On Imre Lakatos", in Cohen, Feyerabend, Wartofsky (eds) (1976), pp. 37–38.

1989 "Rehearsing a Revolution. Karl Popper: A Life – A Section entitled Music and Politics: Karl Popper Meets Arnold Schönberg and the Eislers and Gives up Communism", Regional Meeting of the Mont Pèlerin Society, Christchurch, New Zealand, 27–30 November 1989.

1990 *Unfathomed Knowledge, Unmeasured Wealth. On Universities and the Wealth of Nations*, La Salle, Illinois: Open Court Publishing Company, 1990.

BERKSON, William

1976 "Lakatos One and Lakatos Two: An Appreciation", in Cohen, Feyerabend, Wartofsky (eds) (1976), pp. 39–54.

BERNSTEIN, Richard J.

1983 *Beyond Objectivism and Relativism. Science, Hermeneutics, and Praxis*, Philadelphia: University of Pennsylvania Press, 1983.

BIAGIOLI, Mario

1993 *Galileo, Courtier. The Practice of Science in the Culture of Absolutism*, Chicago–London: University of Chicago Press, 1993.

BIRD, Alexander J.

1991 *Arithmetic, Grammar and Ontology*, Ph.D. dissertation, University of Cambridge, 1991.

2000 *Thomas Kuhn*, Princeton: Princeton University Press, Princeton 2000.

2002 "Kuhn's Wrong Turning", *Studies in History and Philosophy of Science*, 33, 2002, pp. 443–463.

2003 "Kuhn, Nominalism, and Empiricism", *Philosophy of Science*, 70, 2003, pp. 690–719.

2004 "Kuhn, Naturalism, and the Positivist Legacy", *Studies in History and Philosophy of Science*, 35, 2004, pp. 337–356.

BORGES, Jorge L.

1952 *El Aleph*, Buenos Aires: Losada, 1952; English translation by Andrew Hurley, in Jorge L. Borges, *Collected Fictions*, New York: Viking, 1998, pp. 181–288.

1984 *Atlas*, in collaboration with María Kodama, Buenos Aires: Editorial Sudamericana, 1984.

BORRADORI, Giovanna

1991 "Il muro dell'Atlantico", in Giovanna Borradori, *Conversazioni americane. Con W.O. Quine, D. Davidson, H. Putnam, R. Nozick, A.C. Danto, R. Rorty, S. Cavell, A. MacIntyre, Th. S. Kuhn*, Rome–Bari: Laterza, 1991, pp. 3–30; English translation by Rosanna Crocitto, "The Atlantic Wall", in Giovanna Borradori, *The American Philosopher. Conversations with Quine, Davidson, Putnam, Nozick, Danto, Rorty, Cavell, MacIntyre, and Kuhn*, Chicago–London: University of Chicago Press, 1994, pp. 1–25.

BRIDGMAN, Percy W.

1927 *The Logic of Modern Physics*, New York: Macmillan, 1927.

BROWN, Harold I.

1977 *Perception, Theory, and Commitment. The New Philosophy of Science*, Chicago: Precedent Publishing, 1977.

1983 "Incommensurability", *Inquiry*, 26, 1983, pp. 3–29.

Bruner, Jerome S., Postman, Leo J.

1949 "On the Perception of Incongruity: A Paradigm", *Journal of Personality*, XVIII, 1949, pp. 206–223.

Bucciantini, Massimo

2003 *Galileo e Keplero. Filosofia, cosmologia e teologia nell'Età della Controriforma*, Turin: Einaudi, 2003.

Buchdahl, Gerd

1969 *Metaphysics and the Philosophy of Science. The Classical Origins: Descartes to Kant*, Cambridge, Massachusetts–London: The MIT Press, 1969.

Buchwald, Jed Z.

1989 *The Rise of the Wave Theory of Light*, Chicago: University of Chicago Press, 1989.

1992 "Kinds and the Wave Theory of Light", *Studies in History and Philosophy of Science*, 23, 1992, pp. 39–74.

Buchwald, Jed Z., Smith, George E.

1997 "Thomas S. Kuhn, 1922–1996", *Philosophy of Science*, 64, 1997, pp. 361–376.

2001 "Incommensurability and the Discontinuity of Evidence", *Perspectives on Science*, 9, 2001, pp. 463–498.

Buck, Roger C., Cohen, Robert S. (eds)

1971 *PSA 1970. In Memory of Rudolf Carnap*, Dordrecht–Boston: D. Reidel Publishing Company, 1971.

Burtt, Edwin A.

1925 *The Metaphysical Foundations of Modern Physical Science. A Historical and Critical Essay*, London: Kegan Paul & Co., 1925, 1932[2].

Buzzoni, Marco

1986 *Semantica, ontologia ed ermeneutica della conoscenza scientifica. Saggio su T.S. Kuhn*, Milan: Franco Angeli, 1986.

Campelli, Enzo

1999 "Un rapporto *imaginabilis*? Ludwik Fleck e Thomas Kuhn", in Campelli (ed.) (1999), pp. 7–52.

Campelli, Enzo (ed.)

1999 *T.S. Kuhn: come mutano le idee sulla scienza*, Milan: Franco Angeli, 1999.

CARNAP, Rudolf

1928a *Der logische Aufbau der Welt*, Berlin: Weltkreis, 1928; Berlin: Felix Meiner, 1962²; English translation by Rolf A. George in Rudolf Carnap, *The Logical Structure of the World and Pseudoproblems in Philosophy*, London: Routledge & Kegan Paul, 1967, pp. 1–300.

1928b *Scheinprobleme in der Philosophie. Das Fremdpsychische und der Realismusstreit*, Weltkreis, Berlin 1928; Berlin: Felix Meiner, 1962²; English translation by Rolf A. George in Rudolf Carnap, *The Logical Structure of the World and Pseudoproblems in Philosophy*, London: Routledge & Kegan Paul, 1967, pp. 310–343.

1934 *Logische Syntax der Sprache*, Vienna: Julius Springer, 1934; English translation by Amethe Smeaton (Countess von Zeppelin), revised by Olaf Helmer, *The Logical Syntax of Language*, London: Kegan Paul, Trench, Trubner & Co., 1937.

1935 *Philosophy and Logical Syntax*, London: Kegan Paul, 1935.

1936a "Von der Erkenntnistheorie zur Wissenschaftslogik", in *Actes du Congrès international de philosophie scientifique, Sorbonne, Paris 1935*, vol. 1, *Philosophie scientifique et empirisme logique*, Paris: Hermann & C.ie, 1936, pp. 36–41.

1936b "Wahrheit und Bewährung", in *Actes du Congrès international de philosophie scientifique, Sorbonne, Paris 1935*, vol. 4, *Induction et probabilité*, Paris: Hermann & C.ie, 1936, pp. 18–23; English translation by Herbert Feigl, adapted by the author, "Truth and Confirmation", in Herbert Feigl, Wilfrid Sellars (eds), *Readings in Philosophical Analysis*, New York: Appleton-Century-Crofts, 1949, pp. 119–127.

1936–1937 "Testability and Meaning", *Philosophy of Science*, 3, 1936, pp. 419–471 and 4, 1937, pp. 1–40; reprinted as *Testability and Meaning*, New Haven: Yale University Press, 1950.

1946 "Remarks on Induction and Truth", *Philosophy and Phenomenological Research*, 6, 1946, pp. 590–602.

1950a *Logical Foundations of Probability*, Chicago: University of Chicago Press, 1950, 1962².

1950b "Empiricism, Semantics, and Ontology", *Revue Internationale de philosophie*, 4, 11, 1950, pp. 20–40; reprinted in Carnap (1947, 1956), pp. 205–221; revised reprint in Philip P. Wiener (ed.), *Readings in the Philosophy of Science. Introduction to the Foundations and Cultural Aspects of the Sciences*, New York: Charles Scribner's Sons, 1953, pp. 509–522 and 633–634.

1956 "The Methodological Character of Theoretical Concepts", in Herbert Feigl, Michael Scriven (eds), *The Foundations of Science and the Concepts of Psychology and Psychoanalysis*, Minneapolis: University of Minnesota Press, 1956, pp. 38–76.

1963a "Intellectual Autobiography", in Schilpp (ed.), (1963), pp. 3–84.

1963b "Replies and Systematic Expositions", in Schilpp (ed.), (1963), pp. 859–1013.

1966 *Philosophical Foundations of Physics. An Introduction to the Philosophy of Science*, edited by Martin Gardner, New York–London: Basic Books, 1966.

CARRIER, Martin
2002a "Incommensurability and Empirical Comparability: The Case of the Phlogiston Theory", in Peter Gärdenfors, Jan Woleński, Katarzyna Kijana-Placek (eds), *In the Scope of Logic, Methodology and Philosophy of Science. Volume Two of the 11th International Congress of Logic, Methodology and Philosophy of Science, Cracow, August 1999*, Dordrecht–Boston–London: Kluwer Academic Publishers, 2002, pp. 551–564.
2002b "Explaining Scientific Progress: Lakatos' Methodological Account of Kuhnian Patterns of Theory Change", in Kampis, Kvasz, Stöltzner (eds) (2002), pp. 53–71.

CARTWRIGHT, Nancy
1983 *How the Laws of Physics Lie*, Oxford: Oxford University Press, 1983.

CASSIRER, Ernst
1910 *Substanzbegriff und Funktionbegriff. Untersuchungen über die Grundfragen der Erkenntniskritik*, Darmstadt: Wissenschaftliche Buchgesellschaft, 1910.

CEDARBAUM, Daniel G.
1983 "Paradigms", *Studies in History and Philosophy of Science*, 14, 1983, pp. 173–213.

CHEN, Xiang
1997 "Thomas Kuhn's Latest Notion of Incommensurability", *Journal for General Philosophy of Science*, 28, 1997, pp. 257–273.

COFFA, J. Alberto
1991 *The Semantic Tradition from Kant to Carnap. To the Vienna Station*, edited by Linda Wessels, Cambridge: Cambridge University Press, 1991.

COHEN, I. Bernard
1985 *Revolution in Science*, Cambridge, Massachusetts: Harvard University Press, 1985.

COHEN, Robert S., FEYERABEND, Paul K., WARTOFSKY, Marx W. (eds)
1976 *Essays in Memory of Imre Lakatos*, Dordrecht–Boston: D. Reidel Publishing Company, 1976.

COHEN, Robert S., SCHNELLE, Thomas (eds)
1986 *Cognition and Fact. Materials on Ludwik Fleck*, Dordrecht–Boston–Lancaster–London: D. Reidel Publishing Company, 1986.

COHEN, Robert S., WARTOFSKY, Marx W. (eds)

1965 *Proceedings of the Boston Colloquium for the Philosophy of Science, 1962–64. In Honor of Philipp Frank.* Dordrecht–Boston: D. Reidel Publishing Company, 1965, pp. 223–261.

CONIGLIONE, Francesco

1990 *Realtà ed astrazione. Scuola polacca ed epistemologia post-positivista*, Catania: C.U.E.C.M., 1990.

1996 *Nel segno della scienza. La filosofia polacca del Novecento*, Milan: Franco Angeli, 1996.

CONIGLIONE, Francesco, POLI, Roberto, WOLEŃSKI, Jan (eds)

1993 *Polish Scientific Philosophy. The Lvov-Warsaw School*, Amsterdam–Atlanta: Rodopi, 1997.

CORRY, Leo

1993 "Kuhnian Issues, Scientific Revolutions and the History of Mathematics", *Studies in History and Philosophy of Science*, 24, 1993, pp. 95–117.

CORVI, Roberta

1992 *I fraintendimenti della ragione. Saggio su P.K. Feyerabend*, Milan: Vita e Pensiero, 1992.

CROMBIE, Alistair C. (ed.)

1963 *Scientific Change. Historical Studies in the Intellectual, Social and Technical Conditions for Scientific Discovery and Technical Invention, from Antiquity to the Present*, London: Heinemann Educational Books, 1963.

DAHMS, Hans-Joachim

1995 "The Emigration of the Vienna Circle", in Stadler, Webel (eds) (1995), pp. 57–79.

DAVIDSON, Donald

1974 "On the Very Idea of a Conceptual Scheme", *Proceedings and Addresses of the American Philosophical Association*, 47, 1974, pp. 5–20; reprinted in Davidson (1984), pp. 183–198.

1984 *Inquiries into Truth and Interpretation*, Oxford: Oxford University Press, 1984.

DE SWART, Harrie C.M. (ed.)

1988 *Gerrit Mannoury and Dutch Significs*, special issue of *Methodology and Science*, 21, 1988.

DEVITT, Michael
1979 "Against Incommensurability", *Australasian Journal of Philosophy*, 57, 1979, pp. 29–50.
1984 *Realism and Truth*, Oxford: Basil Blackwell, 1984, 1991[2].
2001 "Incommensurability and the Priority of Metaphysics", in Hoyningen-Huene, Sankey (eds) (2001), pp. 143–157.

DEVITT, Michael, STERELNY, Kim
1987 *Language and Reality. An Introduction to the Philosophy of Language*, Cambridge, Massachusetts: The MIT Press, 1987, 1999[2].

DOKIC, Jérôme, ENGEL, Pascal
2001 *Ramsey. Verité et succès*, Paris: Presses Universitaires de France, 2001; English edition, *Frank Ramsey. Truth and Success*, London–New York: Routledge, 2002.

DOPPELT, Gerald
1978 "Kuhn's Epistemological Relativism. An Interpretation and Defence", *Inquiry*, 21, 1978, pp. 33–86; reprinted in Meiland, Krausz (eds) (1982), pp. 113–146.
1986 "Relativism and the Reticulational Model of Scientific Rationality", *Synthese*, 69, 1986, pp. 225–252.
1990 "The Naturalistic Conception of Methodological Standards in Science: A Critique", *Philosophy of Science*, 57, 1990, pp. 1–19.

DUERR, Hans P. (ed.)
1980 *Versuchungen. Aufsätze zur Philosophie Paul Feyerabends*, 2 vols., Frankfurt am Main: Suhrkamp, 1980.

DUHEM, Pierre
1906 *La théorie physique. Son objet et sa structure*, Paris: Chevalier & Rivière, 1906; *La théorie physique: son objet – sa structure*, Paris: Marcel Rivière, 1914[2]; English translation by Philip P. Wiener, *The aim and structure of physical theory*, Princeton: Princeton University Press, 1954.
1996 *Essays in the History and Philosophy of Science*, translated and edited by Roger Ariew and Peter Barker, Indianapolis: Hackett Publications & Co., 1996.

EARMAN, John
1993 "Carnap, Kuhn, and the Philosophy of Scientific Methodology", in Horwich (ed.) (1993), pp. 9–36.

ECO, Umberto
1990 *The Limits of Interpretation*, Bloomington: Indiana University Press, 1990; revised Italian edition, *I limiti dell'interpretazione*, Milan: Bompiani, 1990.

1997 *Kant e l'ornitorinco*, Milan: Bompiani, 1997; English translation by Alastair McEwen, *Kant and the Platypus. Essays on Language and Cognition*, London: Secker & Warburg, 1999.

2003 *Dire quasi la stessa cosa. Esperienze di traduzione*, Milan: Bompiani, 2003.

EDWARDS, Paul (ed.)

1967 *The Encyclopedia of Philosophy*, New York: The Macmillan Company & The Free Press, and London: Collier-Macmillan, London 1967, 8 vols.

ENÇ, Berent

1976 "Reference of Theoretical Terms", *Nous*, 10, 1976, pp. 261–282.

ENGLISH, Jane

1978 "Partial Interpretation and Meaning Variance", *The Journal of Philosophy*, 75, 1978, pp. 57–76.

FARRELL, Robert P.

2003 *Feyerabend and Scientific Values. Tightrope-Walking Rationality*, Dordrecht–Boston–London: Kluwer Academic Publishers, 2003.

FEIGL, Herbert

1969 "The Wiener Kreis in America", in Donald Fleming, Bernard Bailyn (eds), *The Intellectual Migration. Europe and America, 1930–1960*, Cambridge, Massachusetts: Belknap Press, 1969, pp. 630–673.

1970 "The 'Orthodox' View of Theories: Remarks in Defense as well as Critique", in Radner, Winokur (eds) (1970), pp. 3–16.

FEIGL, Herbert, MAXWELL, Grover G. (eds)

1961 *Current Issues in the Philosophy of Science. Symposia of Scientists and Philosophers. Proceedings of Section L of the American Association for the Advancement of Science, 1959*, New York: Holt, Rinehart & Winston, 1961.

FEYERABEND, Paul K.

1955 "Wittgenstein's 'Philosophical Investigations'", *Philosophical Review*, 64, 1955, pp. 449–483; revised reprint in Feyerabend (1981b), pp. 99–130.

1957 "On the Quantum-Theory of Measurement", in Stephen Körner, Michael H.L. Pryce (eds), *Observation and Interpretation. A Symposium of Philosophers and Physicists*, Academic Press: New York, 1957, pp. 121–130.

1958a "An Attempt at a Realistic Interpretation of Experience", *Proceedings of the Aristotelian Society*, 58, 1958, pp. 143–170; reprinted in Feyerabend (1981a), pp. 17–36.

1958b "Complementarity", *Proceedings of the Aristotelian Society*, Supplementary Volume, 32, 1958, pp. 75–104.

1960 "Das Problem der Existenz Theoretischer Entitäten", in Ernst Topitsch (ed.), *Probleme der Wissenschaftstheorie. Festcshrift für Viktor Kraft*, Vienna: Springer, 1960, pp. 35–72; reprinted in Feyerabend (1978a), pp. 40–73; English translation by Daniel Sirtes and Eric Oberheim, "The Problem of the Existence of Theoretical Entities", in Feyerabend (1999a), pp. 16–49.

1961a *Knowledge Without Foundations*, Two Lectures Delivered on the Nellie Heldt Lecture Fund, Oberlin, Ohio: Oberlin College, 1961; reprinted in Feyerabend (1999a), pp. 50–77.

1961b "Comments on Hanson's 'Is There a Logic of Scientific Discovery?'", in Feigl, Maxwell (eds) (1961), pp. 35–39.

1961c "Niels Bohr's Interpretation of the Quantum Theory", in Feigl, Maxwell (eds) (1961), pp. 371–390.

1962a "Explanation, Reduction, and Empiricism", in Herbert Feigl, Grover G. Maxwell (eds), *Scientific Explanation, Space, and Time*, Minneapolis: University of Minnesota Press, 1962, pp. 28–97; reprinted in Feyerabend (1981a), pp. 44–96.

1962b "Problems of Microphysics", in Robert G. Colodny (ed.), *Frontiers of Science and Philosophy*, Pittsburgh: University of Pittsburgh Press, 1962, pp. 189–283; reduced reprint in Sidney Morgenbesser (ed.), *Philosophy of Science Today*, New York–London: Basic Books, 1967, pp. 136–147.

1964 Review of Crombie (ed.) (1963), *The British Journal for the Philosophy of Science*, 15, 1964, pp. 244–254.

1965a "Problems of Empiricism", in Robert G. Colodny (ed.), *Beyond the Edge of Certainty*, Englewood Cliffs, New Jersey: Prentice-Hall, 1965, pp. 145–260.

1965b "On the 'Meaning' of Scientific Terms", *The Journal of Philosophy*, 62, 1965, pp. 266–274; reprinted in Feyerabend (1981a), pp. 97–103.

1965c "Reply to Criticism: Comments on Smart, Sellars, and Putnam", in Cohen, Wartofsky (eds) (1965), pp. 223–261; revised reprint in Feyerabend (1981a), pp. 104–131.

1965d Review of Popper (1963a), *Isis*, 56, 1965, p. 88.

1966 "Herbert Feigl: A Biographical Sketch", in Feyerabend, Maxwell (eds) (1966), pp. 3–13.

1967a "Boltzmann, Ludwig", in Edwards (ed.) (1967), vol. 1, pp. 334–337.

1967b "Heisenberg, Werner", in Edwards (ed.) (1967), vol. 3, pp. 466–467.

1967c "Planck, Max", in Edwards (ed.) (1967), vol. 6, pp. 312–314.

1967d "Schrödinger, Erwin", in Edwards (ed.) (1967), vol. 7, pp. 332–333.

1970a "Consolations for the Specialist", in Lakatos, Musgrave (eds) (1970), pp. 197–230.

1970b "Against Method. Outline of an Anarchistic Theory of Knowledge", in Radner, Winokur (eds) (1970), pp. 17–130.

1970c "Problems of Empiricism, Part II", in Robert G. Colodny (ed.), *The Nature and Function of Scientific Theories. Essays in Contemporary Science and Philosophy*, Pittsburgh: University of Pittsburgh Press, 1970, pp. 275–353.

1975 *Against Method. Outline of an Anarchistic Theory of Knowledge*, London: New Left Books, 1975; London: Verso, 1988[2], 1993[3].

1977 "Changing Patterns of Reconstruction", *The British Journal for the Philosophy of Science*, 28, 1977, pp. 351–369.

1978a *Der Wissenschaftstheoretische Realismus und die Autorität der Wissenschaften. Ausgewählte Schriften*, Braunschweig: Friedrich Vieweg & Sohn, 1978.

1978b "Kuhns Struktur wissenschaftlicher Revolutionen. Ein Trostbüchlein für Spezialisten?", in Feyerabend (1978a), pp. 153–204.

1978c *Science in a Free Society*, London: New Left Books, 1978.

1979 *Erkenntnis für freie Menschen*, Frankfurt am Main: Suhrkamp, 1979, 1980[2].

1981a *Realism, Rationalism, and Scientific Method. Philosophical Papers*, vol. 1, Cambridge: Cambridge University Press, 1981.

1981b *Problems of Empiricism. Philosophical Papers*, vol. 2, Cambridge: Cambridge University Press, 1981.

1987a *Farewell to Reason*, London–New York: Verso, 1987.

1987b "Putnam on Incommensurability", *The British Journal for the Philosophy of Science*, 38, 1987, pp. 75–81; reprinted in Feyerabend (1987a), pp. 265–272.

1989a "Realism and the Historicity of Knowledge", *The Journal of Philosophy*, 86, 1989, pp. 393–406.

1989b *Dialogo sul metodo*, edited by Roberta Corvi, Rome–Bari: Laterza, 1989; English edition in Paul K. Feyerabend, *Three Dialogues on Knowledge*, Oxford: Blackwell Publishers, 1991, pp. 47–159.

1994 Review of Biagioli (1993) and Hoyningen-Huene (1989a/1993), *Common Knowledge*, 3, 1994, p. 173.

1995a "Two Letters of Paul Feyerabend to Thomas S. Kuhn on a Draft of *The Structure of Scientific Revolutions*", edited by Paul Hoyningen-Huene, *Studies in History and Philosophy of Science*, 26, 1995, pp. 353–387.

1995b *Killing Time. An Autobiography*, Chicago–London: University of Chicago Press, 1995.

1999a *Knowledge, Science and Relativism. Philosophical Papers*, vol. 3, edited by John M. Preston, Cambridge: Cambridge University Press, 1999.

1999b *Conquest of Abundance. A Tale of Abstraction versus the Richness of Being*, edited by Bert Terpstra, Chicago–London: University of Chicago Press, 1999.

2006 "More Letters by Paul Feyerabend to Thomas S. Kuhn on Proto-Structure", edited by Paul Hoyningen-Huene, *Studies in History and Philosophy of Science*, 37, 2006, pp. 610–632.

Forthcoming *Physics and Philosophy*, edited by Joseph Agassi and Stefano Gattei, New York: Cambridge University Press, forthcoming.

FEYERABEND, Paul K., MAXWELL, Grover G. (eds)

1966 *Mind, Matter, and Method. Essays in Philosophy and Science in Honor of Herbert Feigl*, Minneapolis: University of Minnesota Press, 1966.

FIELD, Hartry
1973 "Theory Change and the Indeterminacy of Reference", *The Journal of Philosophy*, 70, 1973, pp. 462–481.

FINE, Arthur
1975 "How to Compare Theories: Reference and Change", *Nous*, 9, 1975, pp. 17–32.

FLECK, Ludwik
1935 *Entstehung und Entwicklung der wissenschaftlichen Tatsache*, Basel: Benno Schwabe, 1935; Frankfurt am Main: Suhrkamp, 1980[2]; English translation by Fred Bradley and Thaddeus J. Trenn, *Genesis and Development of a Scientific Fact*, edited by Thaddeus J. Trenn and Robert K. Merton, preface by Thomas S. Kuhn, Chicago–London: University of Chicago Press, 1979.
1983 *Erfahrung und Tatsache. Gesammelte Aufsätze*, Frankfurt am Main: Suhrkamp, 1983.

FREGE, F. L. Gottlob
1892 "Über Sinn und Bedeutung", *Zeitschrift für Philosophie und philosophische Kritik*, 100, 1892, pp. 25–50; English translation by Max Black, "On Sense and Reference", in Gottlob Frege, *Translations from the Philosophical Writings of Gottlob Frege*, edited by Peter Geach and Max Black, Oxford: Basil Blackwell, 1962, 1970[2], pp. 56–78.
1919 "Der Gedanke. Eine Logische Untersuchung", *Beiträge zur Philosophie des deutschen Idealismus*, I, 1918–1919, pp. 58–77; English translation by R. H. Stoothoff, "Thoughts", in Gottlob Frege, *Logical Investigations*, edited with a Preface by Peter T. Geach, Oxford: Basil Blackwell, 1977, pp. 1–30.

FRIEDMAN, Michael
1987 "Carnap's *Aufbau* Reconsidered", *Nous*, 21, 1987, pp. 521–545; reprinted in Friedman (1999), pp. 89–113.
1991 "The Re-evaluation of Logical Positivism", *The Journal of Philosophy*, 88, 1991, pp. 505–519; reprinted as "Introduction" in Friedman (1999), pp. 1–14.
1992a "Philosophy and the Exact Sciences: Logical Positivism as a Case Study", in John Earman (ed.), *Inference, Explanation, and Other Frustrations. Essays in the Philosophy of Science*, Los Angeles: University of California Press, 1992, pp. 84–98.
1992b "Epistemology in the *Aufbau*", *Synthese*, 93, 1992, pp. 15–57; reprinted in Friedman (1999), pp. 114–162.
1993 "Remarks on the History of Science and the History of Philosophy", in Horwich (ed.) (1993), pp. 37–54.
1999 *Reconsidering Logical Positivism*, Cambridge: Cambridge University Press, 1999.

2000 *A Parting of the Ways. Carnap, Cassirer, and Heidegger*, Chicago–La Salle: Open Court Publishing Company, 2000.
2001 *Dynamics of Reason*, Stanford: CSLI Publications, 2001.
2002a "Kuhn and Logical Empiricism", in Nickles (ed.) (2002), pp. 19–44.
2002b "Kant, Kuhn, and the Rationality of Science", *Philosophy of Science*, 69, 2002, pp. 171–190.

FRIES, Jakob F.
1828–1831 *Neue oder anthropologische Kritik der Vernunft*, 3 vols., Heidelberg: Winter, 1828–1831.

FROLA, Eugenio
1947 "La matematica come lingua chiusa e la conoscenza del mondo fisico", in Abbagnano, Buzano, Buzzati-Traverso, Frola, Geymonat, Persico (1947), pp. 91–109.
1964 *Scritti metodologici*, edited by Ludovico Geymonat, Turin: Giappichelli, 1964.

FULLER, Steve W.
2000 *Thomas Kuhn. A Philosophical History of Our Times*, Chicago–London: University of Chicago Press, 2000.
2004 "The Case of Fuller vs Kuhn", *Social Epistemology*, 18, 2004, pp. 3–39.

GALILEI, Galileo
1632 *Dialogo sopra i due massimi sistemi del mondo*, Florence: Giovan Battista Landini, 1632; English translation by Stillman Drake, *Dialogue Concerning the Two Chief World Systems, Ptolemaic & Copernican*, foreword by Albert Einstein, Berkeley: University of California Press, 1953, 1967[2].
1638 *Discorsi e dimostrazioni matematiche intorno a due nuove scienze attenenti alla mecanica et i movimenti locali*, Leyden: Elsevier, 1638; English translation by Henry Crew and Alfonso De Salvio, *Dialogues Concerning Two New Sciences*, with an introduction by Antonio Favaro, New York: The Macmillan Company, 1914.

GALISON, Peter L.
1993 "The Cultural Meaning of *Aufbau*", in Stadler (ed.) (1993), pp. 75–93.
1995 "Context and Constraints", in Jed Z. Buchwald (ed.) *Scientific Practice. Theories and Stories of Doing Physics*, Chicago–London: University of Chicago Press, 1995, pp. 13–41.
1996 "Constructing Modernism: The Cultural Location of *Aufbau*", in Giere, Richardson (eds) (1996), pp. 17–44.

GATTEI, Stefano
1995 "Con Parmenide guardando la Luna", *Reset*, 23, 1995, pp. 28–29.
2000a "Tom a Paul", in Kuhn (2000a), pp. 223–231.

2000b "La filosofia della scienza di Thomas S. Kuhn: una ricostruzione", in Kuhn (2000a), pp. 293–344.

2000c "Nota biografica", in Kuhn (2000a), pp. 345–349.

2002a "The Ethical Nature of Karl Popper's Solution to the Problem of Rationality", *Philosophy of the Social Sciences*, 32, 2002, pp. 240–266.

2002b "The Positive Power of Negative Thinking", *Cladistics*, 18, 2002, pp. 446–452.

2002c "Razionalità senza fondamenti", in Gattei (ed.) (2002), n. 1, pp. 10–14.

2002d "La ragione come libera scelta", in Gattei (ed.) (2002), n. 1, pp. 25–38.

2003 "A Plea for Criticism in Matters Epistemological", in Gattei (ed.) (2003), pp. 161–168.

2004 "Karl Popper's Philosophical Breakthrough", *Philosophy of Science*, 71, 2004, pp. 448–466.

2005a "The Complex Story of Popper's Early Intellectual Development", *Learning for Democracy*, 1, 2005, pp. 88–91.

2005b "Thomas Kuhn and the Legacy of Logical Positivism", in Stephen Rainey, Barbara Gabriella Renzi (eds), *Noesis. Essays in the History and Philosophy of Science, Philosophy of Language, Epistemology and Political Philosophy*, Cambridge: Cambridge Scholars Press, 2005, pp. 31–42.

2005c "Due approcci al problema della razionalità", in Roberta Corvi (a cura di), *Esperienza e razionalità. Prospettive contemporanee*, Franco Angeli Editore, Milano 2005, pp. 60–77.

2006 "Rationality without Foundations", in Ian C. Jarvie, Karl Milford, David W. Miller (eds), *Karl Popper: A Centenary Assessment*, vol. II: *Metaphysics and Epistemology*, Aldershot: Ashgate, 2006, pp. 131–144.

2007 *Introduzione a Popper*, Rome–Bari: Laterza, 2007.

Forthcoming *Karl Popper's Philosophy of Science: Rationality Without Foundations*, London–New York: Routledge, forthcoming.

GATTEI, Stefano (ed.)

2002 *Karl R. Popper, 1902–2002: ripensando il razionalismo critico*, special monographic issues of *Nuova Civiltà delle Macchine*, XX, 1–2, 2002.

2003 *The Kuhn Controversy*, special monographic issue of *Social Epistemology*, 17, 2–3, 2003.

GEYMONAT, Ludovico

1934 *La nuova filosofia della natura in Germania*, Turin: Bocca, 1934.

1935 "Nuovi indirizzi della filosofia austriaca", *Rivista di filosofia*, XXVI, 1935, pp. 146–175.

1936 "Logica e filosofia delle scienze", *Rivista di filosofia*, XXVII, 1936, pp. 250–265.

1945 *Studi per un nuovo razionalismo*, Turin: Chiantore, 1945.

1953 *Saggi di filosofia neorazionalistica*, Turin: Einaudi, 1953.

1957 *Galileo Galilei*, Turin: Einaudi, 1957.

1960 *Filosofia e filosofia della scienza*, Milan: Feltrinelli, 1960.

1963 “Eugenio Frola”, *Atti della Accademia delle Scienze di Torino I. Classe di Scienze Fisiche, Matematiche e Naturali*, 97, 1963, pp. 986–997.

1964 “Il significato del contributo di Eugenio Frola alla rinascita della metodologia della scienza in Italia”, in Frola (1964), pp. 7–33.

GHINS, Michel

1998 “Kuhn: Realist or Antirealist?”, *Principia*, 2, 1998, pp. 37–59.

GIEDYMIN, Jerzy

1968 “Revolutionary Changes, Non-Translatability, and Critical Experiments”, in Lakatos, Musgrave (eds) (1968), pp. 223–227.

1970 “The Paradox of Meaning Variance”, *The British Journal for the Philosophy of Science*, 21, 1970, pp. 257–268.

1971 “Consolations for the Irrationalist?”, *The British Journal for the Philosophy of Science*”, 22, 1971, pp. 39–48.

1973 “Logical Comparability and Conceptual Disparity between Newtonian and Relativistic Mechanics”, *The British Journal for the Philosophy of Science*, 24, 1973, pp. 270–276; reprinted in Giedymin (1982), pp. 196–205.

1974 Review of Ajdukiewicz (1949/1973), *The British Journal for the Philosophy of Science*, 25, 1974, pp. 189–194.

1975 “Antipositivism in Contemporary Philosophy of Social Science and Humanities”, *The British Journal for the Philosophy of Science*, 26, 1975, pp. 275–301.

1977 “On the Origin and Significance of Poincaré’s Conventionalism”, *Studies in History and Philosophy of Science*, 8, 4, 1997, pp. 271–301; reprinted in Giedymin (1982), pp. 1–41.

1978 “Radical Conventionalism, Its Background and Evolution: Poincaré, LeRoy and Ajdukiewicz”, in Ajdukiewicz (1978), pp. xix–lii; reprinted in Giedymin (1982), pp. 109–148.

1980 “Hamilton’s Method in Geometrical Optics and Ramsey’s View of Theories”, in David H. Mellor (ed.), *Prospects for Pragmtism. Essays in Memory of F.P. Ramsey*, Cambridge: Cambridge University Press, 1980, pp. 229–254; reprinted as “The Physics of the Principles and its Philosophy: Hamilton, Poincaré and Ramsey”, in Giedymin (1982), pp. 42–89.

1982 *Science and Convention. Essays on Henri Poincare’s Philosophy of Science and the Conventionalist Tradition*, Oxford: Pergamon Press, 1982.

1991 “Geometrical and Physical Conventionalism of Henri Poincaré in Epistemological Formulation”, *Studies in History and Philosophy of Science*, 22, 1991, pp. 1–22.

1992 “Conventionalism, the Pluralist Conception of Theories and the Nature of Interpretation”, *Studies in History and Philosophy of Science*, 23, 1992, pp. 423–443.

GIERE, Ronald N.

1988 *Explaining Science. A Cognitive Approach*, Chicago–London: University of Chicago Press, 1988.

GIERE, Ronald N., RICHARDSON, Alan W. (eds)
1996 *Origins of Logical Empiricism*, Minneapolis: University of Minnesota Press, 1996.

GILLIES, Donald A.
1992 "The Fregean Revolution in Logic", in Gillies (ed.) (1992), pp. 265–305.
1993 *The Philosophy of Science in the Twentieth Century: Four Central Themes*, Oxford: Blackwell Publishers, 1993; Italian translation by Matteo Motterlini, with additional chapters by Giulio Giorello, *La filosofia della scienza nel XX secolo*, Rome–Bari: Laterza 1995.

GILLIES, Donald A. (ed.)
1992 *Revolutions in Mathematics*, Oxford: Oxford University Press, 1992.

GINZBURG, Carlo
1998 *Occhiacci di legno. Nove riflessioni sulla distanza*, Milan: Feltrinelli, 1998.

GIORDANO, Giuseppe
1997 *Tra paradigmi e rivoluzioni. Thomas Kuhn*, Soveria Mannelli: Rubbettino, 1997.

GIORELLO, Giulio
1976a "Filosofia della scienza e storia della scienza nella cultura di lingua inglese", in Ludovico Geymonat, *Storia del pensiero filosofico e scientifico*, Milan: Garzanti, 1970–1976, vol. VII, pp. 190–298.
1976b "Introduzione", in Imre Lakatos, Alan Musgrave (ed.), *Critica e crescita della conoscenza*, Milan: Feltrinelli, 1976, pp. 7–63.
1977 "Introduzione", in Giorello (ed.) (1977), pp. vii–xl.
1979 "Prefazione", in Paul K. Feyerabend, *Contro il metodo. Abbozzo di una teoria anarchica della conoscenza*, Milan: Feltrinelli, 1979, pp. 1–12.
1986 "Le correnti marxiste e il problema del cambiamento concettuale in scienza", in Agazzi (ed.) (1986), pp. 259–301.
1992 "The 'Fine-Structure' of Mathematical Revolutions: Metaphysics, Legitimacy, and Rigour. The case of the Calculus from Newton to Berkeley and Maclaurin", in Gillies (ed.) (1992), pp. 134–168.

GIORELLO, Giulio (ed.)
1977 *L'immagine della scienza. Il dibattito sul significato dell'impresa scientifica nella cultura italiana*, Milan: il Saggiatore, 1977.

GIORELLO, Giulio, MONDADORI, Marco
1978 "Dinamica della conoscenza scientifica e dialettica", in Umberto Curi (ed.), *La razionalità scientifica*, Padua: Francisci, 1978, pp. 107–149.

1992 "Ludovico Geymonat e la filosofia della scienza", in Corrado Mangione (ed.), *Omaggio a L. Geymonat*, Padua: Franco Muzzio Editore, 1992, pp. 25–41.

GUICCIARDINI, Niccolò

1999 *Reading the* Principia. *The Debate on Newton's Mathematical Methods for Natural Philosophy from 1687 to 1736*, Cambridge: Cambridge University Press, 1999.

GUTTING, Gary (ed.)

1980 *Paradigms and Revolutions. Appraisals and Applications of Thomas Kuhn's Philosophy of Science*, Notre Dame–London: University of Notre Dame Press, 1980.

HACKER, Peter M.S.

1996 *Wittgenstein's Place in Twentieth-Century Analytic Philosophy*, Oxford: Blackwell Publishers, 1996.

HACKING, Ian

1983 *Representing and Intervening. Introductory Topics in the Philosophy of Natural Sciences*, Cambridge: Cambridge University Press, 1983.

1993 "Working in a New World: The Taxonomic Solution", in Horwich (ed.) (1993), pp. 275–310.

HACKING, Ian (ed.)

1981 *Scientific Revolutions*, Oxford: Oxford University Press, 1981.

HACOHEN, Malachi H.

2000 *Karl Popper – The Formative Years, 1902–1945. Politics and Philosophy in Interwar Vienna*, New York: Cambridge University Press, 2000.

HAHN, Hans, NEURATH, Otto, CARNAP, Rudolf

1929 *Wissenschaftliche Weltauffassung. Der Wiener Kreis*, Veröffentlichungen des Vereines Ernst Mach, Artur Wolf, Vienna 1929; partial English translation by Marie Neurath, "The Scientific Conception of the World: The Vienna Circle", in Neurath (1973), pp. 299–318.

HALLER, Rudolf

1995 "'Philosophy – Tool and Weapon'", in Stadler, Webel (eds) (1995), pp. 80–87.

HANSON, Norwood R.

1958 *Patterns of Discovery. An Inquiry into the Conceptual Foundations of Science*, Cambridge: Cambridge University Press, 1958.

1969 "Logical Positivism and the Interpretation of Scientific Theories", in Achinstein, Barker (eds) (1969), pp. 57–84.

Harding, Sandra G. (ed.)
1976 *Can Theories be Refuted? Essays on the Duhem–Quine Thesis*, edited by Sandra G. Harding, Dordrecht–Boston: D. Reidel Publishing Company, 1976.

Hardwig, John
1991 "The Role of Trust in Knowledge", *The Journal of Philosophy*, 88, 1991, pp. 693–708.

Harré, Rom (ed.)
1975 *Problems of Scientific Revolution. Progress and Obstacles to Progress in the Sciences*, Oxford: Oxford University Press, 1975.

Hattiangadi, Jagdish N.
1965 *Truth, Acceptance, and Agreement. A Discussion of Professor Kuhn's Theory of Science*, Master's Thesis, University of London, 1965.

Heilbron, John L.
1998 "Thomas Samuel Kuhn: 18 July 1922 – 17 June 1996", *Isis*, 89, 1998, pp. 505–515.

Hempel, Carl G.
1935 "On the Logical Positivists' Theory of Truth", *Analysis*, 2, 1935, pp. 49–59; reprinted in Carl G. Hempel, *Selected Philosophical Essays*, edited by Richard C. Jeffrey, New York: Cambridge University Press, 2000, pp. 9–20.
1952 *Fundamentals of Concept Formation in Empirical Science*, Chicago: University of Chicago Press, Chicago 1952.
1958 "The Theoretician's Dilemma. A Study in the Logic of Theory Construction", in Herbert Feigl, Michael Scriven, Grover G. Maxwell (eds), *Concepts, Theories, and the Mind-Body Problem*, Minneapolis: University of Minnesota Press, 1958, pp. 37–98; reprinted in Hempel (1965a), pp. 173–226.
1965a *Aspects of Scientific Explanation and Other Essays in the Philosophy of Science*, New York: The Free Press, 1965.
1965b "Fundamentals of Taxonomy", in Hempel (1965a), pp. 137–154.
1966 *Philosophy of Natural Sciences*, Englewood Cliffs, New Jersey: Prentice-Hall, 1966.
1970 "On the 'Standard Conception' of Scientific Theories", in Radner, Winokur (eds) (1970), pp. 142–163.
1983 "Kuhn and Salmon on Rationality and Theory Choice", *The Journal of Philosophy*, 83, 1980, pp. 570–572.

1989 *Oltre il positivismo logico. Saggi e ricordi*, edited by Gianni Rigamonti, Rome: Armando Armando, 1989.
1993 "Thomas Kuhn, Colleague and Friend", in Horwich (ed.) (1993), pp. 7–8.

HEMPEL, Carl G., OPPENHEIM, Paul
1948 "Studies in the Logic of Explanation", *Philosophy of Science*, 15, 1948, pp. 135–175; reprinted, with a postscript, in Hempel (1965a), pp. 245–295.

HERSCHEL, John F.W.
1830 *A Preliminary Discourse on the Study of Natural Philosophy*, London: Longman, Rees, Orme, Brown, Green & Taylor, 1830, 1833[2].

HESSE, Mary B.
1963 Review of Kuhn (1962a), *Isis*, 54, 1963, pp. 286–287.
1983 "Comment on Kuhn's 'Commensurability, Comparability, Communicability'", in Asquith, Nickles (eds) (1983), vol. 2, pp. 704–711.

HILBERT, David
1899 *Grundlagen der Geometrie*, Leipzig: B.G. Teubner, 1899; English translation by Leo Unger, *The Foundations of Geometry*, La Salle, Illinois: Open Court, 1971.

HORWICH, Paul (ed.)
1993 *World Changes. Thomas Kuhn and the Nature of Science*, Cambridge, Massachusetts–London: The MIT Press, 1993.

HOWSON, Colin (ed.)
1976 *Method and Appraisal in the Physical Sciences*, Cambridge: Cambridge University Press, 1976.

HOYNINGEN-HUENE, Paul
1989a *Die Wissenschaftsphilosophie Thomas S. Kuhns. Rekonstruktion und Grundlagenprobleme*, Braunschweig: Friedrich Vieweg & Sohn, 1989; English translation by Alexander T. Levine, *Reconstructing Scientific Revolutions. Thomas S. Kuhn's Philosophy of Science*, Chicago–London: University of Chicago Press, 1993.
1989b "Idealist Elements in Thomas Kuhn's Philosophy of Science", *History of Philosophy Quarterly*, 6, 1989, pp. 393–401.
1990 "Kuhn's Conception of Incommensurability", *Studies in History and Philosophy of Science*, 21, 1990, pp. 481–492.
1997a "Obituary of Thomas S. Kuhn (1922–1996)", *Erkenntnis*, 45, 1997, pp. v–viii.
1997b "Paul K. Feyerabend", *Journal for General Philosophy of Science*, 28, 1997, pp. 1–18.
1997c "Thomas S. Kuhn", *Journal for General Philosophy of Science*, 28, 1997, pp. 235–256.

2000 "Paul K. Feyerabend and Thomas Kuhn", in Preston, Munévar, Lamb (eds) (2000), pp. 102–114.

HOYNINGEN-HUENE, Paul, OBERHEIM, Eric, ANDERSEN, Hanne
1996 "On Incommensurability", *Studies in History and Philosophy of Science*, 27, 1996, pp. 131–141.

HOYNINGEN-HUENE, Paul, SANKEY, Howard (eds)
2001 *Incommensurability and Related Matters*, Dordrecht–Boston–London: Kluwer Academic Publishers, 2001.

HULL, McAllister H., Jr.
1995 "The Influence of the Vienna Circle on American Philosophy of Science", in Stadler, Webel (eds) (1995), pp. 88–96.

IRZIK, Gürol
2000 "Il nuovo Kuhn: la formulazione tassonomica", *Iride*, 31, 2000, pp. 636–645.
2002 "Carnap and Kuhn: a Belated Encounter", in Peter Gärdenfors, Jan Woleński, Katarzyna Kijana-Placek (eds), *In the Scope of Logic, Methodology and Philosophy of Science. Volume Two of the 11th International Congress of Logic, Methodology and Philosophy of Science, Cracow, August 1999*, Dordrecht–Boston–London: Kluwer Academic Publishers, 2002, pp. 603–620.
2003 "Changing Conceptions of Rationality: From Logical Empiricism to Post-Positivism", in Parrini, Salmon, Salmon (eds) (2003), pp. 325–346.

IRZIK, Gürol, GRÜNBERG, Teo
1995 "Carnap and Kuhn: Arch Enemies or Close Allies?", *The British Journal for the Philosophy of Science*, 46, 1995, pp. 285–307.
1998 "Whorfian Variations on Kantian Themes: Kuhn's Linguistic Turn", *Studies in History and Philosophy of Science*, 29, 1998, pp. 207–221.

JACOBS, Struan
2002 "Polanyi's Presagement of the Incommensurability Concept", *Studies in History and Philosophy of Science*, 33, 2002, pp. 105–120.
2003 "Two Sources of Michael Polanyi's Prototypal Idea of Incommensurability: Evans-Pritchard on Azande Witchcraft, and St. Augustine on Conversion", *History of the Human Sciences*, 16, 2003, pp. 57–76.

JAMES, William
1904 "Humanism and Truth", *Mind*, 13, 1904, pp. 457–475.

JØRGENSEN, Jørgen

1951 *The Development of Logical Empiricism*, Chicago: University of Chicago Press, 1951; reprinted in Neurath, Carnap, Morris (eds) (1970), pp. 845–936.

KAMPIS, George, KVASZ, Ladislav, STÖLTZNER, Michael (eds)

2002 *Appraising Lakatos. Mathematics, Methodology, and the Man*, Dordrecht–Boston–London: Kluwer Academic Publishers, 2002.

KANT, Immanuel

1781 *Kritik der reinen Vernunft*, Riga: Hartknoch, 1781, 1787²; English translation by Norman Kemp Smith, *Critique of Pure Reason*, London: Macmillan, 1929; New York: Palgrave Macmillan, 2003³.

1784 "Beantwortung der Frage: Was ist Aufklärung?", *Berlinische Monatsschrift*, IV, 1784, pp. 481–494; English translation by Hugh B. Nisbet, "An Answer to the Question: 'What is Enlightenment'?", in Immanuel Kant, *Kant's Political Writings*, edited with an introduction by Hans Reiss, Cambridge: Cambridge University Press, 1970, pp. 54–60.

KATZ, David

1944 *Gestaltpsychologie*, Basel: B. Schwab & Co., 1944; English translation by Robert Tyson, *Gestalt Psychology. Its Nature and Significance*, New York: Ronald, 1950.

KJANIA-PLACEK, Katarzyna, WOLEŃSKI, Jan (eds)

1998 *The Lvov-Warsaw School and Contemporary Philosophy*, Dordrecht–Boston–London: Kluwer Academic Publishers, 1998.

KOFFKA, Kurt

1935 *Principle of Gestalt Psychology*, New York: Harcourt, Brace & Co., 1935.

KOPPELBERG, Dirk

1987 *Die Aufhebung der Analytischen Philosophie. Quine als Synthese von Carnap und Neurath*, Frankfurt am Main: Suhrkamp, 1987.

KORDIG, Carl R.

1971 *The Justification of Scientific Change*, Dordrecht–Boston: D. Reidel Publishing Company, 1971.

KOYRÉ, Alexandre

1931 "Die Philosophie Emile Meyersons", *Deutsch–Französische Rundschau*, 4, 1931, pp. 187–217.

1939–1940 *Etudes galileénnes*, 3 vols, Paris: Hermann, 1939–1940; English translation by John Mepham, *Galileo Studies*, Atlantic Highlands, New Jersey: Humanities Press, 1978.

KRAFT, Victor

1950 *Der Wiener Kreis: Der Ursprung des Neopositivismus. Ein Kapitel der jüngsten Philosophiegeschichte*, Vienna–New York: Springer, 1950; English translation by Arthur Pap, *The Vienna Circle: The Origins of Neo-Positivism. A Chapter in the History of Recent Philosophy*, New York: Philosophical Library, 1953.

1974 "Popper and the Vienna Circle", in Schilpp (ed.) (1974), vol. I, pp. 185–204.

KRIPKE, Saul A.

1980 *Naming and Necessity*, Oxford: Basil Blackwell, 1980.

1981 "Wittgenstein on Rules and Private Language", in Irving Block (ed.), *Perspectives on the Philosophy of Wittgenstein*, Oxford: Basil Blackwell, 1981, pp. 238–312.

KROON, Fred

1985 "Theoretical Terms and the Causal View of Reference", *Australasian Journal of Philosophy*, 63, 1985, pp. 143–166.

KUHN, Thomas S.

1949 *The Cohesive Energy of Monovalent Metals as a Function of Their Atomic Quantum Defects*, Ph.D. dissertation, Harvard University, 1949.

1950 "An Application of the W. K. B. Method to the Cohesive Energy of Monovalent Metals", *The Physical Review*, 79, 1950, pp. 515–519.

1951a "A Convenient General Solution of the Confluent Hypergeometric Equation, Analytic and Numerical Development", *Quarterly of Applied Mathematics*, IX, 1951, pp. 1–16.

1951b "Newton's '31st Query' and the Degradation of Gold", *Isis*, 42, 1951, pp. 296–298.

1952a "Robert Boyle and Structural Chemistry in the Seventeenth Century", *Isis*, 43, pp. 12–36.

1952b Reply to Marie Boas, "Newton and the Theory of Chemical Solution", *Isis*, 43, 1952, pp. 123–124.

1952c "The Independence of Density and Pore-size in Newton's Theory of Matter", *Isis*, 43, 1952, pp. 364–365.

1955a "Carnot's Version of 'Carnot's Cycle'", *American Journal of Physics*, 23, 1955, pp. 91–95.

1955b "La Mer's Version of 'Carnot's Cycle'", *American Journal of Physics*, 23, 6, 1955, pp. 387–389.

1957 *The Copernican Revolution. Planetary Astronomy in the Development of Western Thought*, Cambridge, Massachusetts–London: Harvard University Press, 1957, 1985[4].

1958a "The Caloric Theory of Adiabatic Compression", *Isis*, 49, 1958, pp. 132–140.

1958b "Newton's Optical Papers", in *Isaac Newton's Papers & Letters On Natural Philosophy and related documents*, edited by I. Bernard Cohen with the assistance of Robert E. Schofield, Cambridge, Massachusetts: Harvard University Press, 1958, 1978², pp. 27–45.

1959a "The Essential Tension: Tradition and Innovation in Scientific Research", in Calvin W. Taylor (ed.), *The Third (1959) University of Utah Research Conference on the Identification of Creative Scientific Talent*, Salt Lake City: University of Utah Press, 1959, pp. 162–167, 169–174; reprinted in Kuhn (1977a), pp. 225–239.

1959b "Energy Conservation as an Example of Simultaneous Discovery", in Marshall Clagett (ed.), *Critical Problems in the History of Science*, Madison: The University of Wisconsin Press, 1959, pp. 321–356; reprinted in Kuhn (1977a), pp. 66–104.

1960 "Engineering Precedent for the Work of Sadi Carnot", *Archives internationales d'Histoire des Sciences*, XIII, 1960, pp. 251–255; also in *Actes du IXᵉ Congrès International d'Historie des Sciences*, Barcelona: Hermann & C.ie, 1960, vol. I, pp. 530–535.

1961a "The Function of Measurement in Modern Physical Science", *Isis*, 52, 1961, pp. 161–193; reprinted in Kuhn (1977a), pp. 178–224.

1961b "Sadi Carnot and the Cagnard Engine", *Isis*, 52, 1961, pp. 567–574.

1962a *The Structure of Scientific Revolutions*, Chicago–London: University of Chicago Press, 1962, 1970².

1962b "Historical Structure of Scientific Discovery", *Science*, 136, 1962, pp. 760–764; reprinted as "The Historical Structure of Scientific Discovery", in Kuhn (1977a), pp. 165–177.

1963a "The Function of Dogma in Scientific Research", in Crombie (ed.) (1963), pp. 347–369.

1963b "Discussion" (on Kuhn (1963a)), in Crombie (ed.) (1963), pp. 386–395.

1964 "A Function for Thought Experiments", in *Mélanges Alexandre Koyré*, vol. II: *L'aventure de l'esprit*, Paris: Hermann, 1964, pp. 307–334; reprinted in Kuhn (1977a), pp. 240–265.

1966 Review of Agassi (1963a), *The British Journal for the Philosophy of Science*, 17, 1966, pp. 256–258.

1967a "The Turn to Recent Science", *Isis*, 58, 1967, pp. 409–419.

1967b Review of Stefan Rozental (ed.), *Niels Bohr. His Life and Work As Seen By His Friends & Colleagues* (Amsterdam: North Holland, 1967), *American Scientist*, 55, 1967, pp. 339A–340A.

1968a "The History of Science", in *International Encyclopedia of the Social Sciences*, David L. Sills, editor, New York: The Macmillan Company & The Free Press, 1968, vol. 14, pp. 74–83; reprinted in Kuhn (1977a), pp. 105–126.

1968b Review of D. ter Haar, *The Old Quantum Theory*, Oxford: Pergamon Press, 1967, *The British Journal for the History of Science*, IV, 1968, pp. 80–81.

1969 "Comment" [on Everett M. Hafner, "The New Reality in Art and Science", James S. Ackerman, "The Demise of the Avant Garde: Notes on the Sociology of Recent American Art", and George Kubler, "Comment"], *Comparative Studies in Society and History*, 11, 1969, pp. 403–412;

reprinted as "Comment on the Relations of Science and Art", in Kuhn (1977a), pp. 340–351.

1970a "Logic of Discovery or Psychology of Research?", in Lakatos, Musgrave (eds) (1970), pp. 1–23.

1970b "Reflections on my Critics", in Lakatos, Musgrave (eds) (1970), pp. 231–278; reprinted in Kuhn (2000b), pp. 123–175.

1970c "Postscript – 1969", in Kuhn (1962a, 1970), pp. 174–210.

1971a "Notes on Lakatos", in Buck, Cohen (eds) (1971), pp. 137–146.

1971b "Les notions de causalité dans le développement de la physique", French translation by Gilbert Voyat, in Mario Bunge, Francis Halbwachs, Thomas S. Kuhn, Jean Piaget, Leon Rosenfeld, *Les théories de la causalité*, Paris: Presses Universitaires de France, 1971, pp. 7–18; reprinted as "Concepts of Cause in the Development of Physics", in Kuhn (1977a), pp. 21–30.

1971c "The Relations Between History and History of Science", *Daedalus*, 100, 1971, pp. 271–304; reprinted as "The Relations between History and the History of Science", in Kuhn (1977a), pp. 127–161.

1972 Review of Martin J. Klein, *Paul Ehrenfest*, vol. 1: *The Making of a Theoretical Physicist* (New York: American Elsevier, 1970), *American Scientist*, 60, 1972, p. 98.

1974a "Discussion" [on I. Bernard Cohen, "History and the Philosopher of Science"] in Suppe (ed.) (1974), pp. 369–370, 373.

1974b "Discussion" [on David Bohm, "Science as Perception-Communication", and Robert L. Causey, "Professor Bohm's View of the Structure and Development of Theories"], in Suppe (ed.) (1974), pp. 409–412.

1974c "Second Thoughts on Paradigms", in Suppe (ed.) (1974), pp. 459–482; reprinted in Kuhn (1977a), pp. 293–319.

1974d "Discussion" [on Frederick Suppe, "Exemplars, Theories and Disciplinary Matrixes"], in Suppe (ed.) (1974), pp. 500–506, 507–509, 510–511, 512–513, 515–516, 516–517.

1975a "Tradition Mathématique et tradition expérimentale dans le développement de la physique", *Annales*, XXX, 1975, pp. 975–998; revised reprint as Kuhn (1976a).

1975b "The Quantum Theory of Specific Heats: A Problem In Professional Recognition", in *Proceedings of the XIV International Congress for the History of Science 1974*, Tokyo: Science Council of Japan, 1975, vol. 1, pp. 170–182.

1975c "Addendum to 'The Quantum Theory of Specific Heats'", in *Proceedings of the XIV International Congress for the History of Science 1974*, Tokyo: Science Council of Japan, 1975, vol. 4, p. 207.

1976a "Mathematical vs. Experimental Traditions in the Development of Physical Science", *The Journal of Interdisciplinary History*, VII, 1976, pp. 1–31; reprinted as "Mathematical versus Experimental Traditions in the Development of Physical Science", in Kuhn (1977a), pp. 31–65.

1976b "Theory-Change as Structure-Change: Comments on the Sneed Formalism", *Erkenntnis*, 10, 1976, pp. 179–199; reprinted in Kuhn (2000b), pp. 176–195.

1976c review of Roger H. Stuewer, *The Compton Effect. Turning Point in Physics*, New York: Science History Publications, 1975, *American Journal of Physics*, 44, 1976, pp. 1231–1232.

1977a *The Essential Tension. Selected Studies in Scientific Tradition and Change*, Chicago–London: University of Chicago Press, 1977.

1977b "The Relations between the History and the Philosophy of Science", in Kuhn (1977a), pp. 3–20.

1977c "Objectivity, Value Judgement, and Theory Choice", in Kuhn (1977a), pp. 320–339.

1978 *Black-Body Theory and the Quantum Discontinuity 1894–1912*, Oxford: Oxford University Press, 1978; Chicago–London: University of Chicago Press, 1987².

1979a "History of Science", in Peter D. Asquith, Henry E. Kyburg (eds), *Current Research in Philosophy of Science*, East Lansing, Michigan: Philosophy of Science Association, 1979, pp. 121–128.

1979b "Metaphor in Science", in Andrew Ortony (ed.), *Metaphor and Thought*, Cambridge: Cambridge University Press, 1979, pp. 409–419; reprinted in Kuhn (2000b), pp. 196–207.

1979c "Foreword", in Fleck (1935/1979), pp. vii–xi.

1980a "The Halt and the Blind: Philosophy and History of Science", *The British Journal for the Philosophy of Science*, 31, 1980, pp. 181–192.

1980b "Einstein's Critique of Planck", in Harry Woolf (ed.), *Some Strangeness in the Proportion. A Centennial Symposium to Celebrate the Achievements of Albert Einstein*, Reading, Massachusetts: Addison-Wesley, 1980, pp. 186–191.

1980c "Open Discussion Following Papers by J. Klein and T.S. Kuhn", in Harry Woolf (ed.), *Some Strangeness in the Proportion. A Centennial Symposium to Celebrate the Achievements of Albert Einstein*, Reading, Massachusetts: Addison-Wesley Publishing Company, 1980, p. 194.

1981 *What are Scientific Revolutions?*, Occasional Paper #18, Center for Cognitive Science, MIT, 1981; reprinted in Kuhn (2000b), pp. 13–32.

1983a "Commensurability, Comparability, Communicability", in Asquith, Nickles (eds) (1983), vol. 2, pp. 669–688; reprinted in Kuhn (2000b), pp. 33–53.

1983b "Response to Commentaries" [on Kuhn (1983a)], in Asquith, Nickles (eds) (1983), vol. 2, pp. 712–716; reprinted as "Postscript: Response to Commentaries", in Kuhn (2000b), pp. 53–57.

1983c "Rationality and Theory Choice", *The Journal of Philosophy*, LXXX, 1983, pp. 563–570; reprinted in Kuhn (2000b), pp. 208–215.

1983d "Foreword", in Bruce R. Wheaton, *The Tiger and the Shark. Empirical Roots of Wave-Particle Dualism*, Cambridge: Cambridge University Press, 1983, pp. ix–xiii.

1984 "Revisiting Planck", *Historical Studies in the Physical Sciences*, 14, 1984, pp. 231–252; reprinted in Kuhn (1978, 1987), pp. 349–370.

1986 "The Histories of Science: Diverse Worlds for Diverse Audiences", *Academe*, 72, 1986, pp. 29–33.

1989a "Possible Worlds in History of Science", in Allén (ed.) (1989), pp. 9–32; reprinted in Kuhn (2000b), pp. 58–86.

1989b "Speaker's Reply" [on Kuhn (1989a)], in Allén (ed.) (1989), pp. 49–51; reprinted as "Postscript: Speaker's Reply", in Kuhn (2000b), pp. 86–89.

1989c "Preface", in Hoyningen-Huene (1989a), pp. 1–3; reprinted as "Foreword", in Hoyningen-Huene (1989a/1993), pp. xi–xiii.

1990 "Dubbing and Redubbing: The Vulnerability of Rigid Designation", in C. Wade Savage (ed.), *Scientific Theories*, Minneapolis: University of Minnesota Press, 1990, pp. 298–318.

1991a "The Road Since Structure", in Arthur Fine, Micky Forbes, Linda Wessels (eds), *PSA 1990. Proceedings of the 1990 Biennial Meeting of the Philosophy of Science Association*, East Lansing, Michigan: Philosophy of Science Association, 1991, vol. 2, pp. 3–13; reprinted in Kuhn (2000b), pp. 90–104.

1991b "The Natural and the Human Sciences", in David R. Hiley, James F. Bohman, Richard Shusterman (eds), *The Interpretive Turn. Philosophy, Science, Culture*, Ithaca, New York–London: Cornell University Press, 1991, pp. 17–24; reprinted in Kuhn (2000b), pp. 216–223.

1992 *The Trouble with the Historical Philosophy of Science*, An Occasional Publication of the Department of the History of Science, Cambridge, Massachusetts: Harvard University, 1992; reprinted in Kuhn (2000b), pp. 105–120.

1993a "Afterwords", in Horwich (ed.) (1993), pp. 311–341; reprinted in Kuhn (2000b), pp. 224–252.

1993b "Introduction" [to Bas C. van Fraassen, "From Vicious Circle to Infinite Regress, and Back Again"], in David Hull, Micky Forbes, Kathleen Okruhlik (eds), *PSA 1992. Proceedings of the 1992 Biennial Meeting of the Philosophy of Science Association*, East Lansing, Michigan: Philosophy of Science Association, 1993, vol. 2, pp. 3–5.

1999 "Remarks on Incommensurability and Translation", in Rema Rossini Favretti, Giorgio Sandri, Roberto Scazzieri (eds), *Incommensurability and Translation. Kuhnian Perspectives on Scientific Communication and Theory Change*, Cheltenham–Northampton, Massachusetts: Edward Elgar, 1999, pp. 33–37.

2000a *Dogma contro critica. Mondi possibili nella storia della scienza*, edited by Stefano Gattei, Milan: Raffaello Cortina Editore, 2000.

2000b *The Road Since* Structure. *Philosophical Essays, 1970–1993, with an Autobiographical Interview*, edited by James Conant and John Haugeland, Chicago–London: University of Chicago Press, 2000.

I-1990 "The Nature of Scientific Knowledge. An Interview with Thomas Kuhn", edited by Skúli Sigurdsson, *Harvard Science Review*, 3, 1, 1990, pp. 18–25.

I-1991 "Paradigmi dell'evoluzione scientifica", in Giovanna Borradori, *Conversazioni americane. Con W.O. Quine, D. Davidson, H. Putnam, R. Nozick, A.C. Danto, R. Rorty, S. Cavell, A. MacIntyre, Th.S. Kuhn*, Rome–Bari: Laterza, 1991, pp. 189–206; English translation by Rosanna Crocitto,

"Paradigms of Scientific Evolution", in Giovanna Borradori, *The American Philosopher. Conversations with Quine, Davidson, Putnam, Nozick, Danto, Rorty, Cavell, MacIntyre, and Kuhn*, Chicago–London: University of Chicago Press, 1994, pp. 153–167.

I-1997a "A Physicist who became a Historian for Philosophical Purposes", a discussion between Thomas S. Kuhn and Aristides Baltas, Kostas Gavroglu, Vasso Kindi, *Neusis*, 6, 1997, pp. 145–200; reprinted as "Discussion with Thomas S. Kuhn", in Kuhn (2000b), pp. 255–323.

I-1997b "Note sull'incommensurabilità", interview by Mario Quaranta, Italian translation by Stefano Gattei, *Pluriverso*, II, 4, 1997, pp. 108–114.

U-1980 "The Natures of Conceptual Change", Perspectives in the Philosophy of Science, University of Notre Dame, Notre Dame, Indiana, 1980.

U-1982 *The Plurality of Worlds. An Evolutionary Theory of Scientific Development*, unfinished manuscript.

U-1984 "Scientific Development and Lexical Change", Thalheimer Lectures, Johns Hopkins University, Baltimore, Maryland, 12–19 November 1984.

U-1987 "The Presence of Past Science", Shearman Memorial Lectures, University College, London, 23–25 November 1987.

U-1990a "An Historian's Theory of Meaning", paper delivered at the Cognitive Science Colloquium, U.C.L.A, 26 April 1990.

U-1990b "A Function for Incommensurability", paper delivered at the Philosophy Colloquium, U.C.L.A., 27 April 1990.

V-1980 *The Crisis of the Old Quantum Theory, 1922–25*, Cambridge, Massachusetts: Science Center, Harvard University, 5 November 1980; video-recording, 120 minutes.

KUHN, Thomas S., HEILBRON, John L.

1969 "The Genesis of the Bohr Atom", *Historical Studies in the Physical Sciences*, 1, 1969, pp. 211–290.

KUHN, Thomas S., HEILBRON, John L., FORMAN, Paul, ALLEN, Lini

1967 *Sources for History of Quantum Physics. An Inventory and Report*, Philadelphia: The American Philosophical Society, 1967.

KUHN, Thomas S., VAN VLECK, John H.

1950 "A Simplified Method of Computing the Cohesive Energies of Monovalent Metals", *The Physical Review*, 79, 1950, pp. 382–388.

LAKATOS, Imre

1968 "Changes in the Problem of Inductive Logic", in Lakatos (ed.) (1968), pp. 315–417.

1969 "Criticism and the Methodology of Scientific Research Programmes", *Proceedings of the Aristotelian Society*, LXIX, 1969, pp. 149–186.

1970 "Falsification and the Methodology of Scientific Research Programmes", in Lakatos, Musgrave (eds) (1970), pp. 91–195; reprinted in Lakatos (1978a), pp. 8–101.

1971a "History of Science and Its Rational Reconstructions", in Buck, Cohen (eds) (1971), pp. 91–135; reprinted in Lakatos (1978a), pp. 102–138.

1971b "Replies to Critics", in Buck, Cohen (eds) (1971), pp. 174–182.

1974 "Popper on Demarcation and Induction", in Schilpp (ed.) (1974), vol. I, pp. 241–273; reprinted in Lakatos (1978a), pp. 139–167.

1976 "Understanding Toulmin", *Minerva*, 164, 1976, pp. 126–143; reprinted in Lakatos (1978b), pp. 224–243.

1978a *The Methodology of Scientific Research Programmes. Philosophical Papers Volume 1*, edited by John Worrall and Gregory P. Currie, Cambridge: Cambridge University Press, 1978.

1978b *Mathematics, Science and Epistemology. Philosophical Papers Volume 2*, edited by John Worrall and Gregory P. Currie, Cambridge: Cambridge University Press, 1978.

1978c "Newton's Effect on Scientific Standards", in Lakatos (1978a), pp. 193–222.

LAKATOS, Imre (ed.)

1967 *Problems in the Philosophy of Mathematics*, Amsterdam: North Holland Publishing Company, 1967.

1968 *The Problem of Inductive Logic*, Amsterdam: North Holland Publishing Company, 1968.

LAKATOS, Imre, FEYERABEND, Paul K.

1995 *Sull'orlo della scienza. Pro e contro il metodo*, edited by Matteo Motterlini, Milan: Raffaello Cortina Editore, 1995; English edition, *For and Against Method*, Chicago–London: University of Chicago Press, 1999.

LAKATOS, Imre, MUSGRAVE, Alan (eds)

1968 *Problems in the Philosophy of Science*, Amsterdam: North Holland Publishing Company, Amsterdam 1968.

1970 *Criticism and the Growth of Knowledge*, Cambridge: Cambridge University Press, 1970.

LAKATOS, Imre, ZAHAR, Elie G.

1976 "Why did Copernicus's Research Programme Supersede Ptolemy's?", in Robert S. Westman (ed.), *The Copernican Achievement*, Los Angeles: University of California Press, 1976, pp. 354–383; reprinted in Lakatos (1978a), pp. 168–192.

LARVOR, Brendan

2003 "Why did Kuhn's *Structure of Scientific Revolutions* Cause a Fuss?", *Studies in History and Philosophy of Science*, 34, 2003, pp. 369–390.

LAUDAN, Larry

1977 *Progress and Its Problems. Towards a Theory of Scientific Growth*, Berkeley–Los Angeles–London: University of California Press, 1977.

1984 *Science and Values. The Aim of Science and Their Role in Scientific Debate*, Berkeley: University of California Press, 1984.
1989 "If It Ain't Broken, don't Fix It", *The British Journal for the Philosophy of Science*, 40, 1989, pp. 369–375.
1990a *Science and Relativism. Some Key Controversies in the Philosophy of Science*, Chicago: University of Chicago Press, 1990.
1990b "Normative Naturalism", *Philosophy of Science*, 57, 1990, pp. 44–59.
1996 *Beyond Positivism and Relativism. Theory, Method and Evidence*, Boulder: Westview Press, 1996.

Le Roy, Edouard
1899–1900 "Science et philosophie", *Revue de métaphysique et de morale*, 7, 1899, pp. 375–425, 503–562, 708–731 and 8, 1900, pp. 37–72.
1900a "Réponse à M. Couturat", *Revue de métaphysique et de morale*, 8, 1900, pp. 223–233.
1900b "La science positive et les philosophies de la liberté", paper delivered at the Congrès International de Philosophie, Paris 1900; report of paper and discussion in *Revue de métaphysique et de morale*, 8, 1900, pp. 575–582.
1901a "Un positivisme nouveau", *Revue de métaphysique et de morale*, 9, 1901, pp. 138–153.
1901b "Sur quelques objections adressées à la nouvelle philosophie", *Revue de métaphysique et de morale*, 9, 1901, pp. 292–327 and 407–432.

Lewis, David
1970 "How to Define Theoretical Terms", *The Journal of Philosophy*, 67, 1970, pp. 427–446.
1984 "Putnam's Paradox", *Australasian Journal of Philosophy*, 62, 1984, pp. 221–236.

Long, Jancis
1987 "The Autonomous Student: A Footnote", *Interchange*, 18, 1987, pp. 26–28.

Lovejoy, Arthur O.
1936 *The Great Chain of Being. A Study of the History of an Idea*, Cambridge, Massachusetts: Harvard University Press, 1936.

Mach, Ernst
1900 *Die Analyse der Empfindungen und das Verhältniss des Physischen zum Psychischen*, Jena: G. Fischer, 1900; English translation by C.M. Williams, revised and augmented by Sydney Waterlow, *The Analysis of Sensations and the Realations of the Physical to the Psychical*, introduction by Thomas S. Szasz, New York: Dover Publications, 1959.

Maier, Anneliese
1949 *Die Vorläufer Galileis im 14. Jahrhundert. Studien zur Naturphilosophie der Spätscholastik*, Rome: Edizioni di "Storia e Letteratura", 1949.

MAJER, Ulrich
1989 "Ramsey's Conception of Theories: An Intuitionistic Approach", *History of Philosophy Quarterly*, 6, 1989, pp. 233–258.

MARCUM, James A.
2005 *Thomas Kuhn's Revolution. An Historical Philosophy of Science*, Continuum International Publishing Group, New York 2005.

MARGOLIS, Howard
1993 *Paradigms and Barriers. How Habits of Mind Govern Scientific Beliefs*, Chicago–London: University of Chicago Press, 1993.

MARGOLIS, Joseph
1991 *The Truth About Relativism*, Oxford: Basil Blackwell, 1991.

MARTIN, Michael
1971 "Referential Variance and Scientific Objectivity", *The British Journal for the Philosophy of Science*, 22, 1971, pp. 17–26.
1972 "Ontological Variance and Scientific Objectivity", *The British Journal for the Philosophy of Science*, 23, 1972, pp. 252–256.

MASTERMAN, Margaret
1970 "The Nature of a Paradigm", in Lakatos, Musgrave (eds) (1970), pp. 59–89.

MEILAND, Jack W., KRAUSZ, Michael (eds)
1982 *Relativism: Cognitive and Moral*, Notre Dame–London: University of Notre Dame Press, 1982.

MELLOR, David H.
1978 "Introduction", in Ramsey (1978), pp. 1–7.
1990 "Introduction", in Ramsey (1990), pp. xi–xxv.
1995 "F.P. Ramsey", *Philosophy,* 70, 1995, pp. 243–262.

MELLOR, David H. (ed.)
1980 *Prospects for Pragmatism. Essays in Memory of F.P. Ramsey*, Cambridge: Cambridge University Press, 1980.

MENGER, Karl
1994 *Reminiscences of the Vienna Circle and the Mathematical Colloquium*, edited by Louise Golland, Brian McGuinness and Abe Sklar, Dordrecht–Boston: Kluwer Academic Publishers, 1994.

METZGER, Hélène
1923 *Les doctrines chimiques en France du début du XVIIe à la fin du XVIIIe siècle*, Paris: Presses Universitaires de France, 1923.

1930 *Newton, Stahl, Boerhaave et la doctrine chimique*, Paris: F. Alcan, 1930.

MEYERSON, Emile

1908 *Identité et realité*, Paris: F. Alcan, 1908, 1926[3]; English translation by Kate Loewenberg, *Identity and Reality*, London: George Allen & Unwin, and New York: The Macmillan Company, 1930.

MILLER, David W.

1997 "Sir Karl Raimund Popper, CH, FBA", *Biographical Memoirs of Fellows of The Royal Society of London*, 43, 1997, pp. 367–409.

MINAZZI, Fabio, PETITOT, Jean

1993 "La connaissance objective comme valeur historique: le néo-illuminisme italien", *Archive de Philosophie*, 56, 1993, pp. 620–660.

MORRIS, Charles W.

1937 *Logical Positivism, Pragmatism, and Scientific Empiricism*, Paris: Hermann & C.ie, 1937.

1938 "Scientific Empiricism", in Neurath, Bohr, Dewey, Russell, Carnap, Morris (1938), pp. 63–75.

1960 "On the History of the International Encyclopedia of Unified Science", *Synthese*, XII, 1960, pp. 517–521.

MOTTERLINI, Matteo

2002 "Professor Lakatos between the Hegelian Devil and the Popperian Deep Blue Sea", in Kampis, Kvasz, Sölzner (eds) (2002), pp. 23–52.

MUNÉVAR, Gonzalo (ed.)

1991 *Beyond Reason. Essays on the Philosophy of Paul Feyerabend*, Dordrecht–Boston–London: Kluwer Academic Publishers, 1991.

MUNZ, Peter

1985 *Our Knowledge of the Growth of Knowledge: Popper or Wittgenstein?*, London: Routledge & Kegan Paul, 1985.

MUSATTI, Cesare L.

1965 "La psicologia della forma", *Aut-Aut*, 89, 1965, pp. 8–38.

MUSGRAVE, Alan E.

1976a "Method or Madness? Can the Methodology of Scientific Research Programmes be rescued from Epistemological Anarchism?", in Cohen, Feyerabend, Wartofsky (eds) (1976), pp. 457–491.

1976b "Why did Oxygen supplant Phlogiston? Research Programmes in the Chemical Revolution", in Howson (ed.) (1976), pp. 181–209.

NAGEL, Ernest
1949 "The Meaning of Reduction in the Natural Sciences", in Robert C. Stauffer (ed.), *Science and Civilization*, Madison: University of Wisconsin Press, 1949, pp. 327–338.
1961 *The Structure of Science*, Harcourt, New York: Brace & World, 1961; London: Routledge & Kegan Paul, 1979[2].

NEIDER, Heinrich
1973 "Memories of Otto Neurath", in Neurath (1973), pp. 45–49.

NEURATH, Otto
1931a "Physicalism: The Philosophy of the Vienna Circle", *The Monist*, 41, 1931, pp. 618–623; reprinted in Neurath (1983), pp. 48–51.
1931b "Physikalismus", *Scientia*, 50, 1931, pp. 297–303; English translation by Robert S. Cohen and Marie Neurath, "Physicalism", in Neurath (1983), pp. 52–57.
1931c "Soziologie im Physikalismus", *Erkenntnis*, 2, 1931, pp. 393–431; English translation by Robert S. Cohen and Marie Neurath, "Sociology in the Framework of Physicalism", in Neurath (1983), pp. 58–90.
1932a "Protokollsätze", *Erkenntnis*, 3, 1932, pp. 204–214; English translation by Robert S. Cohen and Marie Neurath, "Protocol Statements", in Neurath (1983), pp. 91–99.
1932b "Sozialbehaviourismus", *Sociologicus*, 8, 1932, pp. 281–288.
1933 *Einheitswissenschaft und Psychologie*, Vienna: Gerold, 1933.
1934 "Radikaler Physikalismus und 'wirkliche Welt'", *Erkenntnis*, 4, 1934, pp. 346–362.
1935a "Pseudorationalismus der Falsifikation", *Erkenntnis*, 5, 1935, pp. 353–365; English translation by Robert S. Cohen and Marie Neurath, "Pseudorationalism of Falsification", in Neurath (1983), pp. 121–131.
1935b *Le développement du Cercle de Vienne et l'avenir de l'empirisme logique*, Paris: Hermann & C.ie, 1935.
1935c "Einheit der Wissenschaft als Aufgabe", *Erkenntnis*, 5, 1935, pp. 16–22; English translation by Robert S. Cohen and Marie Neurath, "The Unity of Science as a Task", in Neurath (1983), pp. 115–120.
1936a "Une encyclopédie internationale de la science unitaire", in *Actes du Congrès international de philosophie scientifique. Sorbonne, Paris 1935*, vol. II: *Unité de la science*, Paris: Hermann & C.ie, 1936, pp. 54–59; English translation by Robert S. Cohen and Marie Neurath, "An International Encyclopedia of Unified Science", in Neurath (1983), pp. 139–144.
1936b "L'encyclopédie comme 'modèle'", *Revue de synthèse*, 12, 1936, pp. 187–201; English translation by Robert S. Cohen and Marie Neurath, "Encyclopedia as 'Model'", in Neurath (1983), pp. 145–158.
1937a "Unified Science and Its Encyclopedia", *Philosophy of Science*, 4, 1937, pp. 265–277; reprinted in Neurath (1983), pp. 172–182.
1937b "Die neue Enzyklopädie des wissenschaftlichr Empirismus", *Scientia*, 62, 1937, pp. 309–320; English translation by Robert S. Cohen and Marie

Neurath, "The New Encyclopedia of Scientific Empiricism", in Neurath (1983), pp. 189–199.
1938 "Unified Science as Encyclopedic Integration", in Neurath, Bohr, Dewey, Russell, Carnap, Morris (1938), pp. 1–27.
1946 "The Orchestration of the Sciences by the Encyclopedia of Logical Empiricism", *Philosophy and Phenomenological Research*, 6, 1946, pp. 469–508; reprinted in Neurath (1983), pp. 230–243.
1973 *Empiricism and Sociology. With a Selection of Biographical and Autobiographical Sketches*, edited by Marie Neurath and Robert S. Cohen, Dordrecht–Boston: D. Reidel Publishing Company, 1973.
1983 *Philosophical Papers 1913–1946*, edited and translated by Robert S. Cohen and Marie Neurath, with the assistance of Carolyn R. Fawcett, Dordrecht–Boston–Lancaster: D. Reidel Publishing Company, 1983.

NEURATH, Otto, BOHR, Niels, DEWEY, John, RUSSELL, Bertrand, CARNAP, Rudolf, MORRIS, Charles W.
1938 *Encyclopedia and Unified Science*, Chicago: University of Chicago Press, 1938.

NEURATH, Otto, CARNAP, Rudolf, MORRIS, Charles W. (eds)
1955 *Foundations of the Unity of Science. Toward and Encyclopedia of Unified Science*, 2 vols., Chicago: University of Chicago Press, 1955.
1970 *Foundations of the Unity of Science. Toward and Encyclopedia of Unified Science*, vol. II, Chicago: University of Chicago Press, 1970.

NEWTON-SMITH, William H.
1981 *The Rationality of Science*, London: Routledge & Kegan Paul, 1981.

NICKLES, Thomas (ed.)
2002 *Thomas Kuhn*, Cambridge: Cambridge University Press, 2002.

NOLA, Robert
1980a "'Paradigm Lost, or the World Regained' – An Excursion into Realism and Idealism in Science", *Synthese*, 45, 1980, pp. 317–350.
1980b "Fixing the Reference of Theoretical Terms", *Philosophy of Science*, 47, 1980, pp. 505–531.
NOLA, Robert (ed.)
1988 *Relativism and Realism in Science*, Dordrecht–Boston: Kluwer Academic Publishers, 1988.

NOLA, Robert, SANKEY, Howard (eds)
2000 *After Popper, Kuhn and Feyerabend. Recent Issues in Theories of Scientific Method*, Dordrecht–Boston–London: Kluwer Academic Publishers, 2000.

NOTTURNO, Mark A.

1985 *Objectivity, Rationality and the Third Realm: Justification and the Grounds of Psychologism. A Study of Frege and Popper*, Dordrecht–Boston–Lancaster: Martinus Nijhoff Publishers, 1985.

2000 *Science and the Open Society. The Future of Karl Popper's Philosophy*, Budapest: Central European University Press, 2000.

NOVALIS (Friedrich von Hardenberg)

1960 *Schriften*, vol. 2: *Das philosophische Werk I*, edited by Richard Samuel together with Hans-Joachim Mähl and Gerhard Schulz, Stuttgart: W. Kohlhammer Verlag, 1960.

OBERHEIM, Eric

2006 *Feyerabend's Philosophy*, Berlin–New York: Walter de Gruyter, 2006.

OBERHEIM, Eric, HOYNINGEN-HUENE, Paul

1997 "Incommensurability, Realism and Meta-incommensurability", *Theoria*, 12, 1997, pp. 447–465.

O'HEAR, Anthony (ed.)

1995 *Karl Popper: Philosophy and Problems*, Cambridge: Cambridge University Press, 1995.

PAPINEAU, David

1979 *Theory and Meaning*, Oxford: Oxford University Press, 1979.

PARRINI, Paolo

1980 *Una filosofia senza dogmi. Materiali per un bilancio dell'empirismo contemporaneo*, Bologna: Il Mulino, 1980.

1995 *Conoscenza e realtà. Saggio di filosofia positiva*, Rome–Bari: Laterza, 1995; English translation by Paolo Baracchi, *Knowledge and Reality. An Essay in Positive Philosophy*, Dordrecht: Kluwer Academic Publishers, 1998.

2002 *L'empirismo logico. Aspetti storici e prospettive teoriche*, Rome: Carocci, 2002.

PARRINI, Paolo, SALMON, Wesley C., SALMON, Merrilee H. (eds)

2003 *Logical Empiricism. Historical and Contemporary Perspectives*, Pittsburgh: University of Pittsburgh Press, 2003.

PARTEE, Barbara H.

1989 "Possible Worlds in a Model-Theoretic Semantics: A Linguistic Perspective", in Allén (ed.) (1989), pp. 93–123.

PASINI, Mirella, ROLANDO, Daniele (eds)

1991 *Il neoilluminismo italiano: cronache di filosofia, 1953–1962*, Milan: Il Saggiatore, 1991.

PASSMORE, John

1967 "Logical Positivism", in *The Encyclopedia of Philosophy*, Paul Edwards general editor, New York: MacMillan and The Free Press, 1967, vol. V, pp. 52–57.

PERA, Marcello

1981 *Popper e la scienza su palafitte*, Rome–Bari: Laterza, 1981.

1982a *Apologia del metodo*, Rome–Bari: Laterza, 1982, 1996[2].

1982b "Invarianza e cambiamento scientifico. Critica della nuova filosofia della scienza", in Carl R. Kordig, *La giustificazione del cambiamento scientifico*, Rome: Armando Armando, 1982, pp. 7–30.

1984 "Progresso scientifico, storia e valori", in Paul K. Feyerabend, *Scienza come arte*, Rome–Bari: Laterza, 1984, pp. vii–xxvi.

1985 "Dal neopositivismo alla filosofia della scienza", in Adriano Bausola, Giuseppe Bedeschi, Mario Dal Pra, Eugenio Garin, Marcello Pera, Valerio Verra, *La filosofia italiana dal dopoguerra a oggi*, Rome–Bari: Laterza, 1985, pp. 93–173.

1986 "Nascita e trasfigurazione del neopositivismo nella filosofia italiana del dopoguerra", in Agazzi (ed.) (1986), pp. 241–257.

1991 *Scienza e retorica*, Rome–Bari: Laterza, 1991; revised English translation by Clarissa Botsford, *The Discourses of Science*, Chicago–London: University of Chicago Press, 1994.

PERA, Marcello, SHEA, William R. (eds)

1991 *Persuading Science. The Art of Scientific Rhetoric*, Canton, Massachusetts: Science History Publications, 1991.

PETRONI, Angelo M.

1990 *I modelli, l'invenzione e la conferma. Saggio su Keplero, la rivoluzione copernicana e la «New philosophy of science»*, Milan: Franco Angeli, 1990.

POINCARÉ, Jules-Henri

1902 *La science et l'hypothèse*, Paris: Flammarion, 1902; English translation by George B. Halsted, *Science and Hypothesis*, New York: Dover Publications, 1952.

1905 *La valeur de la science*, Paris: Flammarion, 1905; English translation by George B. Halsted, *The Value of Science*, New York: Dover Publications, 1958.

1908 *Science et méthode*, Paris: Flammarion, 1908; English translation by Francis Maitland, *The Value of Science*, New York: Dover Publications, 1952.

1913 *Dernières pensées*, Paris: Flammarion, 1913; English translation by John W. Bolduc, *Mathematics and Science. Last Essays*, New York: Dover Publications, 1963.

POLANYI, Michael

1946 *Science, Faith and Society*, Chicago: University of Chicago Press, 1946, 1964[2].

1951 *The Logic of Liberty*, London: Routledge and Kegan Paul, and Chicago: University of Chicago Press, 1951.

1958 *Personal Knowledge. Towards a Post-Critical Philosophy*, London: Routledge & Kegan Paul, 1958, 1962[2].

1962 "Tacit Knowing. Its Bearing on Some Problems of Philosophy", *Review of Modern Physics*, 34, 1962, pp. 601–616.

1963 "Commentary" (on (1963a)), in Crombie (ed.) (1963), pp. 375–380.

1966 *The Tacit Dimension*, Garden City, New York: Doubleday, 1966.

POPPER, Karl R.

1925 "Über die Stellung des Lehrers zu Schule und Schüler. Gesellschaftliche oder individualistische Erziehung?", *Schulreform*, 4, 1925, pp. 204–208; reprinted in Popper (2006), pp. 3–9.

1927 "Zur Philosophie des Heimatgedankens", *Die Quelle*, 77, 1927, pp. 899–908; reprinted in Popper (2006), pp. 10–26.

1931 "Die Gedächtnispflege unter dem Gesichtspunkt der Selbsttätigkeit", *Die Quelle*, 81, 1931, pp. 607–619; reprinted in Popper (2006), pp. 27–49.

1932 "Pädagogische Zeitschriftenschau", *Die Quelle*, 82, 1932, pp. 301–303, 580–582, 646–647, 712–713, 778–781, 846–849, 930–931; reprinted in Popper (2006), pp. 50–79.

1935 *Logik der Forschung. Zur Erkenntnistheorie der modernen Naturwissenschaft*, Vienna: Julius Springer, 1935; English translation by Karl Popper, Julius Freed and Lan Freed, *The Logic of Scientific Discovery*, London: Hutchinson & Co., 1959.

1940 "What is Dialectic?", *Mind*, New series, 49, 1940, pp. 403–426; reprinted in Popper (1963a, 1989), pp. 312–335.

1945 *The Open Society and Its Enemies*, vol. I: *The Spell of Plato*, vol. II: *The High Tide of Prophecy: Hegel, Marx, and the Aftermath*, London: Routledge & Kegan Paul, 1945, 1966[5].

1949 "Naturgesetze und theoretische Systeme", in Simon Moser (ed.), *Gesetz und Wirklichkeit*, Innsbruck: Tyrolia, 1949, pp. 43–60; English translation as "The Bucket and the Searchlight: Two Theories of Knowledge", in Popper (1972a, 1979), pp. 341–361.

1955 "A Note on Traski's Definition of Truth", *Mind*, New Series, 64, 1955, pp. 388–391; reprinted in Popper (1972a, 1979), pp. 335–340.

1957a "The Aim of Science", *Ratio*, 1, 1957, pp. 24–35; reprinted in Popper (1972a, 1979), pp. 191–204.

1957b *The Poverty of Historicism*, London: Routledge & Kegan Paul, 1957.

1959 "Back to Pre-Socratics", *Proceedings of the Aristotelian Society*, New Series, LIX, 1958–1959, pp. 1–24; reprinted in Popper (1963a, 1989), pp. 136–153.

1962a "Facts, Standards, and Truth: A Further Criticism of Relativism", in Popper (1945, 1962), vol. II, pp. 369–396.

1962b "Julius Kraft, 1898–1960", *Ratio*, 4, 1962, pp. 2–15.

1963a *Conjectures and Refutations. The Growth of Scientific Knowledge*, London: Routledge & Kegan Paul, 1963, 1989[5].

1963b "Truth, Rationality, and the Growth of Scientific Knowledge", in Popper (1963a, 1989), pp. 215–250.

1968a "Remarks on the Problems of Demarcation and Rationality", in Lakatos, Musgrave (eds) (1968), pp. 88–102.

1968b "Epistemology Without a Knowing Subject", in Bob van Rootselaar, Jan F. Staal (eds), *Logic, Methodology and Philosophy of Science III. Proceedings of the Third International Congress of Logic, Methodology and Philosophy of Science*, Amsterdam: North Holland Publishing Company, 1968, pp. 333–373; reprinted in Popper (1972a, 1979), pp. 106–152.

1970 "Normal Science and Its Dangers", in Lakatos, Musgrave (eds) (1970), pp. 51–58.

1972a *Objective Knowledge. An Evolutionary Approach*, Oxford: Clarendon Press, 1972, 1979[2].

1972b "Of Clouds and Clocks. An Approach to the Problem of Rationality and the Freedom of Man", in Popper (1972a, 1979), pp. 206–255.

1972c "Philosophical Comments on Tarski's Theory of Truth", in Popper (1972a, 1979), pp. 319–335.

1974a "Autobiography of Karl Popper", in Schilpp (ed.) (1974), vol. I, pp. 1–181; revised edition, *Unended. Quest. An Intellectual Autobiography*, London: Fontana/Collins, 1976, 1992[4].

1974b "Replies to My Critics", in Schilpp (ed.) (1974), vol. II, pp. 961–1197.

1974c "Kuhn on the Normality of Normal Science, in Popper (1974b), pp. 1144–1148.

1974d "Wisdom on the Similarity between Kuhn and Popper", in Popper (1974b), pp. 1148–1153.

1975 "The Rationality of Scientific Revolutions", in Harré (ed.) (1975), pp. 72–101; reprinted in Hacking (ed.) (1981), pp. 80–106.

1976 "The Myth of the Framework", in Eugene Freeman (ed.), *The Abdication of Philosophy: Philosophy and Public Good. Essays in Honor of Paul Arthur Schilpp*, La Salle, Illinois: Open Court Publishing Company, 1976, pp. 23–48; reprinted in Popper (1994a), pp. 33–64.

1979 *Die beiden Grundprobleme der Erkenntnistheorie*, edited by Troels E. Hansen, Tübingen: J.C.B. Mohr (Paul Siebeck), 1979, 1994[2].

1982a *Postscript to The Logic of Scientific Discovery*, vol. II, *The Open Universe. An Argument for Indeterminism*, edited by William W. Bartley III, London: Hutchinson, 1982.

1982b *Postscript to The Logic of Scientific Discovery*, vol. III, *Quantum Theory and the Schism in Physics*, edited by William W. Bartley III, London: Hutchinson, 1982.

1983 *Postscript to The Logic of Scientific Discovery*, vol. I, *Realism and the Aim of Science*, edited by William W. Bartley III, London: Hutchinson, 1983.

1992 "How the Moon might throw some of her Light upon the two Ways of Parmenides", *The Classical Quarterly*, 42, 1992, pp. 12–19.

1994 *The Myth of the Framework. In Defence of Science and Rationality*, edited by Mark A. Notturno, London–New York: Routledge, 1994.

1998a *The World of Parmenides. Essays on the Presocratic Enlightenment*, edited by Arne F. Petersen with the assistance of Jørgen Mejer, London–New York: Routledge, 1998.

1998b "Beyond the Search for Invariants", in Popper (1998a), pp. 146–222.

2004 *Knowledge and the Body–Mind Problem. In Defence of Interaction*, edited by Mark A. Notturno, London–New York: Routledge, 1994.

2006 *Frühe Schriften*, edited by Troels Eggers Hansen, Tübingen: Mohr Siebeck, 2006.

Preston, John M.

1997a *Feyerabend: Philosophy, Science and Society*, Oxford: Polity Press, 1997.

1997b "Feyerabend's Polanyian Turns", *Appraisal*, 1, Supplementary Issue, 1997, pp. 30–36.

2004 "Bird, Kuhn, and Positivism", *Studies in History and Philosophy of Science*, 35, 2004, pp. 327–335.

2008 *Kuhn's The Structure of Scientific Revolutions, A Reader's Guide*, New York: Continuum Press, 2008.

Preston, John M., Munévar, Gonzalo, Lamb, David (eds)

2000 *The Worst Enemy of Science? Essays in Memory of Paul Feyerabend*, Oxford–New York: Oxford University Press, 2000.

Putnam, Hilary

1965 "How not to Talk about Meaning", in Cohen, Wartofsky (eds) (1965), pp. 205–222.

1972 "The Meaning of 'Meaning'", in Keith Gunderson (ed.), *Language, Mind and Knowledge*, University of Minnesota Press, Minneapolis 1972, pp. 131–193.

1974 "The 'Corroboration' of Theories", in Schilpp (ed.) (1974), vol. I, pp. 221–240; reprinted in Hacking (ed.) (1981), pp. 28–59.

1975 *Mind, Language and Reality. Philosophical Papers, Volume 2*, Cambridge: Cambridge University Press, 1975.

1978 *Meaning and the Moral Sciences*, London–Boston: Routledge & Kegan Paul, 1978.

1981 *Reason, Truth and History*, Cambridge: Cambridge University Press, 1981.

1983 *Realism and Reason. Philosophical Papers, Volume 3*, Cambridge: Cambridge University Press, 1983.
1987 *The Many Faces of Realism. The Paul Carus Letters*, La Salle, Illinois: Open Court Publishing Company, 1987.
1990 *Realism with a Human Face*, edited by James Conant, Cambridge, Massachusetts–London: Harvard University Press, 1990.
1994 "Sense, Nonsense, and the Senses: An Inquiry into the Powers of Human Mind", *The Journal of Philosophy*, 91, 1994, pp. 445–517.

QUINE, Willard V.O.
1951 "Two Dogmas of Empiricism", *Philosophical Review*, 60, 1951, pp. 20–43; revised reprint in Willard V.O. Quine, *From a Logical Point of View*, Cambridge, Massachusetts: Harvard University Press, 1953, 1961[2], pp. 20–46.
1960 *Word and Object*, Cambridge, Massachusetts: The MIT Press, 1960.
1969 *Ontological Relativity and Other Essays*, New York–London: Columbia University Press, 1969.
1970 "On the Reasons for Indeterminacy of Translation", *The Journal of Philosophy*, 67, 1970, pp. 178–183.
1974 *The Roots of Reference*, La Salle, Illinois: Open Court Publishing Company, 1974.
1988 "A Comment on Agassi's Remarks", *Zeitschrift für allgemeine Wissenschaftstheorie*, 19, 1988, pp. 117–118.

RADNER, Michael, Winokur, Stephen (eds)
1970 *Analyses of Theories and Methods of Physics and Psychology*, Minneapolis: University of Minnesota Press, 1970.

RADNITZKY, Gerard, ANDERSSON, Gunnar (eds)
1978 *Progress and Rationality in Science*, Dordrecht–Boston: D. Reidel Publishing Company, 1978.

RAMSEY, Frank P.
1925 "Universals", in Ramsey (1990), pp. 8–30.
1929a "Theories", in Ramsey (1990), pp. 112–136.
1929b "Causal Qualities", in Ramsey (1990), pp. 137–139.
1931 *The Foundations of Mathematics and Other Logical Essays*, edited by Richard B. Braithwaite, with a preface of George E. Moore, London: Kegan Paul, Trench, Trubner & Co., 1931.
1978 *Foundations. Essays in Philosophy, Logic, Mathematics and Economics*, edited by David H. Mellor, with introductions by David H. Mellor, Leonid Mirsky, Timothy J. Smiley and Richard Stone, London–Henley: Routledge & Kegan Paul, 1978.
1990 *Philosophical Papers*, edited by David H. Mellor, Cambridge University Press, Cambridge 1990.

READ, Rupert
2003a "How to Understand Kuhnian Incommensurability: Some Unexpected Analogies from Wittgenstein", in Wilhelm Lütterfelds, Andreas Roser, Richard Raatzsch (eds), *Wittgenstein-Jahrbuch 2001/2002*, Frankfurt am Main: Peter Lang, 2003, pp. 151–171.
2003b "Kuhn: le Wittgenstein des sciences?", *Archives de Philosophie*, 66, 2003, pp. 463–479.
2004 Review of Kuhn (2000b), *The British Journal for the Philosophy of Science*, 55, 2004, pp. 175–178.

REICHENBACH, Hans
1938 *Experience and Prediction. An Analysis of the Foundation of Science*, Chicago: University of Chicago Press, 1938, 1961².
1944 *Philosophical Foundations of Quantum Mechanics*, Berkeley: University of California Press, 1944.

REISCH, George A.
1991 "Did Kuhn Kill Logical Empiricism?", *Philosophy of Science*, 58, 1991, pp. 264–277.
1994 "Planning Science: Otto Neurath and the International Encyclopedia of Unified Science", *The British Journal for the History of Science*, 27, 1994, pp. 153–175.
1995 *A History of the International Encyclopedia of Unified Science*, Ph.D. dissertation, University of Chicago, 1995.

RICHARDSON, Alan R.
1998 *Carnap's Construction of the World. The Aufbau and the Emergence of Logical Empiricism*, Cambridge–New York: Cambridge University Press, 1998.

RORTY, Richard M.
1979 *Philosophy and the Mirror of Nature*, Princeton: Princeton University Press, 1979.

RORTY, Richard M. (ed.)
1967 *The Linguistic Turn. Recent Essays in Philosophical Method*, Chicago: University of Chicago Press, 1967, 1992².

ROSSI, Paolo
1981 "Fatti scientifici e stili di pensiero: appunti intorno a una rivoluzione immaginaria", *Rivista di filosofia*, 21, 1981, pp. 403–428.
1983 "Ludwik Fleck e una rivoluzione immaginaria", in Ludwik Fleck, *Genesi e sviluppo di un fatto scientifico*, Bologna: il Mulino, 1983, pp. 9–42.

SAHLIN, Nils-Eric

1990 *The Philosophy of F.P. Ramsey*, Cambridge University Press, New York 1990.

1997 "'He is No Good for My Work'. On the Philosophical Relations between Ramsey and Wittgenstein", in Matti Sintonen (ed.), *Knowledge and Inquiry. Essays on Jaakko Hintikka's Epistemology and Philosophy of Science*, Amsterdam–Atlanta: Rodopi, 1997, pp. 61–84.

SANKEY, Howard

1990 "In Defence of Untranslatability", *Australasian Journal of Philosophy*, 68, 1990, pp. 1–21; reprinted in Sankey (1997a), pp. 83–109.

1991a "Incommensurability and the Indeterminacy of Translation", *Australasian Journal of Philosophy*, 69, 1991, pp. 219–223; reprinted in Sankey (1997a), pp. 34–38.

1991b "Translation Failure Between Theories", *Studies in History and Philosophy of Science*, 22, 1991, pp. 223–236.

1991c "Incommensurability, Translation and Understanding", *Philosophical Quarterly*, 41, 1991, pp. 414–426; reprinted in Sankey (1997a), pp. 110–122.

1991d "Feyerabend and the Description Theory of Reference", *Journal of Philosophical Research*, XVI, 1991, pp. 223–232.

1993a "Kuhn's Changing Concept of Incommensurability", *The British Journal for the Philosophy of Science*, 44, 1993, pp. 759–774; reprinted in Sankey (1997a), pp. 21–34.

1993b "Kuhn's Model of Scientific Theory Change", *Cogito*, 7, 1993, pp. 18–24.

1994a *The Incommensurability Thesis*, Avebury: Aldershot, 1994.

1994b "Judgement and Rational Theory-Choice", *Methodology and Science*, 27, 1994, pp. 167–182; reprinted in Sankey (1997a), pp. 135–146.

1995a "The Problem of Rational Theory-Choice", *Epistemologia*, 18, 1995, pp. 299–312; reprinted in Sankey (1997a), pp. 125–134.

1995b Review of Hoyningen-Huene (1989a/1993), *Australasian Journal of Philosophy*, 73, 1995, pp. 487–489.

1996a "Rationality, Relativism and Methodological Pluralism", *Explorations in Knowledge*, XIII, pp. 18–36; reprinted in Sankey (1997a), pp. 149–164.

1996b "Normative Naturalism and the Challenge of Relativism: Laudan versus Worrall on the Justification of Methodological Principles", *International Studies in the Philosophy of Science*, 10, 1996, pp. 37–51; reprinted in Sankey (1997a), pp. 165–184.

1997a *Rationality, Relativism and Incommensurability*, Aldershot: Ashgate, 1997.

1997b "Kuhn's Ontological Relativism", in Dimitri Ginev, Robert S. Cohen (eds), *Issues and Images in the Philosophy of Science*, Dordrecht: Kluwer Academic Publishers, 1997, pp. 305–320; reprinted in Sankey (1997a), pp. 42–65.

1997c "Incommensurability: The Current State of Play", *Theoria*, 12, 1997, pp. 425–445.

1997d "Induction and Natural Kinds", *Principia*, 1, 1997, pp. 239–254.

1998 "Taxonomic Incommensurability", *International Studies in the Philosophy of Science*, 12, 1998, pp. 7–16; reprinted in Sankey (1997a), pp. 66–80.

1999 "Incommensurability – An Overview", paper delivered at the international conference *Incommensurability (and related matters)*, Hannover, 13 June 1999.

2000 "Methodological Pluralism, Normative Naturalism and the Realist Aim of Science", in Nola, Sankey (eds) (2000), pp. 211–229.

SANKEY, Howard, HOYNINGEN-HUENE, Paul

2001 "Introduction", in Hoyningen-Huene, Sankey (eds) (2001), pp. vii–xxxiv.

SARTON, George

1952 *A History of Science*, Cambridge, Massachusetts: Harvard University Press, 1952, 2 vols.

SASSOLI, Bernardino

1994 "Struttura e dinamica delle teorie scientifiche nell'epistemologia italiana e francese", in Giulio Giorello (ed.), *Introduzione alla filosofia della scienza*, Milan: Bompiani, 1994, pp. 149–231.

SCHÄFER, Lothar, SCHNELLE, Thomas

1980 "Ludwik Flecks Begründung der soziologischen Betrachtungsweise in der Wissenschaftstheorie", in Fleck (1935, 1980), pp. vii–xlix.

SCHEFFLER, Israel

1967 *Science and Subjectivity*, Indianapolis: The Bobbs-Merrill Company, 1967.

1972 "Vision and Revolution. A Postscript on Kuhn", *Philosophy of Science*, 39, 3, 1972, pp. 366–374.

1996 *On Change and Objectivity in Science: Reflections on Kuhn*, Jerusalem: Monographs from the Mandel Institute, 1996.

SCHILLER, Friedrich J.C.

1795 "Über die ästhetische Erziehung des Menschen. In einer Reihe von Briefen", achter Brief, *Die Horen*, I, 1795, pp. 39–42; English edition and translation by Elizabeth M. Wilkinson and Leonard A. Willoughby, "Eighth Letter", in Friedrich Schiller, *On the Aesthetic Education of Man. In a Series of Letters*, Oxford: Clarendon Press, 1967, pp. 48–53.

SCHILPP, Paul A. (ed.)

1963 *The Philosophy of Rudolf Carnap*, La Salle, Illinois: Open Court Publishing Company, 1963.

1974 *The Philosophy of Karl Popper*, La Salle, Illinois: Open Court Publishing Company, 1974, 2 vols.

SCHLICK, Moritz

1917 "Raum und Zeit in der gegenwärtigen Physik. Zum Einführung in das Verständnis der allgemeinen Relativitätstheorie", *Die Naturwissenschaften*, 5, 1917, pp. 161–167, 177–186; enlarged edition, *Raum und Zeit in der gegenwärtigen Physik. Zum Einführung in das Verständnis der allgemeinen Relativitätstheorie*, Berlin: Springer 1917, 1922[4]; English translation by Henry L. Bose and Peter Heath, "Space and Time in Contemporary Physics. An Introduction to the Theory of Relativity and Gravitation", in Schlick (1979a), pp. 207–269.

1918 *Allgemeine Erkenntnislehre*, Berlin: Springer, 1918, 1925[2]; English translation by Albert E. Blumberg, *General Theory of Knowledge*, New York: Springer Verlag, 1974; reprint, La Salle, Illinois: Open Court Publishing Company, 1985.

1930 "Die Wende der Philosophie", *Erkenntnis*, 1, 1930, pp. 4–11; English translation by Peter Heath, "The Turning-Point in Philosophy", in Schlick (1979b), pp. 154–160.

1934 "Über das Fondament der Erkenntnis", *Erkenntnis*, 4, 1934, pp. 79–99; English translation by Peter Heath, "On the Foundation of Knowledge", in Schlick (1979b), pp. 370–387.

1936 "Meaning and Verification", *Philosophical Review*, XLV, 1936, pp. 339–369; reprinted in Schlick (1979b), pp. 456–481.

1979a *Philosophical Papers. Volume I (1909–1922)*, edited by Henk L. Mulder and Barbara F.B. van de Velde-Schlick, translated by Patrick Heath, Dordrecht–Boston–London: D. Reidel Publishing Company, 1979.

1979b *Philosophical Papers. Volume II (1925–1936)*, edited by Henk L. Mulder and Barbara F.B. van de Velde-Schlick, with translations by Patrick Heath, Wilfrid Sellars, Herbert Feigl and May Brodbeck, Dordrecht–Boston–London: D. Reidel Publishing Company, 1979.

1985 "Lettera inedita di Moritz Schlick a Ludovico Geymonat", in Michele Massafra, Fabio Minazzi (eds), *Il problema delle scienze nella realtà contemporanea. Atti dei Seminari Varesini 1980–1984*, Milan: Franco Angeli, 1985, pp. 298–299.

1986 *Die Probleme der Philosophie in ihrem Zusammenhang. Vorlesung aus dem Wintersemester 1933 und 34*, edited by Henk L. Mulder, Anne J. Kox and Rainer Hegselmann, Frankfurt: Suhrkamp, 1986; English translation by Peter Heath, *The Problems of Philosophy in their Interconnection. Winter Semester Lectures, 1933–34*, Dordrecht–Boston: D. Reidel Publishing Company, 1987.

SCHNELLE, Thomas

1982 *Ludwik Fleck – Leben und Denken. Zur Enstehung und Entwicklung des soziologischen Denkstills in der Wissenschaftstheorie*, Freiburg: Hochschulverlag, 1982.

SEGRE, Michael

1991 *In the Wake of Galileo*, New Brunswick, New Jersey: Rutgers University Press, 1991.

1994 "Peano's Axioms in Their Historical Context", *Archive for History of Exact Sciences*, 48, 1994, pp. 201–342.

2002 "Popper e l'educazione", in Gattei (ed.) (2002), 2, pp. 83–89.

SEILER, Martin, STADLER, Friedrich (eds)

1994 *Heinrich Gomperz, Karl Popper und die Österreichische Philosophie*, Amsterdam–Atlanta: Rodopi, 1994.

SHAPERE, Dudley

1983 *Reason and the Search for Knowledge. Investigations in the Philosophy of Science*, Dordrecht–Boston: D. Reidel Publishing Company, 1983.

SHARROCK, Wes, READ, Rupert

2002 *Kuhn. Philosopher of Scientific Revolution*, Oxford: Polity Press, 2002.

SIEGEL, Harvey I.

1987 *Relativism Refuted. A Critique of Contemporary Epistemological Relativism*, Dordrecht–Boston: D. Reidel Publishing Company, 1987.

1990 "Laudan's Normative Naturalism", *Studies in History and Philosophy of Science*, 21, 1990, pp. 295–313.

SINISI, Vito, WOLEŃSKI, Jan (eds)

1995 *The Heritage of Kazimierz Ajdukiewicz*, Amsterdam–Atlanta: Rodopi, 1995.

SNEED, Joseph D.

1971 *The Logical Structure of Mathematical Physics*, Dordrecht–Boston: D. Reidel, 1971, 1979[2].

1976 "Philosophical Problems in the Empirical Science of Science. A Formal Approach", *Erkenntnis*, 10, 1976, pp. 115–146.

STADLER, Friedrich

1995 "The Vienna Circle and the University of Vienna", in Stadler, Webel (eds) (1995), pp. 44–56.

1997 *Studien zum Wiener Kreis. Ursprung, Entwicklung und Wirkung des Logischen Empirismus im Kontext*, Frankfurt: Suhrkamp, 1997; English translation by Camilla Nielsen, Joel Golb, Sabine Schmidt and Thomas Ernst, *The Vienna Circle. Studies in the Origins, Development, and Influence of Logical Empiricism*, Vienna–New York: Springer, 2001.

STADLER, Friedrich (ed.)

1993 *Scientific Philosophy: Origins and Developments*, Dordrecht–Boston–London: Kluwer Academic Publishers, 1993.

2003 *The Vienna Circle and Logical Empiricism. Re-evaluation and Future Perspectives*, Dordrecht–Boston–London: Kluwer Academic Publishers, 2003.

Stadler, Friedrich, Fischer, Kurt R. (eds)
2007 *Paul Feyerabend. Ein Philosoph aus Wien*, Vienna–New York: Springer, 2007.

Stadler, Friedrich, Webel, Peter (eds)
1995 *Vertreibung der Vernunft. The Cultural Exodus from Austria*, Vienna–New York: Springer, 1995.

Stegmüller, Wolfgang
1973 *Probleme und Resultate der Wissenschaftstheorie und analytischen Philosophie*, Band II, *Theorie und Erfahrung*, zweiter Halbband, *Theorienstrukturen und Theoriendynamik*, Springer, Berlin–Heidelberg: Springer Verlag, 1973; English translation by William Wohlhueter, *The Structure and Dynamics of Theories*, New York–Heidelberg–Berlin: Springer Verlag, 1976.
1975 "Structure and Dynamics of Theories. Some Reflections on J.D. Sneed and T.S. Kuhn", *Erkenntnis*, 9, 1975, pp. 75–100.
1976 "Accidental ('Non-Substantial') Theory Change and Theory Dislodgement: to What Extent Logic Can Contribute to a Better Understanding of Certain Phenomena in the Dynamics of Theories", *Erkenntnis*, 10, 1976, pp. 147–178; reprinted in Gutting (ed.) (1980), pp. 75–93.
1979a *Theory Construction, Structure and Rationality*, New York: Springer Verlag, 1979.
1979b "The Structuralist View: Survey, Recent Developments, and Answers to Some Criticisms", in Ilkka Niiniluoto, Raimo Tuomela (eds), *The Logic and Epistemology of Scientific Change*, Amsterdam: North Holland, 1979, pp. 113–129.

Sterelny, Kim
1983 "Natural Kind Terms", *Pacific Philosophical Quarterly*, 64, 1983, pp. 110–125.

Stern, Joseph P.
1959 *Lichtenberg: a Doctrine of Scattered Occasions. Reconstructed from his Aphorisms and Reflections*, Bloomington: Indiana University Press, 1959.

Stock, Wolfgang G.
1980 "Die Bedeutung Ludwik Flecks für die Theorie der Wissenschaftsgeschichte", *Grazer philosophische Studien*, 10, 1980, pp. 105–118.

STOKES, Geoffrey

1998 *Popper. Philosophy, Politics and Scientific Method*, Cambridge: Polity Press, 1998.

SUPPE, Frederick

1974 "The Search for Philosophic Understanding of Scientific Theories", in Suppe (ed.) (1974), pp. 1–241.

1977 "Afterword – 1977", in Suppe (ed.) (1974, 1977), pp. 615–730.

SUPPE, Frederick (ed.)

1974 *The Structure of Scientific Theories*, Urbana–Chicago–London: University of Illinois Press, 1974, 1977[2].

SZANIAWSKI, Klemens (ed.)

1989 *The Vienna Circle and the Lvov-Warsaw School*, Dordrecht–Boston–London: Kluwer Academic Publishers, 1989.

TAMBOLO, Luca

2007 *L'Oceano della Conoscenza. Il pluralismo libertario di Paul Karl Feyerabend*, Milan: Franco Angeli, 2007.

TARSKI, Alfred

1932 "Der Wahrheitsbegriff in den Sprachen der deduktiven Disziplinen [Zusammenfassung]", *Anzeiger der Akademie der Wissenschaften in Wien: mathematisch-naturwissenschaftliche Klasse*, 69, 1932, pp. 23–25.

1935 "Der Wahrheitsbegriff in den formalisierten Sprachen", *Studia Philosophica*, I, 1935, pp. 261–405; English translation by Joseph H. Woodger, "The Concept of Truth in Formalized Languages", in Alfred Tarski, *Logic, Semantics, Metamathematics. Papers from 1923 to 1938*, Oxford: Clarendon Press, 1956, pp. 152–278.

THOM, René

1990 *Apologie du logos*, Paris: Hachette, 1990.

TOULMIN, Stephen E.

1953 *Philosophy of Science*, London: Hutchinson, 1953.

1961 *Foresight and Understanding. An Enquiry into the Aims of Science*, Bloomington: Indiana University Press, 1961.

1963 "Discussion" (on Kuhn (1963a)), in Crombie (ed.) (1963), pp. 382–384.

1970 "Does the Distinction between Normal and Revolutionary Science Hold Water?", in Lakatos, Musgrave (eds) (1970), pp. 39–47.

1972 *Human Understanding. The Collective Use and Evolution of Concepts*, Princeton: Princeton University Press, 1972.

1977 "From Form to Function: Philosophy and History of Science in the 1950s and Now", *Daedalus*, 106, 1977, pp. 143–162.

UEBEL, Thomas E. (ed.)
1991 *Rediscovering the Forgotten Vienna Circle. Austrian Studies on Otto Neurath and the Vienna Circle*, Kluwer Academic Publishers, Dordrecht–Boston–London 1991.

VENTURI, Franco
1959 "Contributi ad un dizionario storico. I – Was ist Aufklaerung? Sapere aude!", *Rivista storica italiana*, LXXI, 1959, pp. 119–128.

VIDONI, Ferdinando
1993 *Il Positivismo*, Naples: Morano, 1993.

VOELKEL, James R.
2001 *The Composition of Kepler's* Astronomia Nova, Princeton–Oxford: Princeton University Press, 2001.

WAISMANN, Friedrich
1967 *Wittgenstein und der Wiener Kreis. Gespräche, aufgezeichnet von Friedrich Waismann*, edited by Brian F. McGuinness, Oxford: Basil Blackwell, 1967; English translation by Joachim Schulte and Brian F. McGuinness, *Wittgenstein and the Vienna Circle. Conversations Recorded by Friedrich Waismann*, Oxford: Basil Blackwell, 1979.

WATKINS, John W.N.
1958 "Confirmable and Influential Metaphysics", *Mind*, 67, 1958, pp. 344–365.
1970 "Against 'Normal Science'", in Lakatos, Musgrave (eds) (1970), pp. 25–37.
1975 "Metaphysics and the Advancement of Science", *The British Journal for the Philosophy of Science*, 26, 1975, pp. 91–121.
1978 "Minimal Presuppositions and Maximal Metaphysics", *Mind*, 87, 1978, pp. 195–209.
1987 "The LSE 'School' of Philosophy: a Polish Interrogation", *LSE Quarterly*, 1, 1987, pp. 213–226.
2000 "Feyerabend Among Popperians, 1948–1978", in Preston, Munévar, Lamb (eds) (2000), pp. 47–57
2002 "The Propositional Content of the Popper–Lakatos Rift", in Kampis, Kvasz, Stölzner (eds) (2002), pp. 3–12.

WATSON, William H.
1938 *On Understanding Physics*, Cambridge: Cambridge University Press, 1938.

WETTERSTEN, John R.
1985 "The Road through Würzburg, Vienna and Gottingen", *Philosophy of the Social Sciences*, 15, 1985, pp. 487–506.
1987a "On the Unification of Psychology, Methodology and Pedagogy: Selz, Popper, and Agassi", *Interchange*, 18, 1987, pp. 1–14.

1987b "On Education and Education for Autonomy", *Interchange*, 18, 1987, pp. 21–25.
1992 *The Roots of Critical Rationalism*, Rodopi, Amsterdam–Atlanta 1992.
2005 "New Insights on Young Popper", *Journal of the History of Ideas*, 66, 2005, pp. 603–631.

WHORF, Benjamin L.
1956 *Language, Thought and Reality*, edited by John B. Carrol, Cambridge, Massachusetts: The MIT Press, 1956.

WIELAND, Wolfgang
1962 *Die aristotelische Physik. Untersuchungen über die Grundlegung der Naturwissenschaft und die sprachlichen Bedingungen der Prinzipienforschung bei Aristoteles*, Göttingen: Vandenhoeck & Ruprecht, 1962.

WISDOM, John O.
1974a "The Nature of 'Normal' Science", in Schilpp (ed.) (1974), pp. 820–842.
1974b "The Incommensurability Thesis", *Philosophical Studies*, 25, 1974, pp. 299–301.

WITTGENSTEIN, Ludwig
1921 "Logisch-philosophische Abhandlung", *Annalen der Naturphilosophie*, 14, 1921, pp. 185–262; English translation by David F. Pears and Brian F. McGuinness, *Tractatus Logico-Philosophicus*, introduction by Bertrand Russell, London: Routledge & Kegan Paul, 1961.
1953 *Philosophische Untersuchungen. Philosophical Investigations*, translated by G. Elizabeth M. Anscombe, edited by G. Elizabeth M. Anscombe and Rush Rhees, Oxford: Basil Blackwell, 1953.
1969 *On Certainty. Über Gewißheit*, edited by G. Elizabeth M. Anscombe and Georg H. von Wright, Oxford: Basil Blackwell, 1969.
1980 *Bemerkungen über die Philosophie der Psychologie. Remarks on the Philosophy of Psychology*, edited by G. Elizabeth M. Anscombe, Georg H. von Wright and Heikki Nyman, Oxford: Basil Blackwell, Oxford.

WITTGENSTEIN, Ludwig, WAISMANN, Friedrich
2003 *The Voices of Wittgenstein. The Vienna Circle*, transcribed, edited and with an introduction by Gordon Baker, translated by Gordon Baker, Michael Mackert, John Connolly and Vasilis Politis, London–New York: Routledge, 2003.

WITTICH, Dieter
1978 "Die gefesselte Dialektik. Zu den philosophischen Ideen des Wissenschaftshistorikers Th.S. Kuhn", *Deutsche Zeitschrift für Philosophie*, 26, 1978, pp. 785–797.

1981 "Die innere Logik der Philosophie Thomas S. Kuhns", in Kurt Bayertz (ed.), *Wissenschaftsgeschichte und wissenschaftilche Revolution*, Cologne: Pahl-Rugenstein, 1981, pp. 136–159.

WOLEŃSKI, Jan
1989 *Logic and Philosophy in the Lvov-Warsaw School*, Dordrecht–Boston–London: Kluwer Academic Publishers, 1989.
1999 *Essays in the History of Logic and Logical Philosophy*, Kraków: Jagiellonian University Press, 1999.

WOLTERS, Gereon
2003 "Carl Gustav Hempel", in Parrini, Salmon, Salmon (eds) (2003), pp. 109–122.

WORRALL, John
1976 "Imre Lakatos (1922–1974): Philosopher of Mathematics and Philosopher of Science", in Cohen, Feyerabend, Wartofsky (eds) (1976), pp. 1–8.
1988 "The Value of a Fixed Methodology", *The British Journal for the Philosophy of Science*, 39, 1988, pp. 263–275.
1989 "Fix It and Be Damned. A Reply to Laudan", *The British Journal for the Philosophy of Science*, 40, 1989, pp. 376–388.
1990 "Scientific Revolutions and Scientific Rationality. The Case of the 'Elderly Holdout'", in C. Wade Savage (ed.), *The Justification, Discovery and Evolution of Scientific Theories*, Minneapolis: University of Minnesota Press, 1990, pp. 319–354.
1995 "'Revolution in Permanence': Popper on Theory-Change in Science", in O'Hear (ed.), (1995), pp. 75–102.
2000 "Kuhn, Bayes and 'Theory-Choice': How Revolutionary is Kuhn's Account of Theoretical Change?", in Nola, Sankey (eds) (2000), pp. 125–151.
2002 "Normal Science and Dogmatism, Paradigms and Progress: Kuhn 'versus' Popper and Lakatos", in Nickles (ed.) (2002), pp. 65–100.

ZAHAR, Elie G.
1973 "Why did Einstein's Programme Supersede Lorentz's?", *The British Journal for the Philosophy of Science*, 24, 1973, pp. 95–123 and 223–262; reprinted in Howson (ed.) (1976), pp. 211–275.
1982 "The Popper–Lakatos Controversy", *Fundamenta Scientiae*, 3, 1982, pp. 21–54.
1983a "The Popper–Lakatos Controversy in the Light of 'Die beiden Grundprobleme der Erkenntnistheorie'", *The British Journal for the Philosophy of Science*, 34, 1983, pp. 149–171.
1983b "Logic of Discovery or Psychology of Invention?", *The British Journal for the Philosophy of Science*, 34, 1983, pp. 243–261.
1989 *Einstein's Revolution. A Study in Heuristic*, La Salle, Illinois: Open Court Publishing Company, 1989.

2001 *Poincaré's Philosophy. From Conventionalism to Phenomenology*, Chicago: Open Court Publishing Company, 2001.

ZECHA, Gerhard (ed.)
1999 *Critical Rationalism and Educational Discourse*, Amsterdam–Atlanta: Rodopi, 1999.

Index